FEMINIST POLITICAL ECOLOGY

Feminist Political Ecology explores the gendered relations of ecologies, economies, and politics in communities as diverse as the rubber tappers in the rainforests of Brazil to activist groups fighting environmental racism in New York City.

Environmental struggles occur throughout the world from industrial to agrarian societies. Women are often at the centre of these struggles, struggles which concern local knowledge, everyday practice, rights to resources, sustainable development, environmental quality, and social justice.

The book bridges the gap between the academic and rural orientation of political ecology and the largely activist and urban focus of environmental justice movements. The aim is to bring together the theoretical frameworks of feminist analysis with the specificities of women's activism and experiences around the world.

Dianne Rocheleau is Associate Professor of Geography at Clark University; **Barbara Thomas-Slayter** is Professor in the International Development Program at Clark University; **Esther Wangari** is Assistant Professor of Global Studies at Gettysburg College.

INTERNATIONAL STUDIES OF WOMEN AND PLACE

Edited by Janet Momsen, *University of California at Davis*
and Janice Monk, *University of Arizona*

The Routledge series of *International Studies of Women and Place* describes the diversity and complexity of women's experience around the world, working across different geographies to explore the processes which underlie the construction of gender and the life-worlds of women.

FEMINIST POLITICAL ECOLOGY

Global issues and local experiences

Edited by
Dianne Rocheleau,
Barbara Thomas-Slayter,
and Esther Wangari

London and New York

First published 1996
by Routledge
11 New Fetter Lane, London EC4P 4EE

Simultaneously published in the USA and Canada
by Routledge
29 West 35th Street, New York, NY 10001

Typeset in Baskerville by
Florencetype Limited, Stoodleigh, Devon

Printed and bound in Great Britain by
Biddles Ltd, Guildford and King's Lynn

British Library Cataloguing in Publication Data
A catalogue record for this book is available from the British Library

Library of Congress Cataloguing in Publication Data
Feminist political ecology: global issues and local experiences/edited by
Dianne Rocheleau, Barbara Thomas-Slayter, and Esther Wangari.
   p.   cm. – (International studies of women and place)
Includes bibliographical references and index.
   1. Ecofeminism–Political aspects.   2. Human ecology–Political aspects.
3. Feminist theory.   I. Rocheleau, Dianne E.   II. Thomas-Slayter, Barbara P.
III. Wangari, Esther.   IV. Series.
   HQ1233.F48 1996
   305.42'01–dc20   96-425

ISBN 0–415–12026–8
      0–415–12027–6 (pbk)

DEDICATION

For Virginia Swartz, Leona Giroux Rocheleau and Edna Swartz Rocheleau for teaching me about women's worlds in the home, on the shop floor, in the garden and in the forest.

Dianne Rocheleau

For my mother, Virginia Gotaas Pinney, and my aunt, Mary Cecilia Gotaas, whose adventures, accomplishments and struggles in homemaking, mentoring, teaching, travels and research have long been a source of inspiration and pride.

Barbara P. Thomas-Slayter

For my mother, Julia Muthoni, whose spirit is with me, for her commitment to knowledge from her kitchen to the world, her belief in equity and her teaching that things are possible in life. For my father, Gerishon Mutongu, my brothers and sisters – Munene, Gichobi, Wachira, Wangu, Wanjiru, Wanjiku and Wanoi – for their love, and invaluable moral support. For the children, nieces and nephews.

Esther Wangari

CONTENTS

LIST OF ILLUSTRATIONS

PLATES

FIGURES

TABLES

ACKNOWLEDGMENTS

We owe special thanks to many colleagues among the faculty and students at Clark University, where we conducted most of the writing and editorial tasks for the volume. We thank Billie Lee Turner for his efforts on behalf of our project as Director of the Marsh Institute from 1991 to 1994, as well as Jody Emel, Richard Ford, Glenn Elder, Phil Steinberg, Moya Hallstein, and a number of other faculty and students in Geography, Women's Studies, and International Development. Maureen Shaughnessy, Elizabeth Owens, Jean Heffernan, Madeleine Grinkus, and Joan Scott have handled more letters, telephone calls, faxes, and payment vouchers than we can count, with skill and patience. We also wish to acknowledge the invaluable research assistance of David Edmunds and Sunita Reddy during the early part of the project and the diligent, skilled and dedicated work of Viola Harmann as technical editor and Dale Shields, editorial assistant and vigorous critic, who helped bring the volume to completion.

This volume was produced in large part under the auspices of the Gender and Environment Project of the Marsh Institute at Clark University, supported by a grant from the Ford Foundation. We owe a special debt of gratitude to Ms. Cheryl Danley for her enthusiastic promotion of the project and to Dr. Walt Coward for his continuing assistance. Two of our own case studies and editorial tasks were also supported by the Ecology, Community Organization, and Gender (ECOGEN) project of the International Development Program at Clark University, funded through the Systems Approach to Regional Income and Sustainable Resource Assistance (SARSA) Cooperative Agreement by the United States Agency for International Development.

We are also especially grateful for the efforts of Professor Janice Monk and Professor Janet Momsen, series editors, for their helpful suggestions and encouragement throughout and their insightful editorial comments and assistance on the draft manuscript. Tristan Palmer, Matthew Smith and Caroline Cautley of Routledge not only stayed the course with us over a long production process, but kept us on the task, with good humor and patience.

To all of our contributors and co-authors we wish to express our appreciation for providing their original, timely, practical and theoretically significant work for the case study chapters.

We each extend our special acknowledgments to our families and friends who have supported us personally and professionally during our research and writing on this volume: Edwin Munene, Julia Muthoni, Gerishon Mutongu, Rahab Gikunju, Janet Ngugi, Joyce Wanjiku, Louise Wanoi, Tabitha Wanjiru, Bernard Gichobi, Peter Wachira, Daniel Gatabaki, Henry Slayter, Virginia Pinney, Luis, Ramon and Rafael Malaret, Lucy Candib, Richard Schmitt, Andre, Edna, and Marie Rocheleau.

Our thanks to those who worked with us in our field work and on our chapters we have expressed separately in our respective chapters in the volume. However, we wish to acknowledge the many people – in the sites in which these eleven case studies were carried out – who gave generously of their time and energies to recount their experiences and participate in the data gathering and analysis pertaining to key environmental issues in their communities.

INTRODUCTION

Throughout the last decade political ecology and feminist scholarship have developed separate but increasingly convergent critiques of sustainable development. To varying degrees practitioners from both schools of thought have advanced alternative models of environmental policy, practice, and politics. During the last few years an increasing number of political ecology studies have dealt with gender issues and have described the micropolitics of households and communities as well as their linkages to political economies at national and international levels. Likewise feminist scholars have addressed environmental issues within both social theory and case studies. Both groups, in somewhat different tones and contexts, have documented women's social movements, gendered experiences of the environment, and gendered interests in resource management.

This volume constitutes an explicit effort to join feminist and political ecology scholarship from the ground up, based on case studies compiled from collaborative field research or direct participation in the events recounted. We specifically sought to obtain global coverage (Figure I.1) and to transcend the familiar dichotomies between rural and urban environments and industrial and agrarian settings. Based on our own prior work and formal training we identified three themes that encapsulated gender differences in environmental issues: gendered knowledge; gendered rights over resources and environmental quality; and gendered organizations.

Originally we had hoped for an even mix of urban and industrial with rural and agrarian cases in each major geographical region. However, the majority of the European and North American examples already documented and available to us were either urban or industrial or both, while most of the cases studies from Africa, Asia, and Latin America were rural and/or agrarian. This distribution mirrors the current state of research rather than our own perception of the locus and nature of problems by region. The gendered urban and industrial environments of Africa, Asia, and Latin America and the gender issues in rural, resource-based conflicts in Europe and North America constitute important sites for future work in feminist political ecology. Likewise, the eleven case studies presented here did not all treat all three of our themes evenly, but rather emphasized one or two of these topics. We have grouped the cases accordingly by theme, but acknowledge that most, if not all of these stories treat more than one theme.

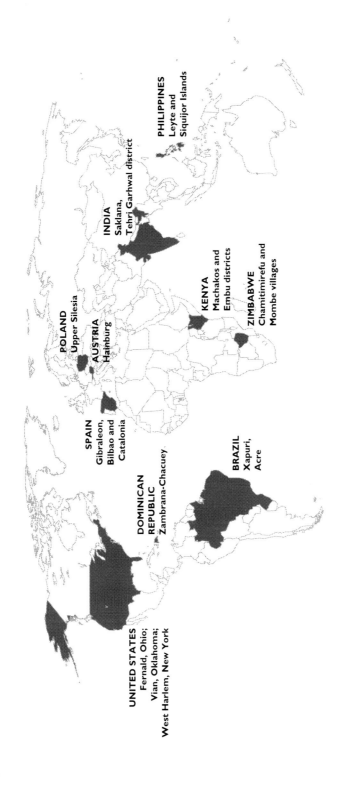

Figure 1.1 Countries and locations represented in case studies

Our overview chapter offers a selective review of the political ecology, feminist and related literatures as well as a conceptual framework for viewing the case studies which follow. We have focused our analysis on gendered knowlege, resource rights and organizations as themes that we have found to be important in our own work. This framework is supported and illustrated by the theoretical, policy, and practical work of authors across several disciplines.

The first group of case studies introduces five very distinct examples of gendered environmental organizations and social movements, with varying degrees of freedom and control for women. The first case study, Chapter 2, presents the experience of the women in rubber tapper communities in the Brazilian Amazon region. Connie Campbell and the Women's Group of Xapuri provide a very different perspective on the struggle of women to gain an equal voice within a well-known forest protection movement already widely documented as a political ecology case study. Chapter 3 chronicles the resistance of the West Harlem community to a major sewage treatment plant, among many other environmental insults. This story illustrates the interaction of race and gender in the course of an African American neighborhood's struggle against environmental racism in New York City. In both of these cases women have risen from the ranks of broad social movements connected to churches and party politics (rubber tappers connected to a liberation theology movement and civil rights activists grounded in African American and liberal protestant churches). In contrast, Chapter 4 on Spain and Chapter 5 on Austria document the dramatic entry of women into local and national politics through site-specific, environmentally focused campaigns that lead them into more formal political arenas. While women in Spain organized a primarily women's movement against toxic waste dumping in their ethnic minority regions, the Austrian case illustrates the combined mobilization of women and men in a resource-based conflict that changed the gender balance and the environmental tone of Austrian national politics.

The second group of case studies focuses on gendered access to resources and control over environmental decisions and technologies in agrarian communities. In each case the gender division of labor and resources is changing in response to commercialization of agricultural production and increasing participation of men in wage labor. Chapter 6 presents the experiences of women in three semi-arid farming communities in Kenya, with an emphasis on the inequitable distribution of land and labor and women farmers' individual and collective responses to chronic poverty and land privatization as well as periodic drought and famine. Chapter 7 discusses the relation between migration, land use change and gendered management of land, crops, trees, and coastal resources, with particular attention to the role of gendered "social capital" within rural communities in the Philippines. The final case study in this section (Chapter 8) presents the agrarian side of the oft heard story of women in the Himalayan foothills of India. While deforestation and erosion do affect women, Manjari Mehta brings into focus the unfavorable terms of on-farm resource control for women under conditions of rapid commercialization and technology change in agriculture.

The third set of case studies addresses two dimensions of gendered knowledge: the process and application of feminist political ecology research and the gender division of knowledge and authority in environmental science and management. The study from Zimbabwe (Chapter 9) treats both issues, by discussing the process and the results of a field study of gendered knowledge and use of land and trees in two rural communities. Louise Fortmann chronicles her experience and her reflections on the ethics of participatory research on gender and environment. Chapter 10 presents a practical application of feminist political ecology to the evaluation and suggested reformulation of a popular reforestation program in the Dominican Republic, based on an analysis of gendered knowledge, work, space, and organizational affiliation. The story of the "Tested Food for Silesia" program (Chapter 11) illustrates the importance of gendered domains of knowledge, both real and popularly perceived. Women's professional duties as environmental engineers and their responsibilities for household food supply and health provided them with both evidence and motivation to organize a movement for a safe food supply. Yet, their popular credibility derived almost solely from the public perception of women's role as mothers. In the case of two rural communities in the midwestern United States (Chapter 12), women mobilized first as housewives and mothers to respond to immediate health threats from toxic waste disposal. They became involved in scientific analysis out of self-defense and were subjected to scorn as practitioners of "housewive's epidemiology." In each of these cases the uneven gender relations of power play a part in the recognition, acquisition and use of environmental knowledge.

The summary and overview (Chapter 13) returns to our original themes and restructures many of our questions in response to the insights from the eleven case studies. Pre-eminent in our review of the case studies is the role of women's organizations and women's groups within broader social movements. Our concluding chapter raises new questions for feminist political ecology and challenges us to build both theory and practice to incorporate the concerns as well as the insights of the people whose stories appear in this volume.

Part I

CONCEPTUAL OVERVIEW

GENDER AND ENVIRONMENT

A feminist political ecology perspective

Dianne Rocheleau, Barbara Thomas-Slayter, and
Esther Wangari

The convergence of interest in environment, gender, and development has emerged under conditions of rapid restructuring of economies, ecologies, cultures, and polities from global to local levels. Global economic, political, and environmental changes have affected both men and women as stakeholders and actors in resource use and allocation, environmental management, and the creation of environmental norms of health and well-being. Some scholars and activists see no gender differences in the ways human beings relate to the environment, except as they are affected by the constraints imposed by inequitable political and economic structures. Others see the gendered experience of environment as a major difference rooted in biology. We suggest that there are *real*, not imagined, gender differences in experiences of, responsibilities for, and interests in "nature" and environments, but that these differences are not rooted in biology per se. Rather, they derive from the social interpretation of biology and social constructs of gender, which vary by culture, class, race, and place and are subject to individual and social change.

In this volume, we explore the significance of these differences and the ways in which various movements, scholars, and institutions have dealt with gendered perspectives on environmental problems, concerns, and solutions. The major schools of feminist scholarship and activism on the environment can be described as:

1 ecofeminist;
2 feminist environmentalist;
3 socialist feminist;
4 feminist poststructuralist; and
5 environmentalist.

Ecofeminists posit a close connection between women and nature based on a shared history of oppression by patriarchal institutions and dominant Western culture, as well as a positive identification by women with nature. Some ecofeminists attribute this connection to intrinsic biological attributes (an essentialist position), while others see the women/nature affinity as a social construct to be embraced and fostered (Plumwood 1993; Merchant 1981, 1989; King 1989; Shiva 1989; Mies and Shiva 1994; Rocheleau 1995). Feminist environmentalism as articulated by Bina Agarwal (1991) emphasizes gendered interests in particular resources and ecological processes on the basis of materially distinct daily work and responsibilities (Seager 1993; Hynes 1989). Socialist feminists have focused on the incorporation of gender into political economy, using concepts of production and reproduction to delineate men's and women's roles in economic systems. They identify both women and environment with reproductive roles in economies of uneven development (Deere and De Leon 1987; Sen and Grown 1987; Sen 1994) and take issue with ecofeminists over biologically based portrayals of women as nurturers (Jackson 1993a and b). Feminist poststructuralists explain gendered experience of environment as a manifestation of situated knowledges that are shaped by many dimensions of identity and difference, including gender, race, class, ethnicity, and age, among others (Haraway 1991; Harding 1986; Mohanty 1991). This perspective is informed by feminist critiques of science (Haraway 1989; Harding 1991) as well as poststructural critiques of development (Escobar 1995; Sachs 1992) and embraces complexity to clarify the relation between gender, environment, and development. Finally, many environmentalists have begun to deal with gender within a liberal feminist perspective to treat women as both participants and partners in environmental protection and conservation programs (Bramble 1992; Bath 1995).

We draw on these views of gender and environment to elaborate a new conceptual framework, which we call feminist political ecology. It links some of the insights of feminist cultural ecology (Fortmann 1988; Hoskins 1988; Rocheleau 1988a and b; Leach 1994; Croll and Parkin 1993) and political ecology (Schmink and Wood 1987, 1992; Thrupp 1989; Carney 1993; Peet and Watts 1993; Blaikie and Brookfield 1987; Schroeder 1993; Jarosz 1993; Pulido 1991; Bruce, Fortmann and Nhira 1993) with those of feminist geography (Fitzsimmons 1986; Pratt and Hanson 1994; Hartmann 1994; Katz and Monk 1993a and b; Momsen 1993a and b; Townsend 1995) and feminist political economy (Stamp 1989; Agarwal 1995; Arizpe 1993a and b; Thomas-Slayter 1992; Joekes 1995; Jackson 1985, 1995; Mackenzie 1995). This approach begins with the concern of the political ecologists who emphasize decision-making processes and the social, political, and economic context that shapes environmental policies and practices. Political ecologists have focused largely on the uneven distribution of access to and control over resources on the basis of class and ethnicity (Peet and Watts 1993). Feminist political ecology treats gender as a critical variable in shaping resource access and control, interacting with class, caste, race, culture, and ethnicity to shape processes of ecological change, the struggle of men and women to sustain ecologically viable livelihoods, and the prospects of any community for "sustainable development."

The analytical framework presented here brings a feminist perspective to political ecology. It seeks to understand and interpret local experience in the context of global processes of environmental and economic change. We begin by joining three critical themes. The first is *gendered knowledge* as it is reflected in an emerging "science of survival" that encompasses the creation, maintenance, and protection of healthy environments at home, at work and in regional ecosystems. Second, we consider *gendered environmental rights and responsibilities*, including property, resources, space, and all the variations of legal and customary rights that are "gendered." Our third theme is *gendered environmental politics and grassroots activism*. The recent surge in women's involvement in collective struggles over natural resource and environmental issues is contributing to a redefinition of their identities, the meaning of gender, and the nature of environmental problems.

GLOBAL PERSPECTIVES FROM LOCAL EXPERIENCE

Until recently, conventional wisdom in international environmental circles suggested that environmental issues in industrialized countries had to do with "quality of life," whereas in Africa, Asia, and Latin America they had to do with survival. If we compare the conservation agenda of wildlife organizations in the United States with the Chipko movement to protect the forests and watersheds of the lower Himalayas, or with women's tree-planting initiatives in Kenya, this view seems accurate. However, there are also wildlife conservation organizations in Africa and citizens' environmental justice movements in the United States. Toxic wastes, contaminated food, and workplace environmental hazards have become more than quality of life issues in many urban and industrial communities as well as in the remote rural areas affected by the same processes.

Perhaps it would be more accurate to recast this dichotomy along different lines, based on a careful analysis of the gender division of rights, responsibilities, and environmental risk in everyday life. While there are several axes of power that may define peoples' access to resources, their control over their workplace and home environments and their definitions of a healthy environment, we focus on gender as one axis of identity and difference that warrants attention. Feminist political ecology deals with the complex context in which gender interacts with class, race, culture and national identity to shape our experience of and interests in "the environment."

Our approach to feminist political ecology examines the very definition of "environment" and the gendered discourse of environmental science, environmental rights and resources, and environmental movements, using feminist critiques of science (Hynes 1989, 1991, 1992; Shiva 1989; Mies and Shiva 1994; Merchant 1982, 1989; Keller 1984; Griffin 1987; Birke and Hubbard 1995; Haraway 1989, 1991; Harding 1986, 1987; Tuana 1989; Hubbard 1990; Zita 1989) as well as the analyses and actions of feminist and environmental movements. For example, Sandra Harding (1986) has raised issues of gender inequities in science as a profession, gender biases and abuses in the practice of science, the myth of gender-neutral objectivity, gendered metaphors employed in scientific explanation and process, and the possibilities for a transformed, socially just science. Donna Haraway (1991) discusses the need to recognize and combine situated knowledges and invokes the "power of partial perspective" as a pathway toward greater objectivity. She advocates a pursuit of scientific knowledge that joins many knowers on the basis of affinities (reaching beyond identities) to build a joint, expanded understanding as part of an explicitly social project.

We also build on the work of socialist feminist scholars such as Nancy Fraser (1987), who has focused on the political discourse of needs and services in social welfare programs in the United States, and Patricia Stamp (1989) who addresses the gendered discourse of "donors and recipients" in international development. We extend their analyses to examine the impact of gender on environmental discourse and its differential effects on women and men (Merchant 1992; Hynes 1989, 1992; Plumwood 1993; Haraway 1991; Harding 1991).

The overview and case studies in this volume draw upon the experience of grassroots environmental movements worldwide, including such diverse situations as the struggle to save old growth forests in Europe, women's initiatives to secure safe food supplies in the industrial core of Poland, community efforts in the United States and Spain to fight toxic waste dumping, women's movements to retain access to land and forest resources in Kenya, and women's participation in the struggles of the rubber tappers' union to protect their forest homes and workplaces in the Brazilian Amazon. Less visible, more diffuse gendered struggles occur at household and community levels in the case study examples from Zimbabwe, the Dominican Republic, the Philippines, and India. The experience of all of these disparate groups provides distinct examples of gendered science, rights, and political organization.

Reviewing these cases we find common threads of concern over:

- survival;
- the rights to live and work in a healthy environment;
- the responsibility to protect habitats, livelihoods, and systems of life support from contamination, depletion (extraction), and destruction; and
- the determination to restore or rehabilitate what has already been harmed.

These common threads surface repeatedly within our varied case studies, which range from urban neighborhoods to arid farmlands to dense rainforests. The commonalities and differences in the relation of gender and environment in these cases both contribute to and challenge prevailing theories and serve to inform policy and practice for environment, development, and women's programs and movements.

THREE THEMES COMMON TO GENDER AND ENVIRONMENT WORLDWIDE

Environmental science and "the international environmental movement" have been largely cast as the domain of men. In fact, while the dominant and most visible structures of both science and environmentalism may indeed be dominated by men, mostly from the wealthier nations, the women of the world – and many men and children with them – have been hard at work maintaining and developing a multiplicity of environmental sciences as well as grassroots environmental movements. And while it is the same few who may lay claim to pieces of the living landscape as private and state property throughout the world, women and many men and children have also been busy maintaining and developing their own places on the planet through the daily management of the living landscape.

The case studies in this volume address the intersection of gender and environment through the lens of three themes: gendered science, gendered rights (over both property and the resource management process) and gendered organizations and political activity. Specific places are treated as culturally and ecologically distinct, but with many shared problems and concerns related to gender and environment in both global and local contexts.

THEME 1: GENDERED SCIENCES OF SURVIVAL

Gendered science can be viewed in terms of the definition of what is science and who does it, in terms of the different possibilities for defining the relation of people and "nature," and in terms of the apparently separate sciences and technologies of production and reproduction, public and private domains, and home, habitat, and workplace spaces. Through the stories of communities involved in a wide range of political and environmental struggles we examine the gender implications of the separation of work and knowledge, science and practice for the gendered science of survival in rural as well as industrial contexts. The case studies presented here illustrate the intersection of rural "local knowledge" with urban and suburban "housewives' epidemiology" and link the gendered knowledge of everyday life in urban and rural, and "north" and "south" contexts.

Our exploration of the convergence of gender, science, and "environment" is informed by several sources, including feminist scholarship, environmental science and policy literature, alternative environmental and development scholarship, women's movements, environmental movements, and alternative "development" movements (including "appropriate technology"). We rely heavily, but not exclusively, on the literature and experience of the last twenty years.

In North America and Europe, feminist health movements and the "housewives" environmentalist and anti-toxics movements have questioned the prevailing paradigm of professional science. They use women's experience to challenge the professionalized definitions of "environment" and ecology, and offer their own alternative perspective on environmental issues related to personal health and the home. Many feminists among the "deep ecologists," social ecologists, and "biocentric" environmentalists have also developed a distinct critique of mainstream environmental science and resource management, with a strong emphasis on the identification of women with nature and the mistreatment of both by a male-dominated, instrumentalist science (Plumwood 1993; Biehl 1991; Merchant 1992). Many advocates of these approaches have been labeled or have begun to call themselves ecofeminists. We suggest that feminist political ecology encompasses much of ecofeminism as well as several related approaches that would not fit that label as currently used.

Many rural women from around the world have also begun to raise their voices internationally to speak of a science of survival largely in the hands of women. Several rural women's movements to protect forests, trees, and water resources in Asia, Africa and Latin America have recently received global recognition and women scholars have in several cases become leaders, advocates and allies of such popular movements (Shiva 1989; Agarwal 1991; Maathai 1989; Seager 1993).

Plate 1.1 Linking environment with health: march in Boston for
Breast Cancer Awareness Month

Source: Lisa Beane

Several common threads have run throughout the scholarship and the movements that address the convergence of gender, science, and environment, but common concerns have often been obscured by the distinct discourses of resistance, critique, and alternative practice. We draw the following points into a common perspective and the authors pursue each of them in the case studies, as appropriate:

1 Women's multiple roles as producers, reproducers, and "consumers" have required women to develop and maintain their integrative abilities to deal with complex systems of household, community, and landscape and have often brought them into conflict with specialized sciences that focus on only one of these domains. The conflict revolves around the separation of domains of knowledge, as well as the separation of knowing and doing, and of "formal" and "informal" knowledge.

2 While women throughout the world under various political and economic systems are to some extent involved in commercial activities (Berry 1989; Jackson 1985), they are often responsible for providing or managing the fundamental necessities of daily life (food, water, fuel, clothing) and are most often those charged with healthcare, cleaning, and childcare in the home, if not at the community level (Moser 1989). This responsibility puts women in a position to oppose threats to health, life, and vital subsistence resources, regardless of economic incentives, and to view environmental issues from the perspective of the home, as well as that of personal and family health.

This does not preclude women from engaging in economic interests, but suggests that they will almost always be influenced by responsibilities for home, health – and in many cases – basic subsistence.

3 Both health and ecology are amenable to feminist and alternative approaches to practice since they do not necessarily require special instrumentation, but rather focus on the "objects" and experience of everyday life, much of which is available through direct observation (Levins 1989). While some aspects of health and ecology have become highly technical, there is much new insight and information to contribute to these disciplines that is still available to observation without specialized instruments beyond the reach of ordinary folk. There is also scope for a feminist practice of ecology that uses specialized tools differently and for different ends.

4 While formal science relies heavily on fragmentation, replication, abstraction, and quantification (Levins 1989), many women have cited the importance of integration and a more holistic approach to environmental and health issues (Candib 1995). Feminist scholars have shown that some women researchers in professional sciences have used distinct approaches based on skills acquired in their socialization as women (Keller 1984; Hynes 1989, 1991, 1992). On a more personal and everyday level, some grassroots women's groups have explicitly stated that "our first environment is our bodies" (Gita Sen, personal communication), calling for a more integrative approach to health, environment, and family planning in development, welfare, and environmental programs.

5 Most feminist or women's environmental movements have incorporated some or all of the elements of the feminist critique of science as summarized by Sandra Harding (1987). The five classes of critique address:

1 inequity of participation and power in science-as-usual;
2 abuse and misuse of science on and about women;
3 assumptions of value-free objectivity and universality in science;
4 use of culturally embedded, gendered metaphors in scientific explanation and interpretation; and
5 development of alternative ways of knowing and ways of learning based on everyday life, women's experience, and explicit statement of values.

Feminist political ecology addresses the convergence of gender, science and environment in academic and political discourse as well as in everyday life and in the social movements that have brought new focus to this issue. In this volume, we explore the critiques of gendered environmental science, as well as the alternative practices of science both within and beyond the current dominant paradigm. And finally, we examine the gendered sciences of survival in a wide range of circumstances, from production systems to responsibilities for health and hygiene.

These sciences occur in several forms, from local environmental knowledge (for example, which plants can cure us and how we can protect them), to recent innovations (new techniques to manage soil, water, and trees; new ways to diagnose exposure to toxic chemicals), to research on the unknown (what is making us sick; or how we can maintain our forest plants in a changing landscape). These various sciences are practiced by diverse groups from rural herbalists and forest farmers to suburban residents, professional nurses, environmental engineers and urban residents and factory workers. While there are many other axes of difference that may shape peoples' experience and understanding of "environment" and their sciences of ecology, feminist political ecology focuses on gender, while including discussions of interactions with class, race, age, ethnicity, and nationality.

THEME 2: GENDERED ENVIRONMENTAL RIGHTS AND RESPONSIBILITIES

Who controls and determines rights over resources, quality of environment, and the definition of a healthy and desirable environment? The question is crucial to the overall debate on gender and environmental rights. Ecofeminism and other feminist critiques of environmental management paradigms have raised questions of gender, power, and paradigms of economic development (Merchant 1981; Hynes 1992; Seager 1990; Shiva 1989), while many feminist critiques of development have focused on access to and control over resources (Agarwal 1991; Deere 1992; Deere and De Leon 1985; Pala Okeyo 1980; Muntemba 1982; Wangari 1991). Although gendered resource tenure has been discussed primarily in the context of rural development, and gendered power over environmental quality has been treated more in terms of urban, industrial sites, the cases in this volume apply and synthesize both approaches in rural and urban contexts across regions.

We recognize gendered environmental rights of control and access as well as responsibilities to procure and manage resources for the household and the community. These rights and responsibilities may apply to productive resources (land, water, trees, animals) or to the quality of the environment. In addition to the gender division of resources, there is a gender division of power to preserve, protect, change, construct, rehabilitate, and restore environments and to regulate the actions of others.

These categories reflect women's and men's often distinct rights and responsibilities in production (subsistence and commercial), their rights and responsibilities to create or maintain a healthy biophysical environment (including chemical aspects), and their rights and responsibilities to determine the quality of life and the nature of the environment. In more abstract terms, we can speak of gendered mandates and terms of control over things, processes, the direction and impact of environmental changes, and over the distribution of those impacts. The rights to control one's own labor and to regulate the actions of others are also highly gendered.

Environmental rights and responsibilities are also gendered spatially. For example, men's or women's domains of access and control are often divided between public and private places, and between home and workplace spaces. Likewise we find gendered spatial categories in different kinds of homes and workplaces, in a continuum of spaces from homestead to cropland; office to factory; suburb to city; indoor to outdoor; and neighborhood to region.

While the specific designation of gendered spaces and the strength and visibility of those divisions may vary dramatically by culture, the existence of gendered spaces is widespread and affects both technocratic and customary systems of resource tenure and control of environmental quality.

Resource tenure

Gendered resource tenure encompasses both rights and responsibilities and can be divided into four distinct domains:

1 control of resources as currently defined;
2 access to resources (de facto and de jure rights; exclusive and shared rights; primary and secondary rights);
3 gendered use of resources (as inputs, products, assets; for subsistence and commercial purposes); and
4 gendered responsibilities to procure and/or manage resources for family and community use.

The recent literature on gendered resource rights in development studies has tended to focus on ownership and use rights in land, trees, water, wildlife, and other rural resources (Hoskins 1982; Fortmann and Bruce 1988; Fortmann 1985; Rocheleau 1988a and b; Bradley 1991; Deere and De Leon 1985; Davison 1988; Carney 1988; Watts 1988; Berry 1989; Peters 1986; Bruce, Fortmann, and Nhira 1993; Leach 1994; Rocheleau and Ross 1995; Schroeder 1993; Jarosz 1993). These resources are often contested, with multiple claimants at different levels: men and women; households of distinct classes; different communities; distinct ethnic groups; and local, national, and international users.

The very notions of property and resources, so often assumed to be fixed, are both variable between groups and places and dynamic in time. Resource values and claims upon them change with human needs, abilities, knowledge, and skills (Rees 1990; Omara-Ojungu 1992) as well as relations of power, based on gender, race, class, ethnicity, locality, and nationality. For example, the land tenure reform in Kenya initiated by the colonial government and later implemented by the newly independent state, excluded women from resources previously available to them through customary rights of use and access. While creating new resource values and property rights for some men, the privatization of land has led to destruction of forests, grasslands, water sources, and soils and the termination of women's access to many areas (Wangari 1991; Pala Okeyo 1980). In Eastern Europe and the Balkans the land tenure reforms spurred by political and economic change have in many cases returned control of rural farmland to traditional patriarchs and male heads of household (LaStarria-Cornhiel 1995).

Similarly, in the Gambia, a land tenure reform and irrigation project specifically intended to benefit women resulted in a redefinition of traditional land and labor rights and destroyed the women's traditional floodplain fields. The seasonal and spatial complementarity of men's and women's cropping systems also broke down, resulting in serious conflicts at the household and community levels over both land and labor resources (Carney 1988;

Watts 1988). Similarly, changes in industrial technology in North America, coupled with simple definitions of land as property, have pitted the value of waste disposal sites for industry (men's domain) against use values of nearby residential property and against the public health of surrounding communities (women's domain).

Types of rights, types of uses, types of resources

The legal standing of resource tenure as well as the kind of tenure tend to reflect gendered relations of power. Environmental rights, especially resource rights, may be either de jure (legal by court precedent or statutory law) or de facto (by practice/custom). Men are often associated with de jure and women with de facto resource rights, which has major implications for the relative strength and security of tenure by gender. In many cases, particularly in Africa and parts of Asia, simultaneous systems of customary and statutory law have exaggerated and distorted the customary gender division of resources. This is particularly true where the customary law of family and marriage is applied to women's claims to household or community resources or environmental rights, while men's claims are settled under "Western" or statutory codes. The ways in which customary rights are distributed are also gendered, though inheritance and marriage laws vary tremendously from one place to another and are constantly changed and renegotiated over time (Mackenzie 1995).

The types of legal and customary rights can also be divided into ownership versus use rights. Rights of exclusive ownership often coincide with dominance by gender as well as class: wealthy men are often owners, while women or poor men are more likely to be users of lands/resources owned by others. Shared uses and multiple user practices are often beyond the legal definitions of property currently recognized, including formal definitions of "common property." The concept of articulated bundles of rights (Fortmann 1985; Riddell 1985; Bruce 1989) provides one tenure framework that lends itself well to gender issues and to rights that pertain to resources and environment, although it has been developed primarily in the context of forestry and rural development. Many forms of customary law incorporate such nested and overlapping rights, while modern legal codes usually do not.

The division between customary rights of control versus rights of use and access has a similar relation to gender (Rocheleau 1988a and b). In many cultures, elder men share authority to allocate resources among themselves and to women and younger men. They exercise control and allocate use rights. Overall, women's rights are often nested within rights controlled by men, or women hold rights to resources that are allocated by men's institutions or organizations (clans, lineages, cooperatives, political committees). This applies equally to "Western" or "Northern" countries but rules are indirectly encoded in the daily practices of political and economic institutions and the disposition of private property rather than explicitly articulated as a gendered legal code. For example, women may encounter difficulty in obtaining credit and home mortgages in their own names or may only be able to receive retirement benefits under their husband's names. At the community level, women may be less likely than men to get elected to planning and zoning boards.

The types of uses enjoyed by men and women also vary. Women often have rights of renewable use (plant crops on soil; harvest leaves from trees; gather dead wood), while men have rights of consumptive use (harvest whole trees; buy and sell land; divert and consume irrigation water). A related question is: rights over what? Men and women may divide use rights or control by the type of resource: land, water, specific animals, plants or their products. These resource categories may also embody a division between resources for use value and resources as commodities.

Responsibilities

Parallel to the gender division of resource rights is an equally important division of responsibilities. They are expressed most concretely at household and community levels, although they may also apply to larger scales of social organization. The most common forms of gendered responsibility for resources include:

1 responsibility to procure particular inputs or products for home use (such as fuelwood, water, milk, and medicinal herbs in rural areas; or bottled water, air filters, pest traps, or disinfectants in urban areas); and
2 responsibility to manage particular resources (such as protection of water sources, maintenance of community forests, and soil conservation in rural areas; or food shopping and meal planning, protection of parks, restoration of neighborhood safety, and detection of home and workplace health hazards in urban and industrial settings).

The relative distribution of resource rights and responsibilities between men and women is far out of balance in many areas (FAO 1988). Women carry a disproportionate share of responsibilities for resource procurement and environmental maintenance, from New York City to the Lower Himalayas, and yet they have very limited formal rights (and limited political and economic means) to determine the future of resource availability and environmental quality. In many cases, the rights of men to extract commodities or to engage in consumptive use have pre-empted women's use of the same resource or the same place, yet women remain responsible for providing the same product or service from another source. The consequences can be serious for the women themselves as well as for the environment. The gender imbalance in environmental rights and responsibilities derives from relations of power based on gender, among other factors.

Relations of power

The relations among resource uses, users, owners, and managers may be relations of conflict, cooperation, complementarity, or coexistence, which raises the issue of power and gender. Throughout the world, as we study gender, environment, and tenure, we find that gendered power relations are expressed in very concrete ways. The case studies in this volume analyze gendered power relations and tenure under shared use situations, as well as under private, state, and formal community ownership of resources, in distinct types of environments. We focus on concrete expressions, rather than explanations of the origins of inequities. We do not promise to resolve the theoretical debates, only to apply relevant insights from the case studies. We also recognize that it is possible to work within, circumvent, ameliorate, or undo the inequities, once they are understood, and we discuss the policy implications of specific approaches within the case studies and in the summary chapter.

Environmental quality

Gendered control over quality of environment encompasses the right to protect, change, or create environmental conditions that meet existing standards of quality (especially with respect to health) and the rights to determine the nature of the environment (land use planning, land use change, structure of homes, neighborhoods, landscape design). In spite of substantial progress in our understanding of gender conflict over resource use and control, and of the link between gendered resource use and environmental change, many areas of interest remain to be explored.

Just as the insights from resource tenure in rural development contexts can inform our understanding of gendered environmental rights in urban and industrial regions, so can the gendered struggles over environmental quality in North America and Europe help us better to comprehend related issues in less industrialized regions. In urban and industrial contexts, for example, conflicts have arisen between grassroots groups, industry and government agencies concerning: rights to use public space; access to and control over clean air and water; and rights to healthy homes and workplaces. Similarly, women in rural areas have a direct stake in the control of pesticide use on commercial crops, and also in the decision to use a given area for commercial rather than subsistence production. Women have been at the forefront of many efforts to address these issues of control over environmental resources and environmental quality. In many cases, their involvement is a response to their prior exclusion from access to resources as well as from the corridors of power where environmental decisions are made in government, industry, and mainstream environmentalist groups.

THEME 3: GENDERED ENVIRONMENTAL POLITICS AND GRASSROOTS ACTIVISM

Our discussion of gendered political participation focuses on the recent wave of women's involvement in collective action for environmental change. For more than a decade, women have been at the forefront of emerging grassroots groups, social movements, and local political organizations engaged in environmental, socioeconomic, and political struggles (Merchant 1992; Seager 1993; Hynes 1992). These phenomena are not localized; they are occurring around the world. They are documented not only by scholars and professionals working in their respective fields, but also by journalists, social critics, politicians, and administrators (Agarwal 1991; Bell 1992; Brown 1991; Collins 1991; Braidotti *et al.* 1994; Dankelman and Davidson 1988; PACA 1990; Freudenberg and Steinsapir 1991; Marcus 1992; Rau 1991; Shiva 1989). We look not only at the reasons for an apparent surge in women's involvement in collective struggles over natural resource and environmental issues, but also at the various forms such activism has taken.

What difference do collective struggles make to sound environmental policy and practice and to "sustainable development?" Three working assumptions are noted below:

1 Given the involvement of women in collective action around the world, there are critical linkages between global environmental and economic processes and the recent surge in women's participation in public fora, particularly in relation to ecological and economic concerns. This surge in women's activism is a response to actual changes in local environmental conditions as well as to discursive shifts toward "sustainable development" in national and international political circles.

2 Applying Gillian Hart's analysis within the Malaysian context (Hart 1991), we transpose her conceptualization of "multiple and interconnected sites of struggle" to an international setting. Different visions of society and differing access to resources and to power are played out according to gender, race, class, ethnicity and nationality, variously connected in complex systems. Pramod Parajuli (1991) provides a similar explanation for the nature of social movements in India.

3 Women are beginning to redefine their identities, and the meaning of gender, through expressions of human agency and collective action emphasizing struggle, resistance, and cooperation. In so doing, they have also begun to redefine environmental issues to include women's knowledge, experience, and interests. While this is a worldwide phenomenon, the process and results in any one place reflect historical, social, and geographical specificity (Alvarez 1990; Egger and Majeres 1992; Friburg 1988; Fraser 1987; Touraine 1988).

Why women? Why now?

When we talk about the environment, we are referring to the ecosystem on which production and reproduction depend. The aspects of a particular ecosystem that are important to the people who live in it vary according to the circumstances of history and the specific demands of their system of production. Regardless of these variations, issues pertaining to the environment are inherently political, and decisions about the environment are not politically neutral. Access to and control of environmental resources are inextricably linked to the positioning of people by gender, race, class, and culture. Environmental issues are central to debates about the nature of the society we live in, the claims that each of us can make on that society, and the realities of justice in distribution. Five considerations are important:

1 Declining ecological and economic circumstances: The increased involvement of women in environmental struggles and in political and social

movements derives from the difficulties they face in ensuring the survival of their families in the face of ecological and economic crises. For many, these difficulties have worsened in the last decade as a result of changes in social and economic relations arising with the spread of capitalism, migration for wage labor, divided families, and the decline in various forms of vertical ties to patrons (Chen 1991; Hart 1991; Kates and Haarmann 1992). Poor households face increased environmental risk, uncertainty, and insecurity, while their entitlements are either precarious or nonexistent.

2 The impact of structural adjustment policies: To these long-term structural changes one must add the immediate implications of the structural adjustment policies of the 1980s and 1990s (Gladwin 1991) and the "retreat of the state" from support of public services, social welfare and environmental regulation in wealthy as well as poor countries. Poor women throughout the world have been severely affected by insufficient food, the rising cost of living, declining services, and eroding economic and environmental conditions. These impacts have spawned not only protest but also strategies for change.

3 Consciousness raising and political awareness: Increasingly people are linking the immediate impact of ecological and economic crises with recognition of a need for structural political changes. Organizations that may have originated from a specific objective, such as India's Chipko movement or the United States' Citizens' Clearinghouse for Hazardous Wastes, have broadened their focus to include the larger social and political systems. In some instances, environmental movements have addressed systems that depress the standard of living for the poor or that emphasize economic growth and military strength to the detriment of environmental safety and personal health.

4 The political marginality of most women: For many women, economic and ecological conditions are potentially catastrophic. They face severe constraints on their livelihood options. They participate little, if at all, in organized politics at the national level. Their activism usually begins locally on matters critical to their own lives, their homes, and their families. It reflects the pressure and distress generated by the system and its impact on family welfare, among people operating "on the edge" both economically and socially. In the last decade, the problems that women face have become increasingly severe. The system does not address their needs, and so they act collectively to secure the necessary conditions to guaranty subsistence, protect the health of their families, and the integrity of the surrounding ecosystem.

5 The role of the women's movement: The women's movement, of which the most recent wave has now been active for over twenty years, has generated international interest in women's issues and women's perspectives. It provides some philosophical moorings for women's activism, while it also derives much of its vitality from the connections between groups that focus on theory and practice, respectively. The United Nations Decade of Women from 1975 to 1985 also contributed to the growing awareness of the distinct

roles and interests of women. The emerging international women's movements have reconfigured the political landscape to address converging issues of gender, race, class, and culture and to treat women's rights as basic human rights. They provide crucial political and ideological underpinnings of support for the increased political activism of women on environmental issues.

Exploring the forms of activism

Women's emerging environmental organizations and movements have had three foci with organizational structures to suit the particular focus:

1 Policy and environmental management issues: Here organizations focus on specific policies, problems, or hazards that are harmful to individuals, households, and communities. Often they start with the intent of documenting an association between the incidence of disease or a health problem and a specific toxic dump site, pesticide spray, workplace hazard, air pollutant or contaminated water source. They may go on to significant victories in legislation and in public information about the specific issue. In the United States the leaders and membership often include significant numbers of women as well as people of color. Environmental racism has become a major topic of concern for many groups. Such organizations, however, are found the world over as people respond to the issues confronting them in daily life. In Bombay, for example, the Society for Promotion of Area Resource Centers (SPARC) is working to demand better living conditions (Bell 1992). In countries of the Caribbean including the Dominican Republic, Dominica, and St. Vincent and the Grenadines, grassroots organizations of informal-sector traders, many of them women, are springing up to claim and work for better working conditions, protection of their rights, and environmental conservation (PACA 1990: 101).

2 Access to and distribution of resources under conditions of environmental decline and resource scarcity: Around the world, local groups are organizing to share the management of resources and to increase their availability. Local-level associations enable people to respond with increased effectiveness to external changes in their environment. They help diminish risk and they create new opportunities. Organizations can provide improved access to land, labor, capital, and information. They may generate exchange opportunities. They may provide access to common property, including resources such as water, forests and communal grazing, or institutions and services such as schools and health clinics.

3 Political change and environmental sustainability: Environmental and economic impoverishment are intertwined and linked to the political structures in which they exist. Organizations may begin with the objective of economic survival, but they often come to a sharp realization of the politics of survival. The Green Belt movement in Kenya, for example, may focus on trees and the rubber tappers' union in Brazil may focus on a search for alternative forest products; but both, along with numerous organizations like

them, find that their strategic interests raise major questions about the political systems in which they operate.

These organizational foci are merely suggestive. Most organizations deal at some time or another with all three categories. Their agendas, as well as the scale of their activities, are purposefully flexible and are continually adjusted as they endeavor to meet both practical needs and strategic, long-term interests.

What difference does women's participation make for women, the environment, and society?

All of these economic and ecological struggles have important implications for the meaning of gender and for the nature of men's and women's roles. These organizations are demanding more equitable development across classes, ethnic groups, castes, gender, and generations. The increased involvement of women is leading to a sense of agency and empowerment. As a result, there are new perceptions of women's roles. Women's visions of their rights, roles, and responsibilities are changing. Increasingly, women are "finding voice" and are being aided in doing so by their participation in groups and organizations (Ronderos 1992: 81).

There are many victories to be claimed by women's environmental action groups around the world. In addition to the cases presented in this volume, we note as examples the widespread planting of trees by the Women's Green Belt movement of Kenya, the protection of a public park in downtown Nairobi by the same group, and the protection of the Himalayan forests from timber concessionaires by the Chipko movement in India. In North America, grassroots movements led by women have prevented the disposal of toxic wastes, as in the case of a landfill in Warren County, North Carolina, and they have pressed legislators and courts in California and Massachusetts to take action on air and water pollution. Recently formed bridging organizations, networks, and coalitions (such as the Women's Congress for a Healthy Planet; WEDO – Women, Environment, and Development Organization; WEDNET – Women, Environment, and Development Network; and Worldwide Network for Women) bring the concerns of these locally based movements to national and international policy fora.

These grassroots organizations, with their significant involvement of women, are stressing the value of all human beings and their rights to satisfy basic human needs, including food security and health (Escobar and Alvarez 1992). They emphasize ecological as well as economic concerns and the needs of future generations as well as those of diverse claimants on existing resources. There is a fundamentally humanitarian, egalitarian, pluralistic, and activist stance to many such organizations, although – as noted by Jackson (1993a and b) – women's organizations are not inherently environmentalist or altruistic.

The myriad of grassroots organizations, with women as well as men involved in them, have begun to blur the distinctions between public and private, productive and reproductive, home and workplace. Such organizations are helping us to reconceptualize and redefine what is political, and what is environmental, as well as what is just and equitable. In the chapters that follow, the authors review gendered political responses to ecological problems exacerbated by economic decline in households and communities around the world. They explore the way in which environmental activism and politics have entered household and community and vice versa. The case studies also document the extensive involvement of women in grassroots organizations in response to declining ecological and economic circumstances in degraded environments or to the magnitude of health and safety problems posed by the "maldevelopment" of previously healthy communities and ecosystems.

REFERENCES

Agarwal, B. (1991) "Engendering the Environment Debate: Lessons from the Indian Subcontinent," CASID (Center for the Advanced Study of International Development) Distinguished Lecture Series, Discussion Paper 8, Michigan State University.

—— (1995) *A Field of One's Own: Gender and Property in South Asia*, Cambridge: Cambridge University Press.

Alvarez, S. E. (1990) *Engendering Democracy in Brazil*, Princeton, New Jersey: Princeton University Press.

Arizpe, Lourdes (1993a) *Cultura y Cambio Global: Percepciones Sociales Sobre la Desforestacion en la Selva Lacandona*, Mexico, D.F.: Centro Regional de Investigaciones Multidisciplinarias, Universidad Nacional Autonoma de Mexico.

Arizpe, Lourdes, Stone, M.P., and Major, D. (eds.) (1993b) *Population and the Environment: Rethinking the Debate*, Boulder, Colorado: Westview Press.

Bath, Paquita (1995) Plenary Address and Comments, "Gender, Communities and Natural Resource Management," Annual Conference of the Center for Latin American Studies, University of Florida, Gainsville, Florida, March 29–April 1, 1995.

Bell, J. Kjellberg (1992) "Women, Environment and Urbanization in Third World Context: A Guide to the Literature," *Women and Environments*, Spring: 12–17.

Berry, S. (1989) "Access and Control and Use of Resources in African Agriculture," *Africa* 59, 1: 41–55.

Biehl, J. (1991) *Rethinking Ecofeminist Politics*, Boston: Southend Press.

Birke, L. and Hubbard, R. (1995) *Reinventing Biology: Respect for Life and the Creation of Knowledge*, Bloomington and Indianapolis: Indiana University Press.

Blaikie, P. and Brookfield, H. (1987) *Land Degradation and Society*, London and New York: Methuen Press.

Bradley, P. (1991) *Women, Woodfuel, and Woodlots*, New York: Macmillan.

Braidotti, R., Charkiewicz, E., Hausler, S., and Wieringa, S. (1994) *Women, the Environment and Sustainable Development: Toward a Theoretical Synthesis*, London: Zed Books.

Bramble, B. (1992) Statement presented to the Earth Summit Meeting, on behalf of The National Wildlife Federation, Rio de Janeiro.

Brown, R. (1991) "Matching Women, Environment and Development Around the World," *Women and Environments*, Winter/Spring: 37–41.

Bruce, J. (1989) *Rapid Appraisal for Resource Tenure Issues*, Rome: Food and Agriculture Organization.

Bruce, J., Fortmann, L., and Nhira, C. (1993) "Tenures in Transition, Tenures in Conflict: Examples from the Zimbabwe Social Forest," *Rural Sociology* 58, 4: 626–42.

Candib, Lucy (1995) *Family Medicine: A Feminist Perspective*. New York: Harper and Row.

Carney, J. (1988) "Struggles over Land and Crops in an Irrigated Rice Scheme: The Gambia," in J. Davison (ed.) *Women and Land Tenure in Africa*, Boulder, Colorado: Westview Press.

—— (1993) "Converting the Wetlands, Engendering the Environment: The Intersection of Gender with Agrarian Change in The Gambia," *Environmental Geography* 69, 4: 329–48.

Chen, M. Alter (1991) *Coping with Seasonality and Drought*, New Delhi: Sage Publications.

Collins, J. (1991) "Women and the Environment: Social Reproduction and Sustainable Development," in R. Gallin and A Ferguson (eds.) *The Women and International Development Annual*, Vol. 2, Boulder, Colorado: Westview Press.

Croll, E. and Parkin, D. (eds.) (1993) *Bush Base: Forest Farm: Culture, Environment, and Development*, London: Routledge.

Dankelman, I. and Davidson, J. (1988) *Women and Environment in the Third World: Alliance for the Future*, London: Earthscan Publications.

Davison, J. (ed.) (1988) *Agriculture, Women, and Land: The African Experience*, Boulder, Colorado: Westview Press.

Deere, C. (1992) "Markets, Machetes Everywhere? Understanding the Cuban Anomaly," *World Development* 20, 6: 825–6.

Deere, C. and De Leon, M. (1985) *Women in Andean Agriculture: Peasant Production and Rural Wage Employment in Columbia and Peru*, International Labor Organization (ILO).

—— (eds.) (1987) *Rural Women and State Policy: Feminist Perspectives on Latin American Agricultural Development*, Boulder, Colorado: Westview Press.

Egger, P. and Majeres, J. (1992) "Local Resource Management and Development: Strategic Dimensions of People's Participation," in D. Ghai and J. M. Vivian (eds.) *Grassroots Environmental Action: People's Participation in Sustainable Development*, London: Routledge.

Escobar, A. (1995) *Encountering Development: The Making of and Unmaking of the Third World*, Princeton: Princeton University Press.

Escobar, A. and Alvarez, S. (eds.) (1992) *The Making of Social Movements in Latin America: Identity, Strategy, and Democracy*, Boulder, Colorado: Westview Press.

FAO (Food and Agriculture Organization) (1988) *Restoring the Balance*, Rome: FAO.

Fitzsimmons, M. (1986) *A New Environmental Politics?* Los Angeles: Graduate School of Architecture and Urban Planning, University of California.

Fortmann, L. (1985) "Seasonal Dimensions of Rural Social Organizations," *The Journal of Development Studies* 21: 377–89.

—— (1988) "Predicting Natural Resource Micro-protest," *Rural Sociology* Vol 54: 357–65.

Fortmann, L. and Bruce, J. (1988) *Whose Trees? Proprietary Dimensions of Forestry*, Boulder, Colorado: Westview Press.

Fraser, N. (1987) *Unruly Practices: Power, Discourse and Gender in Contemporary Social Theory*, Minneapolis: University of Minnesota Press.

Freudenberg, N. and Steinsapir, C. (1991) "Not in our Backyards: The Grassroots Environmental Movement," *Sociology and Natural Resources* 4: 235–45.

Friburg, M. Henne (1988) "Local Mobilization and World Systems Politics," *International Social Science Journal*, No. 117: 341–360.

Gladwin, C. (ed.) (1991) *Structural Adjustment and African Women Farmers*. Gainsville: University of Florida Press.

Griffin, K. (1987) *World Hunger and World Economy: And Other Essays in Development Economics*, New York: Harnes and Lleiner.

Haraway, D. (1989) *Primate Visions: Gender, Race and Nature in the World of Modern Science*, New York: Routledge.

—— (1991) *Simians, Cyborgs and Women: The Reinvention of Nature*, London: Free Association.

Harding, S. (1986) *The Science Question in Feminism*, Ithaca, New York: Cornell University Press.

—— (1987) *Feminism and Methodology: Social Science Issues*, Bloomington: Indiana University Press.

—— (1991) *Whose Science? Whose Knowledge? Thinking from Women's Lives*, Ithaca, New York: Cornell University Press.

Hart, G. (1991) "Engendering Everyday Resistance: Gender, Patronage, and Production Politics in Rural Malaysia," *The Journal of Peasant Studies* 19, 1: 93–121.

Hartmann, B. (1994) *Reproductive Rights and Wrongs: The Global Politics of Population Control and Contraceptive Choice*, New York: Harper.

Hoskins, M. (1982) "Social Forestry in West Africa: Myths and Realities," paper presented at the annual meeting of the American Association for the Advancement of Science, Washington, D.C.

—— (1988) *Restoring the Balance*, Rome: Food and Agriculture Organization.

Hubbard, R. (1990) *The Politics of Women's Biology*, New Brunswick, New Jersey: Rutgers University Press.

Hynes, P. (1989) *The Recurring Silent Spring*, Tarrytown, New York: Pergamon Press.

—— (1991) *Reconstructing Babylon: Essays on Women and Technology*, Bloomington: Indiana University Press.

—— (1992) "The Race to Save the Planet: Will Women Lose?" *Women's Studies International Forum* 14, 5: 473–78.

Jackson, C. (1985) *Kano River Project*, West Hartford, Connecticut: Kumarian Press.

—— (1993a) "Environmentalisms and Gender Interests in the Third World," *Development and Change* 24: 649–77.

—— (1993b) "Doing What Comes Naturally? Women and Environment in Development," *World Development* 21, 12: 1947–1963.

—— (1995) "From Conjugal Contracts to Environmental Relations: Some Thoughts on Labour and Technology," *IDS Bulletin* 26, 1: 33–9.

Jarosz, L. (1993) "Defining and Explaining Tropical Deforestation: Shifting Cultivation and Population Growth in Colonial Madagascar (1896-1940)," *Economic Geography* 69, 4: 366–79.

Joekes, S. (1995) "Gender and Livelihoods in Northern Pakistan," *IDS Bulletin* 26, 1: 66–74.

Kates, R. W. and Haarmann, V. (1992) "Where the Poor Live: Are the Assumptions Correct?" *Environment* 34, 4: 4–11, 25–28.

Katz, C. and Monk, J. (1993a) *Full Circles: Geographies of Women over the Life Course*, London and New York: Routledge.

—— (1993b) "Making Connections: Space, Place and the Life Course," in C. Katz and J. Monk (eds.) *Full Circles: Geographies of Women over the Life Course*, London and New York: Routledge.

Keller, E. Fox (1984) *Reflections on Gender and Science*. New Haven, Connecticut: Yale University Press.

King, Y. (1989) "Ecofeminism," in J. Plant (ed.) *Healing the Wounds*, Philadelphia: New Society Publishers.

LaStarria-Cornhiel, S. (1995) "Impact of Privatization on Gender and Property Rights in Africa," paper prepared for the Gender and Property Rights International E-mail Conference, I.F.R.P.I., May–December.

Leach, M. (1994) *Rainforest Relations: Gender and Resource Use Among the Mende of Gola, Sierra Leone*, Washington, D.C.: Smithsonian Institution Press.

Levins, R. (1989) "Toward the Renewal of Science," paper presented at the Marxism Now/Re-thinking Marxism Conference, University of Massachusetts-Amherst, November.

Maathai, W. (1989) Statement in "The Race to Save the Planet." Video released by the Corporation for Public Broadcasting.

Mackenzie, F. (1995) " 'A Farm is Like a Child who Cannot be Left Unguarded': Gender, Land and Labour in Central Province, Kenya," *IDS Bulletin* 26, 1: 17–23.

Marcus, Frances (1992) "Medical Waste Divides Mississippi Cities," *The New York Times*, June 24, p. 6.

Merchant, C. (1981) "Earthcare: Women and the Environmental Movement," *Environment* 23, 5: 6–13, 38–40.

—— (1982) *The Death of Nature: Women, Ecology and the Scientific Revolution*, London: Wildwood.

—— (1989) *Ecological Revolutions: Nature, Gender and Science in New England*, Chapel Hill: University of North Carolina Press.

—— (1992) *Radical Ecology: The Search for a Liveable World*, New York: Routledge.

Mies, M. and Shiva, V. (1994) *Ecofeminism*, London: Zed Books.

Mohanty, C. (1991) "Under Western Eyes: Feminist Scholarship and Colonial Discourses," in C. Mohanty, A. Russo, and L. Torres (eds.) *Third World Women and the Politics of Feminism*, Bloomington: Indiana University Press.

Momsen, J. (1993a) "Gender and Environmental Perception in the Eastern Caribbean," in *The Development Process in Small Island States*, London and New York: Routledge.

—— (1993b) "Introduction," *Geoforum* 16, 2: 289.

—— (1993c) "Women, Work and the Life Course in the Rural Caribbean," in C. Katz and J. Monk (eds.) *Full Circles: Geographies of Women over the Life Course*, London and New York: Routledge.

Moser, C. (1989) "Gender Planning in the Third World: Meeting Practical and Strategic Planning Needs," *World Development* 17, 11: 1799–1825.

Muntemba, S. (1982) "Women as Food Producers and Suppliers in the Twentieth Century: the Case of Zambia," in *Another Development with Women*, Proceedings of a seminar held in Dakar, Senegal, June 1982, *Development Dialogue* 1, 2.

Omara-Ojungu, P. H. (1992) *Resource Management in Developing Countries*, New York: Halstead Press.

PACA (Policy Alternatives for the Caribbean and Latin America) (1990) *In the Shadows of the Sun: Caribbean Development Alternatives and U.S. Policy*, Boulder, Colorado: Westview Press.

Pala Okeyo, A. (1980) "Daughters of the Lakes and Rivers: Colonization and the Land Rights of Luo Women," in M. Etienne and E. Leacock (eds.) *Women and Colonization, Anthropological Perspectives*, New York: Praeger.

Parajuli, P. (1991) "Power and Knowledge in Development Discourse: New Social Movements and the State in India," *International Conflict Research*, Vol. 43, No. 127: 173–90.

Peet, R. and Watts, M. (1993) "Introduction: Development Theory and Environment in an Age of Market Triumphalism," *Economic Geography* 69: 227–53.

Peters, Pauline (1986) "Community and Intra-Household Issues in Resource Management in Botswana", in Joyce L. Moock (ed.) *Understanding Africa's Rural Households*, Boulder, Colorado: Westview Press.

Plumwood, V. (1993) *Feminism and the Mastery of Nature*, London and New York: Routledge.

Pratt, G. and Hanson, S. (1994) "Geography and the Construction of Difference," *Gender, Place and Culture* 1, 1: 5–29.

Pulido, L. (1991) "Latino Environmental Struggles in the Southwest," Ph.D. Thesis, Department of Urban and Regional Planning, University of California, Los Angeles.

Raintree, J. B. (ed.) (1985) *Land, Trees, and Tenure: Proceedings of an International Workshop on Tenure Issues in Agroforestry, May 1985*, Nairobi: International Council for Research in Agroforestry, and Madison, Wisconsin: Land Tenure Center.

Rau, B. (1991) *From Feast to Famine: Official Lines and Grassroots Remedies to Africa's Food Crisis*, London: Zed Books.

Rees, J. (1990) *Natural Resources: Allocation, Economics and Policy*, 2nd Edition, London: Routledge.

Riddell, J. (1985) "Land Tenure and Agroforestry: A Regional Overview," in J. B. Raintree (ed.) *Land, Trees, and Tenure: Proceedings of an International Workshop on Tenure Issues in Agroforestry, May 1985*, Nairobi: International Council for Research in Agroforestry, and Madison, Wisconsin: Land Tenure Center.

Rocheleau, D. (1995) "Gender and Biodiversity: A Feminist Political Ecology Perspective," *IDS Bulletin* 26, 1: 9–16.

—— (1988a) "Women, Trees and Tenure: Implications for Agroforestry Research and Development," in J. B. Raintree (ed.) *Land, Trees, and Tenure: Proceedings of an International Workshop on Tenure Issues in Agroforestry, May 1985*, Nairobi: International Council for Research in Agroforestry, and Madison, Wisconsin: Land Tenure Center.

—— (1988b) "Gender, Resource Management and the Rural Landscape: Implications for Agroforestry and Farming Systems Research," in S. Poats, M. Schmink, A. Spring, and G. Garing (eds.) *Gender Issues in Farming Systems Research and Extension*, Boulder, Colorado: Westview Press.

Rocheleau, D. and Ross, L. (1995) "Trees as Tools, Trees as Text: Struggles over Resources in Zambrana-Chacuey, Dominican Republic," *Antipode* 27: 407–28.

Ronderos, A. (1992) "Towards an Understanding of Project Impact on Gender Negotiation: Forestry, Community Organization and Women's Groups in Guanacaste, Costa Rica," M.A. Thesis, Program for International Development, Clark University, Worcester, Massachusetts.

Sachs, Wolfgang (1992) *The Development Dictionary: A Guide to Knowledge as Power*, London and New Jersey: Zed Books.

Schmink, M. and Wood, C. (eds.) (1987) *Frontier Expansion in Amazonia*, Gainesville: University Press of Florida.

—— (1992) *Contested Frontiers in Amazonia*, New York: Columbia University Press.

Schroeder, R. (1993) "Shady Practice: Gender and the Political Ecology of Resource Stabilization in Gambian Garden/Orchards," *Economic Geography* 69, 4: 349-365.

Seager, J. (1990) *Atlas Survey of the State of the Earth*, New York: Simon and Schuster.

—— (1993) *Earth Follies: Feminism, Politics, and the Environment*, London: Earthscan.

Sen, G. (1994) "Women, Poverty and Population: Issues for the Concerned Environmentalist," in L. Arizpe, M. P. Stone, and D. C. Major (eds.) *Population and Environment: Rethinking the Debate*, Boulder, Colorado: Westview Press.

Sen, G. and Grown, C. (1987) *Development, Crises, and Alternative Visions*, New York: Monthly Review Press.

Shiva, V. (1989) *Staying Alive: Women, Ecology and Development*, London: Zed Books.

Stamp, P. (1989) *Technology, Gender and Power in Africa*, Ottawa: International Development Research Center.

Thomas-Slayter, B. (1992) "Politics, Class and Gender in African Resource Management: The Case of Rural Kenya," *Economic Development and Cultural Change* 40, 4: 809–28.

Thrupp, L. A. (1989) "Legitimizing Local Knowledge: From Displacement to Empowerment for Third World People," *Agriculture and Human Values* 6, 3: 13–24.

Touraine, A. (1988) *Return of the Actor: Social Theory in Post-Industrial Society*, Minneapolis: University of Minnesota Press.

Townsend, J. (1995) *Women's Voices from the Rainforest*, London: Routledge.

Tuana, N. (1989) *Feminism and Science*, Bloomington: Indiana University Press.

Wangari, E. (1991) "Effects of Land Registration on Small-Scale Farming in Kenya: The Case of Mbeere in Embu District," Ph.D. Thesis, Department of Political Economy, New School for Social Research, New York, New York.

Watts, M. (1988) "Struggles over Land, Struggles over Meaning: Some Thoughts on Naming, Peasant Resistance and the Politics of Place," in R. Golledge, H. Coucelis, and P. Gould (eds.) *A Ground for Common Search*, Santa Barbara, California: Geographical Press.

Zita, J. (1989) "The Premenstrual Syndrome: 'Dis-easing' the Female Cycle," in N. Tuana (ed.) *Feminism and Science*, Bloomington: Indiana University Press.

Part II

GENDERED ORGANIZATIONS

2

OUT ON THE FRONT LINES BUT STILL STRUGGLING FOR VOICE

Women in the rubber tappers' defense of the forest in Xapuri, Acre, Brazil

*Connie Campbell, in collaboration with
The Women's Group of Xapuri*

INTRODUCTION

In Brazil's westernmost state of Acre, the municipality of Xapuri is home to the nationally and internationally recognized movement of the rubber tappers, Amazonian rainforest-dwelling extractive producers who struggled to defend their livelihoods under the leadership of union president Chico Mendes.[1] They have created and sustained a movement whose victories range from grassroots projects of health, education, and cooperative marketing to a national organization that has successfully fought for a different type of land reform, that is, the establishment of extractive reserves (Almeida and Menezes 1994; Allegretti 1979, 1981, 1990, 1994).[2]

When those familiar with this movement in Xapuri reflect on its most powerful images, they may picture a rubber tapper, a solitary man walking the forest trails to extract latex from the rubber trees (*Hevea brasilensis*). An image of the Xapuri union office may come to mind, where Chico and other elected leaders used a strong local base to create a national rubber tappers' organization. Or they may picture an *empate* or standoff, a forest demonstration in which a group of rubber tappers faced ranchers' hired hands and armed police in an attempt to prevent clearing of the forest.

Only in the latter of these images would it be easy to spot a woman. Women have been crucial in the defense of the forest; many women were on the front lines at most *empates*, standing between the chain saws and their forest homes. Women work in the forest, tapping rubber and gathering Brazil nuts.[3] At home, women are responsible for a myriad of domestic activities ranging from childcare to fence building. In the rubber tappers' social movement, women have played critical roles as union members, elected leaders, teachers, church organizers, and rural health agents.

However, the important and varied roles that women have played in the movement have gone largely unnoticed by researchers, environmental advocates and even by the movement and the women themselves to a large degree. Historically, women's social contributions have been recognized by the movement's leaders, but women have remained mostly in the background of the economic and political arenas. This chapter seeks to redress this invisibility by examining women's participation in the movement and exploring how they envision their future.

As Helen Safa argues, Latin American women participate in such movements as an expression of their roles as wives and mothers "redefining and transforming their domestic role from one of private nurturing to one of collective, public protest, in this way challenging the traditional seclusion of women in the private sphere of the family" (Safa 1995: 228). Along with others in various social movements, the women in the Xapuri movement are slowly changing their roles and gaining a stronger voice in their homes, the union office and even in national and international gatherings. As extractive producers, agriculturalists, homemakers, and community leaders, they are actively engaged in moving their community "from protest to production" (Bray 1991: 126), a process that will in large part determine the success of the extractive reserves. Yet without recognizing women's past and present contributions, their future potential to strengthen the movement may not be fully realized.

An important step in the process of recognizing these women and their contributions involves defining their gendered identity. In this chapter, I use the term *mulher seringueira*, or woman rubber tapper, since their identity is distinct from that of their male counterparts, who are traditionally referred to by the generic term of *seringueiro*.[4] The words and experiences of the *mulheres seringueiras* in this chapter convey their identity, dynamism and individuality and dispel the historically static image of the women in this community.[5] To aid in this conveyance, the words and experiences of five women – Sebastiana, Cecilia, Maria, Graça, and Raimunda – will be highlighted throughout this chapter. The identity and image of the *mulheres seringueiras* is crucial in understanding the importance of gender-specific interests and actions in the Xapuri movement and in the establishment and management of the extractive reserves.[6]

The literature on extractive reserves (Allegretti 1990, 1994; Browder 1992; Salafsky *et al.* 1993; Schwartzman 1989; Sobrinho 1992) fails to address fully the heterogeneity of extractive populations, households, and survival strategies which determine access and control over the very resources which are supposed to guarantee the success of the reserves. Field researchers are beginning to document the social organization and gender division of labor in extractive reserves (Almeida 1992; Almeida and Menezes 1994; Campbell 1990). Some have focused on women's roles in production (Campbell 1994a and b; Hecht 1992; Kainer and Duryea 1992) while others have explored women's social and historical roles (Campbell 1990; de Oliveira *et al.* 1994; Simonian 1988; UNIFEM 1990).

The next section of this chapter relates the Xapuri movement to the broader context of the extractive reserves, followed by an examination of social organization in the reserves. The focus then shifts to an exploration of the identity and history of the women in the movement, from participation in the *empates* to their roles in the union and other organizations. The

Plate 2.1 A *mulher seringheira* tapping a rubber tree in Xapuri, Acre, Brazil

Source: Karen Kainer

following section looks at women's present-day roles and their opportunities for more active political and economic participation. Finally, I present these women's visions of their future, as the movement struggles to combine natural resource conservation with economic and social stability.

EXTRACTIVE RESERVES AND THE MOVEMENT IN XAPURI

The rubber tappers' movement in Xapuri and the eventual establishment of extractive reserves was a response to the federal government's Amazon development policy of the 1970s and 1980s. Federal and state governments promoted the expansion of capital-intensive agricultural enterprises in the Amazon, most of which consisted of cattle ranches that enjoyed significant benefits in the form of tax write-offs and reduced-interest financing and which required clearing of large tracts of forest. As the ranchers claimed the old rubber estates, or *seringais*, in the state of Acre, their gunmen and chainsaws drove rubber tapper families from the area. An estimated 10,000 rubber tappers and their families fled to Bolivia, while others took up residence on the periphery of Rio Branco, the capital of Acre. While the BR-364 highway was being paved through the neighboring state of Rondônia on its way to Acre, bringing widespread deforestation and social conflict, the rubber tappers in Xapuri organized themselves to defend the forest and secure their livelihoods. Chico Mendes did not oppose construction of the road, but he organized the rubber tappers to protect their claim to forested areas which lay within the area influenced by the BR-364.

The rubber tappers' most effective tool in defense of the forest was the *empate*. In these demonstrations, women, men, and children would gather at the site of a forest clearing. Unarmed, they approached the ranchers' hired laborers, encouraging them to stop the deforestation by appealing to class interests. The ranchers responded with increasingly violent measures and the rubber tappers were often met by armed police who had been called to the area. In crucial *empates*, women and children formed the front line of defense, placing themselves between the forest and the ranchers' chain saws.

In their struggle for social justice, the rubber tappers sought to avoid eviction and defend their livelihood. They shared a common goal with national and international environmentalists – that of saving the Amazon rainforest. Alliances with key individuals and organizations in the environmental arena strengthened the Xapuri movement. With assistance from various nongovernmental organizations and prominent supporters, the Xapuri rubber tappers worked together with extractive producers throughout the Amazon basin to hold a national meeting in 1985 in Brasília. During that meeting, they created the National Rubber Tappers' Council (*Conselho Nacional dos Seringueiros* – CNS) and put forth the proposal to create a new type of land reform – the establishment of extractive reserves.

The rubber tappers' movement in Xapuri achieved major victories with the establishment of the first extractive settlements in 1987 and the later creation of extractive reserves in 1990. The concept was based loosely on that of "indigenous" reserves, in which peoples recognized by the state as "traditional" are guaranteed usufruct rights on federally protected land. The reserves represented an alternative to other conservation or development initiatives at the federal level because they grew out of a grassroots movement, were based on collective use and did not require division of the land into smaller, individual plots.

The defense of the forest was only one priority for the movement in Xapuri. The movement sought to protect and improve the lives of the forest-dwelling rubber tappers by establishing political and economic autonomy. Through the creation of a rural workers' union, construction of schools and clinics, and establishment of the region's first agro-extractive cooperative, the movement has been able to realize significant life changes for some participants. Families in the area served by the union and the cooperative can sell their products freely to the cooperative or to other traders. These families are released from the conditions of debt-peonage resulting from the patron–client system which was established during the rubber boom in Amazonia. Together with nongovernmental organizations, the movement has improved educational, health, marketing, and employment opportunities in many areas. These achievements are even more significant because of the very low levels of literacy, the sheer physical distances between families in the reserves and their isolation from urban markets. Nevertheless, the rubber tappers in Xapuri still face many challenges, one of which is to recognize and strengthen the roles and distinct interests of women in the movement.

MULHER SERINGUEIRA

Today in the popular press and international environmental circles the image of the *seringueiro* is that of a solitary man. He is visualized leaving home before daybreak to walk the forest's trails, tapping the rubber trees to gather latex. This defender of the forest gathers latex and Brazil nuts, practices slash-and-burn agriculture, and hunts game as his people have done for generations. This chapter examines this picture and explores why we rarely see another image – that of the *mulher seringueira*, or woman rubber tapper. What distinct roles have they played in the movement to defend the rain-forest? How strong are their social, political, and economic voices in the household, in the union hall, in town, in the cooperative, and in the church? How will they integrate their gendered interests and experiences into the struggle to make extractive reserves a viable alternative to other forms of forest use?

The political economy of the household in the rubber tapping culture is fundamental to understanding women's roles in the movement. During the rubber boom of the late 1800s through the 1920s, women were scarce in the *seringal*, or rubber tapping country. Especially in the upper reaches of the Amazon basin, the male *seringueiro* led a very solitary life in the forest (da Cunha 1986). The few women who did live in the *seringal* were the wives of the political and economic elites (Tocantins 1979). Those controlling the latex extraction industry preferred to recruit single males from northeastern Brazil as laborers. Very few *seringueiros* brought, or were allowed to bring, their wives and families with them (Reis 1953). As one observer put it, "[T]he *seringueiro* was a machine to produce rubber. No women dared accompany him to the desolate forest. Women would be dangerous for social stability . . . objects so rare would cause envy and result in covetous crimes of passion" (Tocantins 1979: 166, translation mine). Women's presence would also undermine the bosses' control. Women contributing their labor and services in subsistence agriculture and healthcare would decrease the rubber tappers' purchases of the bosses' dry goods and medicines (UNIFEM 1990).

Rubber tappers in more isolated regions such as Acre requested that their bosses supply them with women. In response, the large commercial trading houses in Belém and Manaus began sending women to the *seringal*. They were delivered in much the same way as dry goods and utensils – at inflated prices which were charged on the *seringueiro*'s account in the boss's store. In one instance, the Governor of the State of Amazonas ordered the Manaus police to round up 150 women from the city's brothels and cabarets. These women were then shipped to and distributed in the Acrean city of Cruzeiro do Sul (Reis 1953). One immigrant, who arrived in Amazonas in 1942, recalled how his uncle made his fortune by taking advantage of the scarcity of women by selling his five sisters to local men (Benchimol 1992: 54).

During the rubber boom, *seringueiros* were prohibited from cultivating their own crops, since this would take time away from rubber tapping and would decrease their purchases at the boss's store. When the price of rubber collapsed on the international market in 1910, bosses were unable to provide their workers with dry goods and so permitted their workers to prepare subsistence agricultural plots (Weinstein 1983). As the bosses' control waned, rubber tappers were able to send back home for their wives and families or create a family through marriage with women in the area. With the presence of female and child labor, the household production system diversified to combine agricultural and extractive activities for subsistence and market production.

The current division of labor in the extractive reserves has not changed dramatically since the collapse of the first rubber boom early this century, although household divisions of labor do vary. Typically, men and older boys are responsible for walking the forest trails to tap the rubber trees and gather latex. Usually all household members participate in gathering Brazil nuts in the forest. Agricultural production normally is shared by all household members. Men care for the larger animals and women are responsible for the chickens and pigs. Women do most of the cooking, cleaning, childcare, water collecting, food processing, and other domestic tasks.

Many *seringueiros* who came to the Amazon during its successive rubber booms were from the northeast of Brazil. Both men and women in the *seringal* cite the northeastern or *nordestino* influence to explain the difficulties that most women face when they participate in activities outside of the home. The male head of household is deemed the family representative in political, social, and economic affairs. Women are publicly recognized as household heads only in the case of widowhood. If the husband is ill or prefers not to be very active in social functions, the wife may assume the responsibility of representing the family. By and large, the family's marketable products (rubber and Brazil nuts) are controlled by the men and cash income passes through the men's hands before being distributed to other household members. Women's work is also devalued because the majority of their tasks, in the agricultural fields or at the house, have no market value.

However, research focusing on women's labor participation in the household's cash crops shows that women do play an important role in production. The few studies that have been done on female participation in latex production indicate that women and girls play a much more important role in rubber tapping than previously thought. In her interviews with elderly *mulheres seringueiras* in the state of Amazonas, Simonian found that it was fairly common for women to tap rubber in the 1920s and 1930s. Today there are still significant numbers of women and children working in the extraction and processing of the latex in this area (Simonian 1988: 9). In Acre, Kainer and Duryea (1992: 422) found that over 64 percent of the women interviewed had cut and collected rubber at some point in their lives while 78 percent reported having regularly collected latex that was tapped by a male household member.[7]

Studies such as these change our understanding of women's labor investments in productive activities widely perceived as "men's work." This knowledge suggests a recognition of women as key economic players. The following section explores how other longstanding roles are changing in the household, the union, and the cooperative as the extractive reserves projects are implemented.

WOMEN IN THE MOVEMENT

Through her extensive experience with rural unions, church communities, and women's groups in Acre, Lucia Ribeiro has observed that there are three typical roles for rural women in these social organizations.[8] The first is to raise the membership numbers; counting women as participants in a church, union, or cooperative gives the impression of a larger organization. Second, women function in their traditional roles as domestic caretakers, staying in the kitchen during meetings and taking care of the home so that the men are free to participate in community activities. Third, women are peacemakers, mediating in potentially violent situations in attempts to avoid escalation of the tension. The men and women interviewed for this chapter agreed that Ribeiro's observation holds true for most *mulheres seringueiras* in Xapuri today. From the home to the forest to the union office, however, women are challenging these three typical roles and gaining a stronger voice for themselves and their daughters.

In order to better study Ribeiro's observation and to explore women's roles in the movement, this chapter presents the voices and experiences of many *mulheres seringueiras*[9] and examines their roles in the movement's initial social organization in the forest.

First steps at organizing

The base communities of the Catholic Church (*Comunidades Eclesiais de Base – CEBs*) were one of the first forms of social organization in the *seringal* and served to teach the rubber tappers about their land rights and to denounce the violent practices of ranchers against them (Bakx 1986; Lima *et al.* 1994). Since women typically lead a fairly isolated life and have infrequent contact with neighbors, the CEBs have been and continue to be important in catalyzing and sustaining women's social interaction. From the formation of the CEBs in the 1970s to the present, women participated actively as monitors of many church groups in the rubber holdings or *seringais* of Xapuri (Boff 1980; Campbell 1990), providing one of the first formal opportunities for women's leadership and power as community leaders (Campbell 1990; de Oliveira, Arruda, and Carneiro 1994). Later we will explore to what degree these initial leadership experiences translated into stronger roles for these and other women in the movement.

The CEBs grew in response to the effects of the federal and state governmental policy of the 1970s that encouraged the expansion of agribusinesses in the region, largely in the form of cattle ranching. Initially, there was little organized resistance by the rubber tappers to the forest clearing and expulsions by ranchers. This was due to the lack of any entities or leaders to whom the *seringueiros* could turn for help. The base communities provided an organizational stepping-stone toward increased unionization of the rubber tappers and other rural workers. Many of those who would go on to become union leaders started off as CEB monitors (Boff 1980; Lima *et al.* 1994), as did Sebastiana and Cecilia. In 1977, the rubber tappers formed a rural workers' union in Xapuri (*Sindicato dos Trabalhadores Rurais* – STR) to assist rubber tappers in moving away from their traditional debt-peonage relationships with their patrons and to better resist the ranchers who were intent on "clearing the forest," often by using violent tactics to drive the rubber tappers from their homes.

From the outset, the union was seen as a social and political space for men. Union records show that of the first 455 members to sign up, 90 percent were men. The women who did join as founding members of the union were largely already accepted in their communities as de jure or de facto heads of household. The latter is the case of Cecilia. Cecilia tapped rubber herself and her husband tended to stay close to home as he was often ill. As she assumed the administrative, marketing, and social duties for their household, she also became the only union member from her family.

Sebastiana joined the union in 1978. She was one of the very few young, single women to do so. Of the first forty-one rural women to join the union, 35 percent were widows, 34 percent were the only ones listed for their household (indicating that they had sole responsibility for representing the household) and 10 percent registered at the same time as their husband, son or brother (indicating that they shared the head of household responsibilities). This means that close to 80 percent of the first women to join the union already had primary or shared responsibility as head of household when it came to political representation. Only one of these forty-one women signed up with her husband. The union and its members saw no need for two people to sign up; the man could speak for everyone in the household. Men in general saw it as a threat to their head-of-household status (both public and private) if their wife was also a union member. Sebastiana recounts how and why women participated in the union in its early days:

> It was mostly widows and the other women, they all played the role of men in the *seringal*. Their role was to represent the family but . . . they didn't have a very active participation in the union. Mostly it was like this; go to the meeting but nobody says anything, nobody contests anything or suggests anything. This persists until today too.
>
> (Sebastiana)

Aside from being very few in number in the union's ranks, the number of women who participated in the community-level union meetings held in the *seringal* was also very low. They tended to gather in the kitchen or listen in from the corridor without actually participating in the meeting. It was reported that union delegates did not permit their wives to participate in meetings because it was not an appropriate place for women.

These early women members did serve a pivotal role in legitimizing the presence of women in the union and in encouraging other women to join. Since the union did not specifically make an effort to increase women's participation, the few early female members served as influential role models and mentors. In a 1994 group interview, Rosa stated that not a single woman had participated in founding the union.

> At first they didn't want women to go; it was really difficult for a woman to go to a meeting. It wasn't until Cecilia joined and then she started explaining to us what the union was. A few women participated in the meetings and some of them had good ideas and gave their opinions. Then Cecilia was elected [by her community] to be union delegate and it was then that the men decided that women could participate in the meetings.
>
> (Rosa)

Cecilia's legitimacy as a de facto head of household and CEB leader allowed her the space and authority to slowly bring other women into the movement.

Sebastiana was the exception to the rule in that she was a single young woman in the union. In 1981, when the first president of the Xapuri union stepped down, Sebastiana was nominated to serve as president to fill out the rest of his term. Sebastiana remembers that her candidacy and election,

> had an incredible impact at the time. Many people were against our slate of candidates because it had a woman on it. But it was interesting in the sense of the work that we carried out. We had a consistent relationship with people at the grassroots and we were able to overcome the feeling of inferiority that people had about me. The union has never had such a cohesive directorship. Cecilia served as Fiscal Officer under me.
>
> (Sebastiana)

Having proven her capability, Sebastiana was selected to be one of the first teachers in an innovative educational effort in the *seringal* – the *Projeto Seringueiro* or Rubber Tapper Project. The union worked together with advisors beginning in 1981 to build schools and clinics and to establish a cooperative. Prior to this project, schools were practically nonexistent in the *seringais* because the bosses did not want the rubber tappers to be able to read and calculate their monthly accounting sheets, which almost always showed a debt due to the boss (Allegretti 1979, 1981). The schools successfully provided basic literacy skills and consciousness-raising in the reserves. They also served to stimulate more community action and, today, continue to be a central focus of the movement (Campbell 1989, 1990).

Women's participation in the *Projeto Seringueiro* paralleled that of women in the union; there were a few exceptional women leaders but most were effectively excluded. In 1988, 32 percent of the teachers were women. There were very few adult women in the classroom, even though the project's main goal was to increase adult literacy. In that same year, 59 percent of the women surveyed in the *seringal* were illiterate. Women who wanted to study found it almost impossible to do so because of their childcare responsibilities, distance to the schools and the class schedule which was designed to accommodate the rubber tapper's daily routine. Most of the women interviewed stated that they really had no need to learn to read and write because their husbands did all of the marketing and dealt with whatever documents the family needed. However, they were adamant in their desire to secure schooling for their children (Campbell 1990). Maria, who currently serves as an elected member of the union council declared in a 1994 interview:

If there was one thing that I could buy, I'd buy more memory because I think that my intelligence is very low ... as a little girl I studied only until the second grade. My dad took me out of school because I couldn't study because I had to cut rubber. ... Then I got married and I was stuck. I wanted to study but my husband wouldn't let me and then we had kids right away.

(Maria)

Her educational experience parallels that of many women and men in the *seringal*. While the literacy and numeracy skills and the political conscious-ness-raising provided by the project were key elements in strengthening household marketing and community action, the direct benefits of the project largely were not accessible to women.

For the most part, women's traditional roles at home were reproduced in community gatherings, schools, and the union hall. Aside from the excep-tions of widows and de facto female heads of household, women's participation in the union was minimal. In the meetings the women were largely muted and served to increase the number of members without addressing the particular needs and contributions of women in the move-ment. The union, schools, and the cooperative were seen as men's spaces. It was the women's job to take care of the home so that husbands and fathers could take part and represent the family. The exceptions were women like Cecilia and Sebastiana who could break the stereotypes to a certain degree and open doors for other women. Exceptional men, like Cecilia's husband, also broke stereotypes and supported their wives' new social roles.

The large majority of *seringueiros*, and Chico Mendes himself, were against moving away from established social roles, even though the movement's ideology promoted partnership or *companheirismo* and radical changes in social relations of production. Men and women interviewed for this chapter stated that, in the early years of the movement, most men held the *machista* atti-tude quoted earlier that women were incapable of making sound contributions to the movement's organizations or of holding leadership posi-tions. This attitude translated into a very low level of self-esteem among the *mulheres seringueiras* and represents one of the greatest challenges to organizing women and changing some of their traditional roles. The next section exam-ines how the movement made effective use of women's perceived traditional roles in the *seringal* and in Brazilian society at large in defending the forest.

WOMEN AND CHILDREN FIRST IN DEFENSE OF THE FOREST

History and strategy of the *empates*

While it would have been a challenge to find even one woman in one of the adult literacy classes or actively participating in a union meeting, it was easier to find women on the front lines of the *empates* or forest demonstra-tions. Women's participation in the *empates* offers an interesting parallel to the role of women in general in the movement. At first there were very few, if any, women in the *empates*. Not wishing to break traditional expectations, women stayed home. As their participation in the union slowly grew, women did take part in the forest demonstrations but it was along the lines of their union roles; a few strong women proving themselves in leadership positions while the majority of women served to increase the numbers without having a strong voice.

In the early 1980s, it was exceptional women who took part in the *empates*, de facto female heads of household who broke the traditional social mold and who had to prove themselves worthy and capable of participating in a male-dominated defense of the forest. In later *empates* women were out in force and out in front face-to-face with armed police, paralleling their increased participation in the union. Their role in the *empates* was highly visible but remained fairly superficial; only a few women participated in the negotiations and dealings with the ranchers and the authorities. Several years later, key women had a more powerful role in decision-making at the *empates* and the sheer number of women on the front lines was stronger than ever. Following the key *empates* of 1988, both men and women took the floor at the Third Municipal Women's Meeting to laud the key role that women and children had played in defending the forest. These speakers also emphasized that, although it was still very difficult for men to accept women in these new roles, such changes were necessary for the women themselves, for the success of the movement and for the defense of the forest (Campbell 1990).

In 1994, the union still encouraged women to play their usual role by being on the front lines in an acknowledged violent climate with armed federal police. However, several *mulheres seringueiras* were key decision-makers in this particular action. Again, this parallels women's roles in the union. Today, while most women listed as union members are largely inactive and serve mostly to increase the membership numbers, several dynamic women have gained elected leadership positions and are striving to expand women's spaces in the union. This section explores the changing nature of the *empates* and women's roles in them, while at the same time drawing some parallels between these experiences and the changing participation of women in the union.

These forest demonstrations were the rubber tappers' nonviolent means of defending the forest from clearing and preventing the evictions of rubber tapper families. Chico Mendes recalled that the rubber tappers of Xapuri and the neighboring municipality of Brasiléia carried out forty-five *empates* between 1975 and 1988:

> These have led to about 400 arrests, forty cases of torture and some of our comrades have been assassinated, but our resistance has saved more than 1,200,000 hectares of forest (5 per cent of the area of the United Kingdom). We've won fifteen and lost thirty of the *empates* but it was worth it.

(Mendes 1989: 79)

Empates were rapidly organized when a forest tract was threatened. Large numbers of rubber tappers and their families would gather at the site of a clearing and would converse with the ranchers' workers in an attempt to peacefully persuade them to set aside their chain saws. Often this tactic consisted of appealing to common class interests, since both the rubber tappers and the ranchers' hired hands (who were mostly small farmers who had lost their land) depended on access to the intact forest for their livelihood. When necessary, the rubber tappers would dismantle the workers' camps to force them out (Allegretti 1994; Mendes 1989). Chico Mendes emphasized that, "when we organize an *empate*, the main argument we use is that the law is being flouted by the landowners and our *empate* is only trying to make sure the law is respected" (Mendes 1989: 66). In later *empates*, the ranchers used their ties with local law officials to bring armed police out to the areas that they sought to clear. This led the rubber tappers' leaders to adopt a different tactic – that of putting women and children up front on the *empate* lines (Mendes 1989).

Cecilia exemplifies one of the exceptions to women's traditional social roles in recounting her experience in one of the earlier *empates* in 1981. Cecilia was a teacher and the elected union delegate for her community. One Sunday morning during a CEB meeting, a union runner appeared with a message from Sebastiana, the union president, saying, "We need you there at the *empate*." She left the meeting right then and there without hesitation. She said:

> I had to go because, earlier, when the community was deciding who should be the new union delegate, there was an old man who spoke up against my nomination. He told everyone that they better choose a man because when a man gets called, he'd go right away and wouldn't hide behind the tree stumps. So I had to go just to prove that old man wrong.
>
> (Cecilia)

She took her 11-year-old daughter with her and used a jar with some kerosene in it as a lamp to light their way on the forest paths. They arrived at the house of the *empate* at 1:30 in the morning. At daybreak, she prepared to go to the clearing site but her daughter insisted on going with her.

> We both got arrested that day along with everyone else. There were seven women and 112 men. They took us all into town. After an hour or so, they let us women go but they kept the men in jail. We left the jail and went looking for some food to take to the jail for them.
>
> (Cecilia)

Her experience parallels that of the few women leaders in the early years of the rubber tappers' movement. She was recognized in the community as the political head of the family and had gained status and leadership responsibilities after having served as a church monitor and a teacher. Many times, those who are literate end up doing double or triple duty in community positions because there are so few who know how to read and write (Campbell 1990). Her experience is also typical in that she was a lone woman who had to prove herself to the skeptics in her community. On the other hand, her experience contrasts sharply with that of the large majority of women who rarely participated in social gatherings outside the home.

At another contested forest site several years later, the *empate* was somewhat different. Just as women's participation in the union had grown slowly, so women's roles in the *empate* changed. At this site, there were three series of *empates*. One participant recalled:

At the first *empate*, my husband went with the other men and I stayed home to wait for him. We [women] didn't go because we didn't have any experience with what the *empates* were all about and we had that fear of contributing and going together with our *companheiros*. At the second one, we were right out in front. We said that we shouldn't let our *companheiros* go alone. We got a bunch of men, women, and children together. There was a bunch of police with carbines all running toward us, to surround us and take our arms away but no one carried any guns. Our *empate* is just to collaborate by talking with them. It's peaceful. It's not to provoke anyone by shooting or that kind of thing. We want peace; what would you carry a gun for? We talked with the police and told them what we wanted, to stop the clearing. I was out in front holding the Brazilian flag.

(Graça)

Another woman remembered the key role that women played: "There were lots of women there. They put us out front because the police were armed there. Sending women to the front was a way of asking for peace" (Daniela). Another recalled:

If women hadn't participated, it would've been weak – we would've had only half the number of people there. Women have more of a calming influence. . . . The police came running down the hill pointing their guns at us. It was us women and the children in the front – and the men all stayed behind us. I wasn't afraid. If one dies, we all die.

(Gabriela)

Yet another woman remembered: "We were told to keep quiet. They [the men] went and talked to the police" (Rosi). Later, when the conflict was moved into town for further demonstrations, two *mulheres seringueiras* participated in the negotiations at the forestry office. These women were peers of Cecilia, older women who had been very vocal in representing their families and serving as CEB leaders.

This series of *empates* was different from the earliest ones in that women formed a very visible and crucial front line. The union learned to employ perceived "traditional" roles by putting women up front, knowing that their presence would defuse the tension and potential for violence. They also used nationalism and motherhood in the *empates*, having women carry their children, hold the Brazilian flag, or sing the national anthem in defense of their homeland and livelihood. Chico Mendes recalled the process that led to the decision to put women up front during an interview in 1986:

At first our strategy was to place ourselves between the forest and the rancher's hired hands to force them to stop clearing. But the rancher ran to the police and then we had to decide how to confront them. We didn't have enough force to go armed and face them and, besides, our primary goal was to avoid conflict; to have a pacifist movement. So we called the women and children to go up front because we knew that the police would think twice before shooting them.[10]

This strategy was employed again in other *empates*, including the crucial demonstration in the Cachoeira area. As a result of this *empate* and subsequent negotiations with federal officials, the government expropriated the area and declared Cachoeira to be an extractive settlement. This victory for the rubber tappers' movement came at a very high price. Darli Alves was the rancher who had purchased the area and who was later indemnified by the government. In retaliation for losing his rights to the Cachoeira area, he issued a death threat against Chico Mendes – a threat that was carried out on December 22, 1988.

As in other *empates*, women played crucial roles in the defense of Cachoeira which eventually led to the creation of an extractive reserve. Most of the women played an invisible role that was responsible for the success of the *empate*; they stayed at home taking care of the children, livestock, and agricultural fields so that their husbands and sons could participate in the prolonged demonstration. Other women took on the burdensome task of housing and feeding over 200 rubber tappers, many of whom brought their families with them. One woman recalled that she watched as the main beam in her house sagged every night as the rubber tappers slung their hammocks from the rafters. She was surprised that the roof did not collapse under the weight. Another participant remembered:

> [A]t first it was mostly men but the *empate* lasted so long that it became a family affair. The people in charge of organizing the *empate* [Chico Mendes, other union leaders and two women from the movement] held a meeting out there and put it to a vote what was the best way to go ahead once the police had arrived. They came up with the idea of putting women in front and put it to a vote. Everyone agreed. The objective of putting the women up front was to show that it was a peaceful thing.
>
> (Alba)

In a 1994 group interview, fourteen of the Cachoeira women recalled their experience. The women said that the outcome of the *empate* would have been much different if they had not been out on the front lines. Even though the rubber tappers always tried to be pacifistic, the women were convinced that it would have been very violent had they not participated in such a visible and effective manner:

> They would've shot at us because they were there with their machine guns. It was then [when we started singing and Chico called some of the police to one side] that they decided to go into town and negotiate there.
>
> (Suely)

Women's active participation was crucial, therefore, in defusing tension during the *empate* out in the forest. When the standoff moved into the town of Xapuri for negotiations, however, women did not participate in such strong numbers and the ranchers responded violently to the rubber tappers' demonstrations. In town, the *seringueiros* occupied the yard of the local Forest Service in an attempt to force local officials to address their protest against the forest clearing. Ilzamar Mendes (Chico's widow) later said:

> The men were in the yard and some of them were laying down on the brick wall that runs between the office yard and the street. I went down there [their house was a few blocks away] and brought the kids who were there over to the house to sleep. A little while later that night, they drove by and shot at them.
>
> (Ilzamar)

Oloci and Darci Alves, two sons of the Alves family, drove by and fired into the crowd, hitting two teenage boys with shotgun blasts. The victims survived and the Alves sons each received twelve-year sentences. The shooting drew the attention of national officials and spurred the discussions which eventually led to the expropriation of the area and the declaration of the Cachoeira extractive settlement.

Judging from earlier *empates*, it would be reasonable to assume that, had the *mulheres seringueiras* participated in the negotiations and the occupation of the Forest Service office, the violent attack might not have occurred. The women from Cachoeira who were interviewed for this chapter stated that a team of three men met with the police and with federal officials during the negotiations, before and after the shooting. They could not recall any women participating in these discussions.

The union in Xapuri still follows its policy of putting the women on the front lines of the *empates*. In 1994, the union became involved in defending a rubber tapper's forest holding. The president instructed a union member to go to the area and talk with the rubber tappers who had gathered there. Due to the very heavy police presence at the site of the clearing and the state of the legal proceedings that were under way, the president's instructions were an attempt to dissuade the rubber tappers from proceeding with forest demonstrations and confronting the police.

What was strikingly different about this defense of the forest was that younger women were actively involved in determining the course of action. At meetings held out in the forest where they discussed whether to go ahead with the *empate* and face the police, Raimunda recalled: "I put in my ideas for a plan of action and I warned my husband that he'd lose if he voted against me. We voted and my plan won out. He should've known better than to vote against my plan!" Interestingly, her plan advocated a non-confrontational approach of waiting to hear from the lawyers involved in the case. The other plan would have placed the *seringueiros* in a very tense situation, confronting armed federal police in direct violation of a judge's order to stay clear of the area while litigation was in process. Not only did women's physical presence at several key *empates* defuse potentially violent situations, but their negotiation skills, in the few cases that they were called upon, effectively reduced the chance for violent confrontation.

In the *empates*, women's participation, by and large, follows the three traditional roles outlined by Ribeiro. Women participated in order to increase the number of people involved; they were used to ensure a pacific outcome in very tense and potentially violent situations; and they were very busy in the kitchens and at home, maintaining the household so that their husbands and sons could participate in the demonstrations. In more recent *empates*, women gained a stronger voice in the decision-making process but still were employed as pacific symbols for the movement. This role is changing slowly as women are translating their defense of the forest into new livelihood strategies that allow them more solid economic and political strength. In the following section, we will move from the forest into the union hall and explore how women's roles changed in the union and the cooperative.

MAKING SPACE FOR WOMEN IN THE MOVEMENT

In the 1970s and 1980s significant changes took place in unions and women's groups around the nation that impacted the organization of women rural workers. Nationwide, Brazil "witnessed the emergence and development of what is arguably the largest, most diverse, most radical, and most successful women's movement in Latin America" (Alvarez 1990: 3). Unionization debates raged about men's and women's roles and the sponsorship of women's meetings and groups (Lavinas 1992) as women in unions across the country began to fight for the right to unionize and for their own space in local and national union organizations (CUT 1991). This section explores the impact of the women's movement and union changes on the Xapuri movement and the Xapuri women's struggle to define and claim a space for themselves.

Most women's roles in the early years of the Xapuri movement followed the three traditional roles outlined by Ribeiro earlier. This profile of women in the union did not change much until the late 1980s. One of the movement's leaders recalled:

> The first one to begin talking about women in the union was Chico [Mendes]. As President, he participated in several union meetings in other states and learned about their work with women. He started talking about women's role in the union during the assemblies and other meetings [in Xapuri]. He was interested in the issue but didn't really know where to go with it. In the beginning, the conversations about women in the union weren't very clear.[11]

It was not until several women from Rio Branco began to conduct research and hold women's meetings in the *seringais* and in town that women were actively drawn into meetings and discussions about the movement. One of the organizers recalled in a 1994 interview:

> In 1986–7 . . . we were the only women who went with the union delegates to the meetings [in the *seringal*]. People thought we were really strange because we stayed in the front room for the meeting while all the other women were in the corridor or back in the kitchen. Afterwards we'd go back into the kitchen and the women would ask us things – about the meeting, their problems, health and so on. So we talked to them and decided to have a meeting in town at the end of the year.[12]

This and other gatherings eventually led to the First Municipal Meeting of Women in Xapuri in December of 1987. At that meeting, the women decided to create a women's group informally linked to the union.

> After the first meeting [in December 1987], we went to talk to Chico about the importance of working with women and what our ideas were. I remember that he asked us, "How can the women's movement help the union? I'd like to know." For us, the question should have been vice versa – "How will the union help the women's movement?"[13]

Although his wife wanted to participate actively in the women's activities, Chico was not in favor. She later recalled:

> Many times when I expressed my interest in joining the women's group or the union, Chico was against it because he thought that I should be taking care of my duties as wife and mother, caring for the children, washing the clothes and cooking the food.[14]

Chico's attitude toward his own wife's participation reflected the greatest challenge for the women's group.

The organizers of the women's group in Xapuri had to deal with domestic violence, social pressure, illiteracy and low self-esteem. They were challenging the status quo by suggesting new ideas about women's rights. One of them recalled that she herself was unfamiliar with the new wave of feminism at the time. She had simply gone out and started talking to the women about their lives in the forest. Later when she went back to the city and read about all the work being done with rural women's groups in northeastern and southern Brazil, she realized that what she had been talking about with women in Xapuri was part of a whole world of feminist theories and practices. She admits that some of the ideas that she and others introduced to the women in the *seringal* were quite shocking. "Here these women thought that their lives were all set and then we showed up and started questioning things that for them were unquestionable".[15]

As the women's group slowly grew, so did the movement as a whole through union growth, expanded services offered by the *Projeto Seringueiro* and the formation of a cooperative. The union had been introduced to the national and international environmental movement and was receiving increasing numbers of visitors and foreign journalists. In 1987, Chico Mendes took part in meetings at the Inter-American Development Bank (Arnt and Schwartzman 1992) and received international awards from the United Nations and the Better World Society. There were approximately 1,000 names in the union's registration book in early 1991. Of these, 120 were women who constituted 12.5 percent of the total union membership. The education component of the *Projeto Seringueiro* had nineteen schools in operation in 1988 (Campbell 1990). On June 30, 1988, the rubber tappers founded the Agro-Extractive Cooperative of Xapuri (Cooperativa Agro-Extrativista de Xapuri – CAEX), the first such association formed by producers in the region without external financial support. Of the thirty founding members, there were three women: two widows who joined along with Cecilia, who was elected Vice President (of these three, only one was still listed as a member in 1994; the other two, including Cecilia, were replaced by their sons). In May of 1988, the women's group held the Second Municipal Meeting of Women Workers in which 169 women participated, 109 of them from the rural areas of Xapuri. Of these 109 women, 26 percent were registered union members and 63 percent were illiterate (STR-Xapuri 1988). The leaders of the women's group met several times in 1988 with Chico and other union leaders to discuss the feasibility of establishing a women's secretariat as part of the union's constituted structure.

In a strange twist, it may be that Chico Mendes' dynamic and charismatic leadership skills served indirectly to lower the number of women members in the union. One of the organizers of the women's group recalled that, in several other rural workers' unions in Acre, factional infighting led to campaigns designed to register women as members, thus increasing the votes for a particular group. The Xapuri union did not have that disruptive history of infighting; it may be that Chico Mendes' leadership style decreased the occurrence of such factionalism, which resulted in women being overlooked as active, voting union members.[16]

In September of 1988, I participated in a meeting at the Xapuri union office with Chico Mendes and several others. We were discussing how to strengthen the role of women in the union. The phone rang in the other room and Chico was told that the call was for him. When he returned to the meeting, someone asked him who had been on the phone. He replied, "It was another death threat." Chico was killed several months later.

It is not hard to imagine why the women's group was not seen as a priority for the union during these times. In a very violent environment, the union was struggling with a myriad of complex issues including the extractive reserves, *empates*, scarce financial resources, increasing international attention and visits from foreign journalists, the day-to-day survival of its members and continued clearing of the forests. The women's group, and its demands for a stronger voice, were seen as something that could wait until after the crisis of the extractive reserves was over. Unfortunately, that crisis led to Chico's death.

The answer to Chico Mendes' question, "What can the women's group do for the union?" was given in 1989, after his death. As the women met in his house with his widow, one of the women said: "Chico has died defending the union and our struggle. He strove to make the movement grow. The women can't let this fail" (Graça). In an interview in 1989, one of the organizers described the women's decision to go forward: "The women decided that, from then on, it wasn't just a movement for the men but that the women also had to carry the struggle forward. They even decided to begin like this: working together with the union but not as part of the union."[17]

The Xapuri women proved to the movement's leaders and members that they were seriously dedicated to the struggle to defend the forest and that they had no desire to divide the *seringueiros'* movement, which some leaders initially thought was the women's intent. Union delegates agreed to have members of the women's group accompany them to union meetings in their respective areas. An organizer recalled in 1989:

> It was approved in the union's general assembly without any dissenting votes. The women could go together with the delegates. Up until now it's been very difficult to hold a meeting in the areas [in the reserves] just with women. Whenever there's a meeting of the cooperative or the base community [of the Church], we can always get a few minutes to talk about women, women's situation, the role of women in the union and so on, but there isn't space for a meeting just about women out there. We can get space for a smaller group in town, like the coordination group, to talk about women, the domination of women and women in the union, but in terms of our work out in the areas, this hasn't been possible. The men are very jealous. They question us, 'What is this?' It's something strange for them. That's why I think this work with the women will have to go very slowly. The women know we have to go slowly to win them over.[18]

The women followed up on their earlier conversations with Chico Mendes and were able to establish a women's secretariat as part of the union's formal structure in the 1989 elections. The women's group met every two months to reflect on: the importance of their participation in the struggle; their role in the union;

the historical domination of women by men; and the value of women, their health and sexuality. The group addressed all of these points in hopes of attaining a higher degree of consciousness and self-esteem for the *mulheres seringueiras*. They also held several fairs where women could sell artisanal goods, the proceeds of which were applied to union dues. In order to sell anything at the fairs, women had to first join the union. This economic incentive was the first effective means of getting active women members into the union.

Even though there was a women's secretariat as part of the union's Council, the women elected to this post had no training or resources with which to continue these activities. They relied solely on the assistance and volunteered time of female organizers from Rio Branco. These organizers were able to put together the Third Municipal Women's Meeting in 1989 which resulted in two craft fairs being held later that year. However, aside from the municipal-wide meeting in town, there no longer were women's meetings out in the reserves. The women's group became almost a "top–down" affair. The few activities that took place were due to the initiative and administration of the group's organizers who were women working at affiliated nongovernmental organizations and who struggled to find time and resources to maintain the Xapuri group. Women from the reserves also faced difficulties in maintaining the group, even though the level of interest was high. They had few leadership skills, resources or organizing capacity to sustain smaller, rural activities. In 1989 and 1994, the rural and urban organizers told me of the difficulties of getting space and support for the women's group in the union. Scheduling and carrying out the municipal meetings was a very difficult task. The resistance they encountered was similar to that from the union leadership in 1987–88, when they first began the group. Following the second craft fair, the women's group faced a lack of support from the union leadership, the absence of the initial organizers and other difficulties and was paralyzed for over two years in 1990–3. There were several attempts during these years to reactivate the women's group by meeting with union leaders and with the women, but to little avail.

Union records show that, even though the women's secretariat and the smaller women's groups out in the reserves were basically inactive, their efforts did serve to bring a new generation of women into the union. From 1977 to 1991, widows constituted 21 percent of women members. Between 1991 and 1994, this fell to 5 percent. The percentage of single women rose from 25 percent to 52 percent of the female membership over the same time period. It is also interesting to note the relative importance of women in the union itself. In the union's first fourteen years (1977–91), women constituted 12.5 percent of the total membership. From 1991 to 1994, women constituted only 3.4 percent of those registering as union members. Even though the number of female members increased due to the women's group and its economic incentives to join the union, female membership as a whole declined relative to male membership in the past three years, so that women in the union today constitute a very small, but increasingly vocal minority. We will see in the next section how a core group of active women is again struggling to promote the women's group and gain more space for its constituents.

WOMEN IN THE MOVEMENT TODAY

The rubber tappers in Xapuri successfully built an autonomous political force which achieved many victories, including saving tracts of forest from clearing, gaining federal protection for the extractive reserves, building schools and clinics, and establishing a cooperative which is able to offer some economic alternatives for some rainforest-dwelling families. Yet, for all of its emphasis on *companheirismo* and the communal struggle, the rubber tappers' movement has been largely a male-dominated one. While some women have been able to strengthen their roles to a certain degree, most have remained, literally, in the kitchen during the union meetings. Many women have gone to the front at the *empates*, when asked, but went back to the kitchen during the negotiations following these dramatic confrontations. Jelin offers useful insight into the difficulties that women and men face in attempting to increase and diversify women's participation in movements such as the rubber tappers. She notes:

> [T]he organization of the family and the sexual division of labor hinder women's public participation because of their responsibilities and the ideo-logical burden of being female. It would seem, therefore, that women more frequently participate in protest movements at critical moments than in long-term, formal, institutionalized organizations that imply taking on responsibilities, dedicating time and effort to the organization and also – why not say it? – the opposition of men. ... Consequently, women can either enter the public domain by adopting masculine codes, behaving like men – demanding equality – or they can set out to transform this domain by incor-porating the knowledge and experience of their own sex, an historically difficult task.
>
> (Jelin 1990b: 186)

This section explores how women in Xapuri are working on both fronts in the movement, demanding equality with men in certain arenas and also attempting to change aspects of the movement by bringing in their own distinct talents and insights.

An important gender dynamic of these efforts is the generational differ-ence among women in the movement. While older women participated in the *empates* and many of them broke a path in the community and the union for younger women to follow, there is a generational difference in their discourse. Older women tend to speak of social and political difficulties, such as not being allowed to leave the house, conduct market transactions, partic-ipate in meetings, or have the opportunity to voice their interests. Today there is a younger, perhaps less radical, generation of women that seeks to work within the household, the union, and the cooperative and effectively to incorporate their own interests and abilities. They have social and polit-ical demands but their economic concerns are in the forefront. They have defended the forest, built schools, established and maintained the base communities of the Catholic church, and are invested in the movement and the extractive reserves. They are increasingly aware of their rights, and their daughters' rights, to access and control over the forest resources that they have defended – rights that may conflict with the rights and interests of their husbands and sons.

These generational differences parallel the dual approaches outlined above by Jelin: women can act like men and demand equality in the movement, or they can take the perhaps more difficult road and incorporate themselves into the organizations and attempt to transform them with their gendered experiences. The older women, or the first generation of women in the movement opened doors by taking the first approach; the widows and de facto women community leaders took on roles in the movement equal to those of men such as base community leader, union delegate, school teacher, union president, or cooperative vice president. These positions were success-fully filled by women, but without particular attention to women's distinct needs and interests; the women carried out their duties as if they were men in a certain sense. The second generation is taking the historically more difficult route, that of working within the movement in an attempt to trans-form it by bringing in their unique histories and experiences as women. By activating an effective women's secretariat in the union, gaining votes in the cooperative, and electing women to leadership positions with the intent of addressing specific women's interests, women in the Xapuri movement today hope to create and maintain a gendered space for themselves and their daughters in which they can promote themselves and the movement as a whole.

Economic changes in the extractive reserves

At the Third Municipal Meeting of Women Rural Workers in 1989, the participants identified their top three priorities: economic opportunities, polit-ical participation, and health/education issues. One might have expected that women would focus on health, which was the main topic of most of their earlier discussions. However, as shown by the 1989 craft fairs, women were primarily interested in ways to earn and control their own income. For several years prior to the 1991 meeting, the price for rubber had plum-meted, sending the rubber tapping community into a severe economic crisis. In declaring their top priority to be economic opportunities, the women were seeking alternative sources of moneymaking to secure their families' livelihoods, an effort similar to the political one they took on as they stood on the front lines of the *empates*. Their economic needs are linked to the larger dilemma of the extractive reserves, that of finding the means to ensure economic and environmental stability.

While their interest and need for their own economic activities has grown, women in Xapuri find that their opportunities remain limited. At one small group discussion during the Third Municipal Meeting, nine out of thirteen women stated that they harvest rubber, but that their husbands demanded half of the cash from the sale of this latex because, in the men's words, "the land is mine." Continuing barriers to women's economic participation were related even several years later by representatives from Xapuri at the ECO-92 Global Forum in Rio de Janeiro:

> [A]side from the problems common to all the *seringueiros*, the women have various, specific problems. They work together with their husbands, but most of the time it is the men who control the money. The stories of the women

Plate 2.2 Vania Mirela and her father, Jose, cracking Brazil nuts at their house in the Cachoeira extractive reserve. Vania and her mother, Celia (not pictured), share processing duties on one of the two nutcrackers so that Vania can attend school

Source: Connie Campbell

from Xapuri reveal that, for women to cut a trail of latex trees or to sell some of the small livestock, they need the permission of their husbands.

(EMAM 1992: 33, translation mine)

Not only do *seringueiro* families still rely mainly on latex and Brazil nuts for their cash income, but these products remain largely within the domain of the male head of household. Even those women who are able to sell some of the products identified as women's business, find very weak markets. Without access to their own income streams, many women find it difficult to keep up with union dues, which in October of 1994 came to US$ 0.85/month. The cooperative's initial dues of 100 kilos of latex (valued at US$ 106 in September 1994) are also far beyond the reach of almost all women and some men in the reserves.

The cooperative in Xapuri is promoting an activity which could address these issues by providing a viable and sustainable economic alternative in the reserves and an economic opportunity for women – the decentralized drying and shelling of Brazil nuts in the extractive reserves. Traditionally, Brazil nuts were gathered on each family's forested tract and sold to the boss who controlled a particular forest estate, or to an itinerant trader. In 1988, the rubber tappers in Xapuri established a cooperative which sold rubber and Brazil nuts to a variety of buyers. The Xapuri cooperative later established a factory in town in which Brazil nuts are shelled, dried and packaged for shipping to domestic and international purchasers.

In the past several years, the cooperative has established three community-level and five household-level mini-processing centers in Xapuri with others being constructed. The decentralized processing reduces the cooperative's transportation and other operational costs, provides employment in the reserves, and adds value locally to an established forest product.[19] Most of the people working as processors are women, many of whom declared that this is the first time that they have had access to and control over their own cash income. For most women in the project, making visible contributions to household income has increased their decision-making power. As the husband of one of the women said, "Women have more of a voice now" due to the project.

Women employed in the decentralized project sites reported that they were satisfied with the project and with their new ability to help with the family's expenses.[20] To date, most of their income has gone for purchases of household goods and clothing. As one woman said: "Sure, it's more work. But before, we were working all the time anyway and not getting paid for it" (Luiza). The control that these women have over their income varies.

In one case, a young man was working with a family as a *meeiro* or rubber sharecropper, a common arrangement in which he tapped rubber on the family's trails and gave them half of the income. In 1994, he switched from tapping rubber to processing Brazil nuts at the community project. Instead of half of his income from rubber going to the male head of household, he now gives part of his Brazil nut income to the female head of household to help with cooking and household expenses. Thus, even indirectly, women's income and control of that income has increased because of the project.

There is some social stratification resulting from the project. Some families are hiring others to care for their agricultural fields or assume childcare/housekeeping duties, including cases of teenage girls working for relatives or neighbors. These girls are earning their own income, but it is not as much as if they themselves were employed by the project. In none of the families that I visited had this situation resulted in adolescent girls dropping out of school, since they had either already completed the four grades available in the *Projeto Seringueiro* schools, or had already stopped attending school. The same is true for families with adolescent sons and daughters working in the Brazil nut project, either at a home-based or community-level site.[21]

The self-confidence and respect gained by most of the women involved in the Brazil nut project has encouraged them to state in a louder voice their desires for a stronger role for women in the union and the cooperative. As Raimunda put it when referring to administrative problems at the cooperative such as delayed paychecks: "[r]ight now they tell us at the office that we can't say anything because we're not members. I want to pay the dues and join the cooperative so that I can go into town and complain." Suely wants to become a member of the cooperative so that she can set up an account in her own name – "that autonomy is very important to me." This new self-confidence extends beyond wanting a stronger political voice or a new economic role in the cooperative. Raimunda is also taking on a new leadership role by starting up a project to market Brazil nut soap that she and several other women are producing with the nuts left over from processing.

The Brazil nut project represents an important opportunity for a few women, but the majority of families in the reserves do not have access to

Plate 2.3 Maria das Dores Almeida and her grandmother, Dona Emilian, shelling beans from their family's agricultural plot

Source: Connie Campbell

the project. On a larger scale, most of the families in the reserves do not participate in the cooperative at all, although the movement is attempting to broaden the cooperative's services and establish new producer associations. The process of building local institutions capable of addressing basic social and economic needs is a "time-consuming and elusive one" (Schmink 1992), but is essential for successful implementation of the extractive reserves and for meeting the particular challenges of marketing the non-timber forest products on which the reserves depend (Schwartzman 1994). The extractive reserve populations rely on such institutions for economic alternatives and, as seen by the Brazil nut processing project, the gender sensitivity of these institutions determines women's access to the benefits they can provide.

Since the Brazil nut project currently reaches a relatively very small number of families in the reserves, the majority of women see no opportunity on the horizon to earn their own income and use that income for their families or for cooperative or union dues. In a 1994 interview, an elected official at the cooperative said about their plans to work with women: "[A]t present we haven't yet laid out a work plan, a policy or any thoughts directed toward women; we don't have one for the men either. She sells [to the cooperative] of her own free will" (José), implying that the cooperative is a gender-neutral space. This is an attitude similar to that of the union leaders in defining how to deal with the women's movement. They saw the goal of the movement as assisting the whole family, but without acknowledging or even being aware of the benefits that result from addressing the specific gender needs or interests of women or adolescents.

Today there are 222 members registered in the cooperative office. Only three of these are women; the number of women members has not changed since the day the cooperative was founded in 1988. One of the three women in the registration book in 1994 was a founding member and the other two joined in 1991 and 1993. The barrier to women joining the cooperative is the initial quota or dues of 100 kilos of rubber (or its equivalent in another product). In a 1994 interview, Raimunda stated that, "[income from] rubber and Brazil nuts always ends up in the man's pocket," indicating that women will have to find another means of paying their quota.

While women who participate in the Brazil nut project represent real exceptions, it is important to note that economic roles for women in general are slowly changing. One woman noted in a 1994 conversation:

> Before, the husband used to do business or make a deal and wouldn't even say anything to his wife. Not today. Before he buys or sells something he arranges it with his wife and kids. If the wife doesn't like it, she'll tell him. Women go to the stores and strike their own bargains. She goes in and talks to the owner. It's a big difference from the days of my mother-in-law.
>
> (Rosi)

While her observation may be somewhat optimistic and generalized, the women that I visited indicated that things are getting better for them in this regard and that women are becoming increasingly aware of their rights and are becoming less afraid to exercise them.

Social and political spaces for women

Learning about one's rights and responsibilities can be a confusing and frightening process as traditional gender roles change. One of the elected women leaders in the union told me:

> Women think that they have their rights. Lots of them confuse rights with liberty. I see it like this – that rights are one thing and liberty is another. At no time are they equal. Women think that they have rights and liberty. They think that they are equal to men in that they can go out drinking and partying, when in fact if she does that, she'll end up separated from her husband. We women have women's rights – to participate, to one day have the possibility of going out and defending those who live in the forest. This is very impor-tant – women senators, state deputies, town councilwomen – these are the women's rights that I'm talking about. The right to do her own marketing, to help her husband at home, to buy stuff at the store. . . . And we have the right to have our glass of wine too . . . we sometimes have meetings with pizza and beer but we don't overdo it.
>
> (Maria)

The women's group was a space wherein women could learn about their rights and about the responsibilities that go with them. Unfortunately, from early 1990 to 1993 the women's group was basically inactive. Meetings with union leaders and with *mulheres seringueiras* did not yield positive results. Several women from Xapuri did participate in statewide meetings in 1992, but there were no rural or urban activities held in Xapuri. This was due to a lack of support from union leaders, a lack of skills and resources within the women's secretariat at the union, dissolution of rural groups and the departure of the women who had initially organized the group (STR-Xapuri 1993).

Despite the lack of organized women's activities, individual women made great personal strides in transforming their social, economic and political roles. Some have been able to translate their new economic role in the Brazil nut project into a stronger political voice outside the household. For example, Raimunda was invited to represent the women of her community by speaking at a local political rally. Another woman related her union experience in a conversation in 1994:

> I went to the meetings all the time before. I spoke up a lot and I paid attention, because it's like the lyrics in that song – if the woman just stays at home listening to what the husband says, then you can forget it – she won't know how to speak up at all.[22]

(Bernadete)

Women in Xapuri now face the task of transforming these personal victories into organizational changes in the union, the cooperative and their political party. One victory in the public sphere for the women was that they overcame all of the challenges they had faced in organizing earlier meetings and successfully held the Fourth Municipal Women's Meeting in Xapuri in June of 1993 with twenty-eight rural women participating. This particular victory was significant because, over the years, the women were able to occupy a public space during a meeting in the front room of their house or in a general assembly of the union and transform that experience into a public, gendered space for their women's meeting. Equally important was that they were able to bring men into their traditional space – the kitchen. At the June 1993 women's meeting, the participants considered it a major victory that there were two men in the union kitchen doing the cooking for the two-day meeting.

For many women, however, their lives have not changed since the movement began. One woman observed during our conversation in 1994: "In this movement, the *seringueiro* got free of the boss man, but the *mulher seringueira* didn't get free from her boss – her husband."

Aside from the social repression of women, their inability to participate can impact the family's economic and political opportunities as well. In one case, a young wife wanted to return as an active participant in her community's CEB since she had served as a leader prior to getting married and having children. Her husband adamantly insisted that she not participate in the Sunday morning gatherings, to the point of physical violence. As she could not be a regular participant in the CEB, she was automatically disqualified from being considered for employment in the Brazil nut project, a job which she eagerly sought. The community had decided earlier that the selection criteria would include the requirement that those who got jobs in the project had to be active community members. Due to her husband's strict control over her, the family lost out on stronger social links with their neighbors and an economic opportunity to increase family income, not to mention the mental and physical suffering that she endured.

A few women argue that husbands are not entirely at fault for women's repression and low levels of participation. However, women would need a safe environment in which to begin to express such self-confidence and resistance to a domineering husband. The majority of women in the extractive reserves have not had the opportunities that Raimunda and others have experienced by participating in the women's group, the union and the co-operative's Brazil nut project.

Changes in the union

Most women in the reserves could identify readily with Zelinda – 54 years old, the mother of six children, ages 5 to 21. During a conversation in 1991, I asked her husband, Francisco, if he was a member of the union. He responded, "Yes, I am. I joined the day the union was founded." I asked the same question of Zelinda. "No, I'm not in the union," she replied. "I don't know how to read or write. I don't know how to think." Her response to my question is typical; not only are the majority of rural women illiterate, but because they are illiterate they believe that they cannot think or contribute to formal social organizations (Campbell 1992).

At the 1993 women's municipal meeting, a union official exhorted women to join up and take the floor right away – otherwise, they would be more useful if they just stayed home. But Raimunda remembers her experience in the movement as being quite different:

> I just wish that women would get involved in the meetings, the *empates*, whatever – just to give their support to the union. I want at least that the women get organized to participate in everything, even if they don't take on anything – just get involved. Because in the beginning, I just started participating and now I'm elected to the union council. I didn't used to understand a lot of things. I didn't know what was going on but I kept going to the meetings and now I give my opinion so that I'm not sitting there without saying anything.
> (Raimunda)

To a certain degree, it is difficult to fault the movement for inattention to women's interests. The movement was not intended to improve the situation of women in the *seringal* and some believe that it is wrongly blamed for its treatment of women. One of the women's group organizers recalls that when they began meeting with Chico Mendes in 1987 and 1988 about the group, it was not clear where the women's interests came in the line of priorities for the union.

What was difficult for the union leaders to see and for the women to express were the benefits that the movement would gain if women were allowed to participate fully in meetings, *empates*, negotiations, the cooperative and household decision-making. Everyone clearly understood the benefits to having the women out in front at the *empates*, but it was less clear what would be gained by breaking with the husband's sole representation of the family and sharing that right and responsibility with women and adolescents.

The road from the kitchen to the union hall to the speaker's podium at an assembly meeting is a scary and difficult one for women who have very low self-esteem, husbands or mothers-in-law who may be adamantly opposed to their participation, and a community that does not value their voices and opinions. It took many years of long conversations for Chico Mendes and others to convince the *seringueiros* to form a union. It took a year and a half of meetings in the *seringal* before thirty rubber tappers joined together to form the Xapuri cooperative. So it takes time and patience to change engendered customs and ideas that are embedded in the social fabric of the rainforest.

Even women who are able to attend union meetings or other gatherings outside the home, face structural and ideological barriers to participating as active and effective social change agents. In part, this is because people see the movement, its history and the opportunities that it offers for women in very different lights. At a women's meeting, a union official spoke of the important role that women have played throughout the unionization movement, but his version of history differed from that written or spoken elsewhere. He stated that, "the women decided to go in the front [of the *empates*] with their children" and that "the women helped in the discussions about creating the cooperative and now many of them are members" (STR-Xapuri 1993). But an organizer of the women's group sees it differently:

> [T]he union is still a man's place today. The majority of women don't identify themselves as a *mulher seringueira* or a *colona* [colonist farmer] or a rural laborer. They identify themselves as housewives and if their husband is a union member, there's no need for the woman to join. There is a lot of non-identification of themselves as rural workers. This will take a long time to change.[23]

By comparing some data on the participants in the 1988 and 1993 Municipal Women's Meetings, we see a new trend in women's unionization (STR-Xapuri 1988, 1993). In 1988, 34 percent of the participants were union members. At the 1993 meeting, 71 percent were registered union members. Unionization among women participating at these meetings increased by 92 percent over the years while unionization among female adolescents increased by 18 percent. If it may seem that women are joining the union in droves, however, that is not the case. The number of women participating in the 1993 meeting was just over a third that of 1988. There is a new generation of women in the union. This new generation may be smaller in number but is more active in terms of being registered members and of serving on the union's leadership council. While the women's group may be serving a smaller total number of women, as shown by the lower number of participants in the 1993 meeting, those who do participate constitute a more active core.

In putting together a list of candidates for the most recent election, union members were encouraged by the national union federation to have at least 30 percent of their leadership positions filled by women. As one member of the elected council recalled in 1994:

> We kept this 30 percent suggestion in mind. We wanted to incorporate as many women as possible but didn't want to put a woman in just because she was a woman. We looked for the *best people for the positions*.[24]

Of the candidates that were elected to the leadership council, 33 percent were women, exceeding the national directive's suggestion. However, a council member stated: "There are four women who speak in the meetings. The rest don't say anything. They vote with the majority" (Carlos). One of the women council members explained:

[The women on the council], they're mostly quiet. Because of the system of the husbands – that the women always trust that the men have to have their turn to speak. So many women are timid because of the custom with their husbands. I'm one of them. I go to the meetings. I understand what's going on but I don't say much because of the system of my husband and that's the way that I was brought up. One of the most difficult things is that he wants to raise our children this way – within this fear. This is going to take a lot of work to change. I was always afraid. It's tough to take the fear away and because of this we don't speak up very much – remembering the fear. To be singled out, to have a *companheiro* say that you're wrong – imagine the shame if he were to say that you'd said something wrong. This is why I think women don't speak up.

(Maria)

In the early years of union organization, it was just as difficult for most men from the *seringal* to speak up at community meetings. Many men told me of the years and years of meetings that they went to without saying a word. This process of gaining a confident voice is even more difficult for women who face the societal constraints in the household and in public spaces described above. The movement itself, and the men and women who participate in it, cannot expect women to step right up and take the floor.

The women who were elected to leadership positions have been able to begin transforming these organizations but there is much work to be done. These and other women have proven that they are capable of making valuable individual contributions to the movement. The organizational resilience of the women's group has proven that women are dedicated to the movement and to their own particular interests and needs. More recently, the participation of Xapuri women in various events throughout the nation has shown the movement's leaders the importance of women's representation. *Mulheres seringueiras* strengthened their voices and their resolve by participating in events which included state-level women's workers meetings, national congresses in São Paulo, the ECO-92 conference in Rio de Janeiro, the Caravan of Women Rural Workers to Brasília and an international congress in Italy.

Mulheres seringueiras from Xapuri who participated in the Global Forum, an NGO event which paralleled the United Nations ECO-92 conference, worked together with other women farmers and forest product gatherers on a policy statement of their goals. They stated that they wanted to remain in the rural sector, free from the misery encountered by those families who migrate to urban areas, but also free of rural violence from land conflicts. They saw the extractive reserves as a solution to land conflicts which also permits natural resource management and conservation. They also declared the need to break down the isolation of women due to illiteracy and lack of access to health services; address the devaluation of women's work within the household and their double duty/double identity as housewives and workers; and revise the policies of banks, cooperatives, and agro-industries which largely ignore women's labor and economic roles (EMAM 1992).

In a way, this parallels the situation of the Xapuri *seringueiros*, who first received recognition and support in international circles before Brazil's national media, NGOs, and environmental movement knew about them.

Plate 2.4 Dona Antonia and Dona Maria Lino meet with Maria da Concepcao de Barros (far right) and other women at the union hall in Xapuri. Dona Maria's two daughters (second and fourth from right) also participate actively in the women's group

Source: Connie Campbell

Here too, the women had to go outside of Xapuri to Rio de Janeiro and Brasília to gain recognition and legitimacy. The question remains as to whether they will be able to translate this "stamp of approval" into action on behalf of women's distinct interests or if the movement leadership will try to co-opt or even dampen the women's momentum.

LOOKING TOWARD THE FUTURE: "THE PATH FOR OUR DAUGHTERS"

The women in the Xapuri movement have achieved significant changes in the three stereotypical roles explored earlier in this chapter. Yet, the political, social and economic challenges that they and their *companheiros* face remain daunting. Just as the women gained visibility in the household, in the movement, and even in national and international audiences, so too must the issues of domestic, economic and social change be put on the table for discussion in order to achieve the basic goals of the extractive reserves.

Others have documented the critical roles of women in natural resource management in other tropical rainforest areas (Kelkar and Nathan 1991; Rodda 1991; Shiva 1989). Yet, researchers dedicated to analyzing the viability of extractive reserves pay insufficient attention to critical social factors such as gender division of knowledge, labor, and authority. This chapter has shown that the *mulheres seringueiras* have crucial roles to play in the economic growth, political representation, social stability, and natural resource conservation of the extractive reserves. Along with their husbands, sons, and daughters, *mulheres seringueiras* are "daily managers of the living environment" (Dankelman and Davidson 1988, cited in Rocheleau 1991: 157). As Kainer and Duryea have argued, "[T]he success of the extractive reserves as a development model therefore depends on the recognition and positive exploitation of differences in gender" (1992: 424).

The challenges that *seringueiro* communities, their representative organizations, researchers, planners, and project implementors face in successfully implementing the extractive reserves are just as daunting as those faced by many *mulheres seringueiras* who "are prisoners in their own homes," as a rubber tapper leader put it. Many of the *mulheres seringueiras* whose voices comprise this chapter were able to identify and exploit opportunities to slowly change their lives and strengthen the Xapuri movement. In a similar vein, when researchers and others regard the future of Amazonia, there is "room for cautious optimism. The challenge is to identify and exploit those limited but potentially significant degrees of freedom that permit new directions" (Schmink and Wood 1992: 355). Successful implementation of the extractive reserves is one such new direction that joins those on the rubber trails in the forest, in the kitchen, and on the front lines at *empates* with others in union offices, universities, and policy-makers' offices. There is much to learn from one another in this common cause.

ACKNOWLEDGMENTS

I wish to thank Lucia Ribeiro and Eliete Braga, of the *Rede Acreana de Mulheres e Homens* in Rio Branco, for collaboration on an earlier version of this chapter. Special thanks are due to Marianne Schmink for her guidance and support throughout the research and writing. I am also grateful to Eduardo Romero, Marianne Schmink, and Amelia Simpson for their very helpful and insightful reviews.

NOTES

1   Chico Mendes was assassinated by a local cattle ranching family in December of 1988 because of his successful mobilization of rural populations to prevent clearing of forested lands by ranchers.

2   Extractive reserves are federally owned and protected areas in which resident populations are given usufruct rights to practice non-degradatory and small-scale extractive activities. The reserves were designed along the lines of indigenous reserves and provide an alternative to government-sponsored colonization projects in the Amazon. In 1987, the first extractive settlement projects were established under the administration of the National Colonization and Land Reform Agency. In 1990, a presidential decree established extractive reserves under the administration of the Brazilian Environmental Institute (IBAMA). Today, both the extractive settlement projects and reserves are commonly referred to as extractive reserves. There are seventeen extractive reserves in varying stages of legalization and demarcation, totaling 3,700,600 hectares. The federal government is responsible for monitoring resource use in the reserves while associations of residents in the reserve are responsible for drawing up management plans which must meet federal approval. Within the reserves, land is held in a form of common property in that each household has control over its forest tract of 200–500 hectares but the individual tracts are not demarcated. The head of household must register with the federal government to sign a usufruct agreement which states that the land may be sold or passed on only to another extractive producer.

3 The extraction of wild Brazil nuts from the forest is a major source of cash income for many Amazonian households.

4 In Portuguese, a male rubber tapper is referred to as a *seringueiro* while a female rubber tapper is called a *mulher seringueira*, a term used in Xapuri to avoid confusion with the word *seringueira*, which refers to the rubber tree.

5 Mohanty (1991) warns against the creation of static images of third world women in general.

6 This chapter draws on my research in the town of Xapuri and surrounding forests from 1988 through 1994. Extended visits over the years provided opportunities to create and sustain close personal relationships with numerous families and leaders in the movement. From our friendship and trust emerged the many long and intimate conversations that form the basis of this chapter. I agree with other anthropologists that "the imaginative and hard work of learning about others' lives" greatly enriches and differentiates the texts that are generated (Gudeman and Rivera 1990: 4). As the women in this chapter seek to define their own identities in their homes and the movement, it is "crucial to build up a micro-history based on the retrieval of popular recollections and the recollections of the actors themselves and of their own movements, which has so much to do with the process of establishing identities" (Jelin 1990a: 8).

7 In my 1994 household interviews in two extractive reserves in Acre, I encountered one family in which the woman was a full-time rubber tapper while her husband was employed as a schoolteacher.

8 Lucia Ribeiro was born in Xapuri and has worked for many years with local workers' unions and the *Rede Acreana de Mulheres e Homens*, a nongovernmental organization based in Rio Branco which implements training courses and various projects with producer groups and their representative organizations.

9 In participating in the interviews for this chapter, women and men in Xapuri understood that their voices would be published. In a similar publication, Townsend (1995: 4) and her collaborators reflect that, "[t]he information and life stories obtained were explicitly given to be published, but would the givers now approve of the extracts which we are publishing? We cannot know. For this reason, the people and communities . . . are identified only by fictional names." For the same reason, I have chosen to use fictional names for the *mulheres seringueiras*.

10 Lecture given by Chico Mendes at the Department of Geography, University of São Paulo (Piracicaba) 1986. Videotape.

11 Interview with the Regional Coordinator of the CNS, November 7, 1994, Rio Branco.

12 Interview with the Project Coordinator, *Rede Acreana de Mulheres e Homens*, October 25, 1994, Rio Branco.

13 See note 8.

14 Interview with Ilzamar Gadelha Mendes, *Comitê 8 de Março Informativo*, March 1989, Rio Branco.

15 See note 8.

16 See note 8.

17 Interview with the Project Coordinator of the *Centro dos Trabalhadores da Amazônia*, June 29, 1989, Rio Branco.

18 See note 10.

19 Cooperatives and producer associations in other extractive reserves have undertaken similar efforts to add value locally to the forest products gathered in the reserves. These processing efforts include Brazil nuts and heart-of-palm in the eastern Amazon.

20 In 1994, I held many long conversations and extended interviews in the reserves with the men and women participating in the Brazil nut project. Many of these people had been key participants in earlier research relating to the *Projeto Seringueiro* in 1988 and 1989. Through another research project and consultancies, I had visited regularly with these families since before the Brazil nut project began in certain pilot sites in 1991 through my 1994 field work. Over the years, I have interviewed all of

the families involved in the Brazil nut project in two of the three CAEX pilot sites. This field work involved extended stays with these families in the reserves.

21 In a brief field visit to another project site, Hecht (1995: 14) explored this same issue and found that "adult women soon ceased working" in the Brazil nut project which now "employs adolescents and children at the expense of their schooling." From my extended conversations with project personnel and members of neighboring communities, I found that this situation is due to a longstanding dispute between the family controlling the Brazil nut project and the family of the school teacher; the children's absences from school are not due to the project. In fact, the family that operates the Brazil nut project would probably make more money if their children were in school or at least not involved in the project because available adult labor would yield a higher return than the children achieve for the family managing the site. Interviews with the Brazil Nut Project Coordinator, July–December 1994, Xapuri.

22 In reference to a popular song sung frequently at women's meetings: "*Olhe mulher rendeira, olhe mulher renda, se a mulher não sai de casa, nunca vai se libertar.*" "Look lacemaker, look seamstress, if a woman never leaves the house, she'll never be free."

23 See note 8.

24 See note 7.

REFERENCES

Allegretti, M. H. (1979) "Os Seringueiros: Estudo de Caso em un Seringal Nativo do Acre," unpublished M.A. thesis, Universidade de Brasilia.

—— (1981) "Projeto Seringueiro: Cooperativa, Educaçao e Saude para Seringueiros de Xapuri-Acre," mimeograph, Rio Branco, Acre.

—— (1990) "Extractive Reserves: An Alternative for Reconciling Development and Environmental Conservation in Amazonia," in A. Anderson (ed.) *Alternatives to Deforestation*, New York: Columbia University Press.

—— (1994) "Reservas Extrativistas: Parametros para uma Politica de Desenvolvimento Sustentável na Amazônia," in R. Arnt (ed.) *O Destino da Floresta: Reservas Extrativistas e Desenvolvimento Sustentável na Amazônia*, Instituto de Estudo Amazônicos, Rio de Janeiro: Relume-Dumará.

Almeida, M. W. (1992) "Rubber Tappers of the Upper Jurua River: The Making of a Forest Peasantry," unpublished Ph.D. thesis, University of Cambridge.

Almeida, M. W. and Menezes, M. A. (1994) "Acre-Reserva Extrativista do Alto Juruá," in R. Arnt (ed.) *O Destino da Floresta: Reservas Extrativistas e Desenvolvimento Sustentável na Amazônia*, Instituto de Estudo Amazônicos, Rio de Janeiro: Relume-Dumará.

Alvarez, S. E. (1990) *Engendering Democracy in Brazil: Women's Movements in Transition Politics*, Princeton: Princeton University Press.

Arnt, R. A. and Schwartzman, S. (1992) *Um Artifício Orgânico: Transição na Amazônia e Ambientalismo (1985–1990)*, Rio de Janeiro: Rocco.

Bakx, K. (1986) *Peasant Formation and Capitalist Development: The Case of Acre, South-west Amazonia*, unpublished Ph.D. thesis, University of Liverpool.

Benchimol, S. (1992) *Romanceiro da Batalha da Borracha*, Manaus: Imprensa Oficial.

Boff, C. (1980) *Deus e o Homem no Inferno Verde: Quatro Meses de Convivência com as CEBs do Acre*, Petrópolis, Rio de Janeiro: Editora Vozes.

Bray, D. (1991) "'Defiance' and the Search for Sustainable Small Farmer Organizations: A Paraguayan Case Study and a Research Agenda," *Human Organization* 50, 2: 125–35.

Browder, J. O. (1992) "Social and Economic Constraints on the Development of Market-Oriented Extractive Reserves in Amazon Rain Forests," in D. C. Nepstad and S. Schwartzman (eds.) *Non-Timber Forest Products from Tropical Forests: Evaluation of a Conservation and Development Strategy*, Advances in Economic Botany, Volume 9, Bronx, New York: New York Botanical Garden.

Campbell, C. E. (1989) "Community Mobilization for Education and Conservation: A Case Study of the Rubber Tappers in Acre, Brazil," *The Latinamericanist* 24, 2: 2–7.

—— (1990) "The Role of a Popular Education Project in the Mobilization of a Rural Community: A Case Study of the Rubber Tappers of Acre, Brazil," unpublished M.A. thesis, University of Florida.

—— (1992) "Rural Women's Groups in the Western Amazon," paper presented at the University of Florida's Women's Studies Program Conference "Women and Politics in the 1990s," March 20–21, Gainesville, Florida.

—— (1994a) "Gender Issues and Non-Timber Forest Products: Some Preliminary Observations from On-Going Research in the Brazilian Amazon," paper presented at the Latin American Studies Association, XVII International Congress, March 10-12, Atlanta, Georgia.

—— (1994b) "A Socioeconomic Analysis of Forest Product Processing Projects in the Western Brazilian Amazon," paper presented at the Annual Meeting of the Society for Applied Anthropology, April 13–17, Cancún, Mexico.

CUT (Commissão sobre a Mulher Trabalhadora) (1991) *Mulheres Trabalhadores Rurais: Participação e Luta Sindical*, Publicação do Departamento Nacional dos Trabalhadores Rurais, Rio de Janeiro: Centro Ecuménico de Documentaçao e Informaçao (CEDI).

da Cunha, E. (1986) *Um Paraiso Perdido: Ensaios, Estudos e Pronunciamentos sobra a Amazonia*, Rio de Janeiro: Jose Olympio Editora and the Fundação de Desenvolvimento de Recursos Humanos, da Cultura e do Desporto do Governo do Estado do Acre.

Dankelman, I. and Davidson, J. (1988) *Women and Environment in the Third World: Alliance for the Future*, London: Earthscan.

de Oliveira, J., Arruda, M.A., and Carneiro, V.C. (1994) "A Mulher Seringueira: Casa, Trabalho e Politica," mimeo, Rio Branco, Acre: Universidade Federal do Acre, Departamento de Historia.

Dugelby, B. and Terborgh, J. (1993) "Can Extractive Reserves Save the Rain Forest?" *Conservation Biology* 7, 1: 39–52.

EMAM (1992) Escola de Mulheres para Eduacaçao Ambiental, *Com Garra e Qualidade – Mulheres em Economias Sustentáveis: Agricultura e Extrativismo*, Rio de Janeiro: Publicação Rede Mulher.

Gudeman, S. and Rivera, A. (1990) *Conversations in Colombia: The Domestic Economy in Life and Text*, Cambridge: Cambridge University Press.

Hecht, S. B. (1992) "Extractive Communities, Biodiversity and Gender Issues in Amazonia," in L. M. Borkenhagen and J. N. Abramovitz (eds.) *Proceedings of the International Conference on Women and Biodiversity*, Washington, D.C.: World Resources Institute.

—— (1995) "Surplus Extraction? Women, Extraction and Rural Industrialization in Acre, Brazil," submitted to *Journal for Peasant Studies*.

Jelin, E. (1990a) "Introduction," in E. Jelin (ed.) *Women and Social Change in Latin America*, London: Zed.

—— (1990b) "Citizenship and Identity: Final Reflections," in E. Jelin (ed.) *Women and Social Change in Latin America*, London: Zed.

Kainer, K. and Duryea, M. (1992) "Tapping Women's Knowledge: Plant Resource Use in Extractive Reserves, Acre, Brazil," *Economic Botany* 46, 4: 408–25.

Kelkar, G. and Nathan, D. (1991) *Gender and Tribe: Women, Land and Forests in Jharkhand*, New Delhi: Kali for Women.

Lavinas, L. (1992) "Mulheres trabalhadoras rurais: abrindo novos rumos no sindicalismo," in *Mulheres daqui e de lá: Diálogo entre as trabalhadoras do Brasil e do Quebec*, CUT, Commissão sobre a Mulher Trabalhadora.

Lima, J. S., Martins, J. S., Souza, E. S., de Souza, M. C., da Silva, M. L. and da Silva, R. C. (1994) "A Igreja Católica do Vale do Acre e Purus na Organização dos Trabalhadores Rurais: Seringueiros de Xapuri," mimeo, Rio Branco: Universidade Federal do Acre.

Mendes, C. (1989) *Fight for the Forest: Chico Mendes in His Own Words*, London: Latin American Bureau.

Mohanty, C. (1991) "Under Western Eyes: Feminist Scholarship and Colonial Discourses," in C. Mohanty, A. Russo and L. Torres (eds.) *Third World Women and the Politics of Feminism*, Bloomington: Indiana University Press.

Reis, A. C. (1953) *O Seringal e o Seringueiro*, Rio de Janeiro: Serviço de Informação Agrícola, Ministerio da Agricultura.

Rocheleau, D. E. (1991) "Gender, Ecology and the Science of Survival: Stories and Lessons from Kenya," *Agriculture and Human Values* 8, 1/2: 156–65.

Rodda, A. (1991) *Women and the Environment*, London: Zed Books.

Safa, H. I. (1995) "Women's Social Movements in Latin America," in C. E. Bose and E. Acosta-Belén (eds.) *Women in the Latin American Development Process*, Philadelphia: Temple University Press.

Salafsky, N., Dugelby, B. L. and Terborgh, J. W. (1993) "Can Extractive Reserves Save the Rain Forest? An Ecological and Socioeconomic Comparison of Nontimber Forest Product Extraction Systems in Petén, Guatemala and West Kalimantan, Indonesia," *Conservation Biology* 7, 1: 39–52.

Schmink, M. (1992) "Building Institutions for Sustainable Development in Acre, Brazil," in K. H. Redford and C. Padoch (eds.) *Conservation of Neotropical Forests: Working from Traditional Resource Use*, New York: Columbia University Press.

Schmink, M. and Wood, C. H. (1992) *Contested Frontiers in Amazonia*, New York: Columbia University Press.

Schwartzman, S. (1989) "Extractive Reserves: The Rubber Tappers' Strategy for Sustainable Use of the Amazon Rain Forest," in J. O. Browder (ed.) *Fragile Lands of Latin America: Strategies for Sustainable Development*, Boulder, Colorado: Westview Press.

——(1994) "Mercados para Produtos Extrativistas da Amazonia Brasileira," in R. Arnt (ed.) *O Destino da Floresta: Reservas Extrativistas e Desenvolvimento Sustentável na Amazônia*, Instituto de Estudos Amazônicos, Rio de Janeiro: Relume-Dumará.

Shiva, V. (1989) *Staying Alive: Women, Ecology and Development*, London: Zed Books.

Simonian, L. T. L. (1988) "Mulheres Seringueiras na Amazônia Brasileira: Uma Vida de Trabalho e de Silencio," mimeo, Brasília: Secretaria de Açoes Sócio-Culturais do Ministerio da Cultura and Ministerio do Desenvolvimento e Reforma Agraria.

Sobrinho, P. V. (1992) *Capital e Trabalho na Amazônia Ocidental*, São Paulo: Cortez Editora.

STR-Xapuri (1988) Sindicato dos Trabalhadores Rurais – Xapuri, "Segundo Encontro das Mulheres Trabalhadoras de Xapuri-AC: Relatório," mimeo, Xapuri, Acre.

——(1993) Sindicato dos Trabalhadores Rurais – Xapuri, "4º Encontro Municipal de Mulheres Trabalhadores Rurais de Xapuri: Relatório," mimeo, Xapuri, Acre.

Tocantins, L. (1979) *Formação Histórica do Acre: Volume I*, Rio de Janeiro: Editora Civilização Brasiliera.

Townsend, J. G. (1995) *Women's Voices from the Rainforest*, London: Routledge.

UNIFEM (1990) "A Vida e Trabalho da Mulher Seringueira," video, Rio Branco: Fundação Cultural do Governo do Estado do Acre.

Weinstein, B. (1983) *The Amazon Rubber Boom: 1850–1920*, Stanford: Stanford University Press.

Westwood, S. and Radcliffe, S. A. (1993) "Gender, Racism and the Politics of Identities in Latin America," in S. A. Radcliffe and S. Westwood (eds.) *'VIVA': Women and Popular Protest in Latin America*, London: Routledge.

3

FEMINIST POLITICS AND ENVIRONMENTAL JUSTICE:

Women's community activism in West Harlem, New York

Vernice Miller, Moya Hallstein, and Susan Quass[1]

INTRODUCTION

Since the founding of the United States, African-Americans and other marginalized peoples have had to struggle for their survival against ever-changing adversities. The newest affront to the existence of these culturally diverse communities is manifested in the disparate placement of environmentally hazardous waste and industrial sites in their neighborhoods. Because these communities are often greatly disadvantaged economically and politically, local and state political institutions place very little value on them and subsequently allow or actively pursue environmentally racist practices. In recent decades, the environmental assaults on these communities of color have multiplied, as have the number of affected communities (Anthony 1990; Bryant and Mohai 1992; Bullard 1990; Lee 1987; Martinez 1991; *National Law Journal* 1992; Shepard 1994). In response, community activists have begun networking throughout the country in opposition to policies and practices which encourage the continual degradation of their quality of life. The growing environmental justice movement has directly challenged the value systems and resultant actions which have been imposed on these communities of color for decades (Anthony 1990; Bryant and Mohai 1992; Bullard 1990, 1992; Durning 1989; Grossman 1992; Lee 1992; Miller 1993; Shepard 1994).

Community-based struggles aimed at combating acts of environmental racism and the ensuing disparities are often gendered, with much of the work carried out by women outside the realm of the visible, "valued" social and political spheres. West Harlem Environmental Action (WHE ACT)[2] formed in 1988, is one example of a local initiative where women's interests, concerns, experiences, and knowledge have shaped the direction and significance of community-based struggles against environmentally hazardous sites. WHE ACT was created through the initiative of two women who rose to confront the City of New York about its calculated decision to place the North River Sewage Treatment Plant (the NRSTP or North River) in West Harlem, primarily an African-American neighborhood with a growing number of immigrants from the Dominican Republic. The efforts of WHE

ACT to increase community awareness and rally opposition to the NRSTP have revealed the intersection of race and gender within West Harlem. From the supportive networks of African-American women uniting to preserve their quality of life to the controversial and often unsupportive political spheres of City and District elected officials, race and gender have come together in various overlapping and at times conflicting ways.

Although initial responses to the sewage treatment plant were racially based, the impetus for much of the subsequent legal, political, and community actions has been driven by strong "behind the scenes" networks of women who were often confronted with a combination of both racial and gender biases. Women in WHE ACT have drawn upon and continued a tradition of African-American women assuming the unseen, "behind the scenes" leadership roles to mobilize and facilitate activities that maintain and promote the cultural and social survival of their communities. Especially active in WHE ACT is a generation of senior citizens, largely women, who have a strong commitment to maintaining their quality of life and have the skills and experience from previous struggles against multiple threats to the survival of their community.

The North River Sewage Treatment Plant is not the first environmentally hazardous facility to be located in West Harlem – it is the fourteenth! Other sites include a solid waste marine transfer station and seven of the eight Manhattan municipal bus depots (*Amicus Journal* 1994). This long-standing precedent of using West Harlem as a "dumping ground" has worn thin, and residents have mobilized against the multiplicity of threats to their quality of life, community survival, and rights of self-determination. WHE ACT has been able to employ West Harlem's community-level strengths to address issues ranging from health concerns, community empowerment, and employment to local control of open spaces, land use, and zoning. The traditional base of women's neighborhood activism has been fundamental to its formation and subsequent successes.

WHE ACT was initially created as a means to confront the City of New York and the North River Sewage Treatment Plant effectively. In the process, it has challenged limits to women's political leadership roles, strengthened community networks, and lessened dependence on outside organizations by developing new environmental research skills within West Harlem. In recent years, WHE ACT has broadened its scope far beyond the sewage treatment plant to join the national environmental justice movement in challenging the limited, and often discriminatory, definitions of environment employed by both governments and mainstream environmental groups. Members of WHE ACT have broadened their own definition: "our environment is where we live, work and play" (Miller 1994). Narrow conceptions of "environmental" issues have allowed mainstream institutions to ignore the irresponsible practice of locating hazardous waste sites in communities of color, across class, yet predominantly in "nonaffluent" communities (*Amicus Journal* 1994; Anthony 1990; Bryant and Mohai 1992; Bullard 1990, 1992; Durning 1989; Grossman 1992; Lee 1987, 1992; Martinez 1991; *National Law Journal* 1992; Shepard 1994). Through examining the history of community action against the North River Sewage Treatment Plant and the evolution of WHE ACT, this chapter will both address women's environmental activism in West Harlem and provide a local perspective on the national environmental justice movement.

THE WEST HARLEM COMMUNITY

Harlem has long existed as a shifting ethnic mosaic, with a rich cultural heritage which also bears the imprint of discrimination toward its kaleidoscope of communities. In the 1800s, Harlem began to attract affluent white families anxious to relocate away from the center of New York City. West Harlem, in particular, soon encompassed many of the most prestigious neighborhoods in the city, including the luxurious mansions of wealthy landowners such as Alexander Hamilton. Although much of the architecture and housing stock still clearly reflect Harlem's early bourgeois days, by the end of the 1920s Harlem had changed faces. It resonated with a new ethnic, social, and economic diversity while becoming increasingly distinct from the rest of New York City.

Discriminatory housing policies led to the clustering of New York's new African-American inhabitants. By the late 1920s/early 1930s, with the advent of what has become known as the Harlem Renaissance, Harlem had become an important center for the development and perpetuation of African-American tradition and culture. The following decades produced some of the most innovative African-American music, literature, and art of the twentieth century. During this time, Harlem presented itself as a cultural and economic haven to African-Americans. Throughout these changes, West Harlem continued to house much of the upper and middle class of this now African-American population. Caught in the midst of this uneasy intersection of a flourishing culture and racial discrimination, Harlem emerged as the capital of African America.

The pattern of environmental discrimination toward this community also began decades ago. In the 1930s, New York City's commissioner of parks designed 255 parks to be built over the next few decades, only one of which was planned for Harlem (Caro 1974). Harlem is a significant section of the city in upper Manhattan, stretching east to west from the Harlem/East River to the Hudson River. Riverside Park, considered by many to be one of the most beautiful in New York, is one such park that could have included Harlem, but did not. The park extends along much of the Hudson River's bank until it abuts Harlem and the eyesore of industrial use, now dominated by the North River Sewage Treatment Plant.

Whether in housing policies, provision of social services, or siting of public disamenities, Harlem has continually been regarded as a low-value area. Land uses and the public policy process in New York City have consistently reflected an intentional denigration of the Harlem community (Shepard 1994). These longstanding zoning policies have set the stage for an unyielding process of disinvestment and environmental racism which ultimately permitted the construction of the North River Sewage Treatment Plant in this historic community.

THE NORTH RIVER SEWAGE TREATMENT PLANT
MEETS WEST HARLEM

North River Sewage Treatment Plant is designed to treat 170 million gallons of raw sewage daily.[3] The catchment area far exceeds that of West Harlem, encompassing the entire west side of Manhattan with a residential population of approximately 600,000. However, since the service area includes most of Manhattan's commercial and business districts the daytime population surpasses one million. With only 78,000 residents, West Harlem produces just a small portion of the total waste water processed at the NRSTP. Thus, this community is host to an environmentally hazardous facility which provides an undeniably necessary environmental quality control. Ironically the vast majority of those serviced by the plant remain insulated from its noxious attributes (Gold 1991a and b; Miller 1994; Public Hearing 1968; Engineering News Record 1984).

The West Harlem community first learned of the North River Sewage Treatment Plant in 1968 when the New York State Assembly passed the "Open Meetings Law" which required that meetings regarding governance at the state or municipal levels be open to the public (Miller 1994: 715). Prior to this new law the City Planning Commission could determine land use, siting, and city development policy behind closed doors, conveniently avoiding opposition to any controversial decisions. However, by the time the North River Sewage Treatment Plant became public knowledge, its location had been fully approved and construction was imminent (Public Hearing 1968).

Empathizing with the West Harlem community in its opposition to the sewage treatment plant, John Lindsay, elected mayor of New York City in 1966, offered to mitigate the situation with decorative water fountains for the top of the plant (Miller 1994; New York City 1984). While the people of West Harlem rejected Lindsay's idea, his support signaled to them that they could ameliorate some of the negative impacts of the plant through proposals for alternate uses for the top of the NRSTP. In an effort to appease the West Harlem community, they were given two years to design a more appropriate and much needed recreational park (Gold 1991a). In 1970, the state of New York committed several million dollars to fund the construction of the River Bank State Park (RBSP). Every year from 1970 to 1992 members of the West Harlem community fought with the state to maintain the funding line in the annual budget appropriations. Construction of the park finally began in 1989, while the plant itself was still under construction. The River Bank State Park opened on May 27, 1993 (Holloway 1993).

Long before its construction, however, the park was a controversial issue for those concerned with the health of their community and of the children who would use the park daily. To date, there have been only limited studies of hazardous emissions and other potential sources of adverse health effects from the sewage treatment plant, although there is substantial evidence of possible links with symptoms of respiratory disease including bronchitis and asthma, among other illnesses. The park has proven to be a mixed blessing, offering much needed recreational facilities while simultaneously placing its users at an unknown health risk due to its location over the sewage treatment plant (Miller 1994; Public Hearing 1968; Ryan 1968; Severo 1989; Shepard 1994).

Since 1968, some members of the West Harlem community have been actively building opposition to the North River Sewage Treatment Plant through a series of locally based initiatives (King 1968). It was not until the mid-1980s, however, when the plant was almost complete, that a larger segment of the community began to strengthen their position to make their voices heard. As the plant materialized, becoming a very real obtrusion in the West Harlem landscape, men, in particular, realized that the plant was being constructed without their involvement; in fact, not one minority contractor had been hired (Engineering News Record 1984; Shepard 1994). The more than $1.1 billion construction of the NRSTP was one of the largest public works projects of a nonmilitary nature in the United States during the latter half of this century (Gold 1991a; Ryan 1968). There were legitimate concerns over how much of this capital would filter into the community. Serious questions were raised about who was going to work in the plant and how many people of color, especially those who live in West Harlem, would be hired. Residents of West Harlem decided that it was time to engage in a long overdue dialogue with the City of New York.

At the time, a newly formed independent political organization of the New York County Democratic Party offered the most viable avenue through which to advocate, develop, and strengthen their cause. In 1985, residents of West Harlem formed the West Harlem Independent Democratic Club (WHID) to address the issues of political empowerment and community control more effectively. Through creating the first reform Democratic club in Harlem, members of West Harlem took a strong progressive stance in breaking away from the larger centralized party structure. While the club still functions within the context of the Democratic Party, it is organized and run on a collective basis. Its efforts are generally still targeted at party politics through selectively influencing party agendas, supporting specific candidates, and also creating a forum for issues and concerns relevant to the community which otherwise might not be addressed.

Within the context of complex and volatile Harlem politics characterized by class, race, gender, and neighborhood divisions, the formation of WHID was in many respects an affront to existing party structures and local elected officials. It exacerbated tensions between West, Central, and East Harlem. Over the years a strong schism has developed in the Harlem community in which West Harlem has been disparagingly perceived as the "Black bourgeoisie." Black elected officials representing Harlem have tended to be from Central Harlem and to many West Harlem residents their elected officials have deliberately turned a blind eye to much of what transpires in West Harlem. WHID emerged as an independent grassroots organization within this atmosphere of political contention.

Much of WHID's success in rallying community support can be attributed to the commitment that women and senior citizens – especially senior women – have shown in opposition to the North River Sewage Treatment Plant and in the larger political structure. When the NRSTP began to operate it became immediately clear that jobs at the plant was only one of the numerous issues which the North River presented to West Harlem. The health-related implications of the pervasive odor of sewage from the malfunctioning plant were immediately apparent with the abrupt increases in asthma, bronchitis, and other ailments (Dockery *et al.* 1993; Ford 1993; Goldstein and Izeman 1990; Severo 1989; Shepard 1994). Women, and senior women in particular, played a significant role in broadening WHID's scope to include these far-reaching concerns about community health, the Harlem environment, and overall quality of life.

At the expense of the West Harlem community, the North River Sewage Treatment Plant began to work in 1986 before it was fully constructed or operational. It went directly into full capacity primary treatment, bypassing the standard test phase in which unforeseen problems are addressed and remedied (Gold 1991a and b; Miller 1994; Muzynski 1993; Severo 1989). The premature start-up of the NRSTP was the combined result of legally imposed completion mandates and construction delays. In 1976, the United States Environmental Protection Agency brought a lawsuit against the City of New York for failing to satisfy the four-year completion deadline it had been given after the passage of the U.S. Clean Water Act in 1972. The court mandated a timetable for compliance, requiring that primary treatment be fully operational by 1986, secondary treatment by 1991. Subsequently, to avoid substantial federal fines, plant construction was continued under an expedited schedule which led to numerous alterations and shortcuts (Ash 1992; New York City 1984; Engineering News Record 1984; Miller 1994; Officer 1993; Severo 1989).

The noxious odors emanating from the exceptionally large, uncovered settling tanks were immediately apparent (Browne 1994; Gold 1991a; Severo 1989; WHE ACT 1993). At thirty feet, the tanks at North River are the deepest in the United States, far surpassing the standard depth. The size of the tanks presents a considerable problem in that the treatment process at the NRSTP entails two processes: primary, in which waste is physically treated through screening, settling, and skimming; and secondary, a bio-organic process achieved through oxygenation in which microorganisms digest most of the contaminants (Commoner 1987; Gold 1991a and c; New York City 1984). The tanks at North River are so deep that the sewage cannot aerate, subsequently turning septic before this secondary process can be completed (Commoner 1987; Gold 1991c). This produces, among other effects, the noxious odors which filter into the West Harlem neighborhoods (Gold 1991c). Compounding the situation, the sophisticated odor control system found in most sewage treatment plants was not installed until 1991–92. Until then, sixteen stacks at the northern end of the plant and a large stack at the southern end emitted toxic and noxious gases into the air and operated without the requisite "cleansing" digester system. Digester systems are important because they process the ambient emissions before releasing them so that the noxious odors and toxins are removed or greatly reduced. The southern stack burned pure methane for five years before it was turned off. Aside from the hydrogen sulfide produced by the oil-burning engines, there is very little data on other toxic emissions[4] (New York City 1992; Muzynski 1993; Severo 1989). Neither the State nor the City Departments of

Environmental Protection have released comprehensive studies or reports of any ambient emissions (Engineering News Record 1984; Miller 1994; New York City 1984). Meanwhile, the West Harlem community is subject to the adverse effects of long-term exposure to unknown contaminants (Commoner 1987; Goldstein and Izeman 1990; Muzynski 1993; Severo 1989).

IMPACT ON WEST HARLEM

Since 1968, when dialogue with city officials began, West Harlem residents have been assured that North River Sewage Treatment Plant would be a "good neighbor." Because it was the newest sewage treatment facility in the United States, city officials claimed it would be "state of the art" with all the latest technology (Engineering News Record 1984; Severo 1989). Instead, what West Harlem has received is a physical monstrosity in its frontyard and a pattern of increasing numbers of complaints about adverse health conditions. It was not until the plant began to operate, and immediately malfunctioned, that the community realized the magnitude of the problems that the NRSTP would produce. The all-pervasive, inescapable quality of the noxious odors invaded people's lives, permeating their homes, while revealing the true nature of this new neighbor and providing much of the impetus for the ensuing struggle.

Shortly after the NRSTP began to operate, general health conditions in West Harlem began to decline with an increased prevalence of bronchitis, allergies, and respiratory difficulties, especially asthma. Skin rashes and constant tearing of the eyes also have been clear indicators that the environment in West Harlem has changed (Commoner 1987; Ford 1993; Dockery *et al.* 1993; Goldstein and Izeman 1990; Severo 1989; Shepard 1994). Children and the elderly, often the most vulnerable in any community, have suffered the greatest impact of these adverse health conditions, at times having to be hospitalized[5] (Ford 1993; Shepard 1994). Residents have noted yet other repercussions of the alterations to the environment and quality of life in West Harlem, including a year-round mosquito population and an apparent increase in both the population and average size of sewer rats.

Also of concern was the effect of ambient air emissions from North River on residents who suffer from sickle-cell anemia. Methane, a natural gas produced by raw sewage, is a common by-product in most sewage treatment plants. Prolonged exposure to methane can exacerbate existing sickle-cell conditions in those who have the disease. Sickle-cell anemia is a genetic, hereditary condition which occurs predominantly in African-Americans. It is characterized by cell deformities resulting from defective hemoglobin in which cells suffer from insufficient oxygen[6] (Severo 1989). Originally, the methane produced by the 170 million gallons of sewage treated daily at NRSTP was to be recaptured and used to power the plant. This process was delayed, however, and pure methane was burned and released into the air twenty-four hours a day for five years (Commoner 1987; Goldstein and Izeman 1990; Severo 1989).

Ironically, the odors which served as the catalyst to community awareness and action turned out to be not as significant as the daily release of less noticed toxins into the ambient air from the plant's stacks. Had it not been for these noxious odors emanating from the plant, residents might not have immediately realized what was happening in their "front yard." As it became unmistakably clear that North River Sewage Treatment Plant was not the "good neighbor" it had been promised to be, women and senior citizens came to the forefront of the struggle in an attempt to uncover the truth and mitigate the situation. They were, and continue to be, the same women who have traditionally been the motivating force behind community-based activities, initiatives, and struggles. The women who had served on the local community board, been involved with the churches, and led community garden and cleanup efforts, had now become aware of the deplorable and insulting conditions created by the NRSTP. They initiated their campaign against the plant within the community forum provided by the creation of WHID. Among numerous issues, they brought with them a concern about declining health conditions and a determination to confront and halt the legacy of discrimination toward West Harlem.

Coupled with these health-related concerns was yet another contention with the NRSTP. Not only was the plant constructed in flagrant disregard for the integrity of West Harlem, it was built along prestigious Riverside Drive, whose occupants are primarily senior citizens with a lifetime of investment in their community, and an increasing number of young professionals who also have a strong commitment to maintaining the quality of their Riverside Drive neighborhood. The construction of North River was an affront to their lifestyle, obstructing what had previously been "their" valued, scenic, and unobstructed view of the Hudson River. This added insult brought a group of people into the debate who might otherwise not have had so clear an investment in opposing the North River Sewage Treatment Plant.

As more people joined the community effort in opposition to the NRSTP, meetings became increasingly animated with vigorous discussions about ways to rectify the situation. WHID cosponsored a "town meeting" with Franz Lichter, the local State Senator, at the end of October 1986 at the Riverview Terrace Coop on Riverside Drive. The residents of Riverview Terrace were particularly disturbed as, prior to North River, their twenty-two story building had an unobstructed view of the Hudson River. It was also one of the earlier low-income cooperative apartments built in Manhattan, and its inhabitants were middle- and working-class people who had struggled hard to buy into the "American Dream." With very little publicity, WHID was able to attract close to 300 people to this historic meeting. It was clear that the NRSTP was having a notable impact on the West Harlem community and this time residents were not going to accept another hazardous facility without a fight. The construction of the NRSTP, and the magnitude of the difficulties it had created, served to illuminate the harsh reality that West Harlem had become host to a grossly disproportionate number of the environmentally hazardous and noxious public facilities in the Borough of Manhattan.

The unprecedented community response at the October WHID-sponsored town meeting led to the formation of the Community Coalition Against North River (CCANR), an organization designed to address the disamenities produced by the NRSTP. Holding monthly meetings over the next few years, CCANR carried on the struggle with city officials to address, mitigate, and compensate for the construction of the NRSTP in West Harlem. Residents continued to assert that deteriorating health conditions in West Harlem were directly correlated to the plant's emissions. Though they could not see or touch the emissions that were continually released through the seventeen stacks protruding above the plant – and to that date, there had been no information made available to the public – it was unmistakably clear that the air the West Harlem community was breathing had changed (Commoner 1987; Goldstein and Izeman 1990; Muzynski 1993; Severo 1989). When confronted, both the New York City Department of Environmental Protection and the Mayor's office repeatedly responded that there were no problems with the NRSTP, and complaints about adverse health conditions resulting from noxious odors and emissions were dismissed as "figments of their imagination." The members of CCANR quickly realized that they were going to be ridiculed rather than supported by the administration of Mayor Ed Koch (Schultz 1989; Severo 1989), so they turned to the Manhattan Borough President, David Dinkins, who at the time had been a resident of Harlem for three decades.

Dinkins, the highest ranking African-American elected city official, showed immediate interest in the concerns of CCANR and was extremely supportive of its struggle against the NRSTP. In 1987, he authorized $15,000 from his discretionary budget enabling CCANR to contract an independent assessment of ambient emissions and related processes at the NRSTP. Barry Commoner's Center for Biology of Natural Systems, an independent organization based at Queens College, undertook the study and within a year released a comprehensive document endorsed by David Dinkins. The report revealed that for every measurable standard North River Sewage Treatment Plant far exceeded New York State air quality levels (Commoner 1987; Severo 1989). The West Harlem community, with the support of CCANR and David Dinkins, could now begin to substantiate their claims against the plant, which had thus far been met with vehement denial by city officials. The struggle, however, continued to be a local effort because sewage treatment plants are not within the jurisdiction of the Federal Clean Air Act. Fortunately, state and municipal environmental laws are quite stringent, a fact from which CCANR was able to benefit.

While the New York City administration continued to reject accusations of racism as a decisive factor in the location of the NRSTP (Shepard 1994), the United Church of Christ's (UCC) Commission for Racial Justice published its landmark report *Toxic Wastes and Race in the United States* (Lee 1987). In this report the UCC posits the concept of environmental racism: the phenomenon of targeting specific communities for the location of environmentally hazardous facilities and waste sites because of their ethnic, racial, and socioeconomic composition (Bryant and Mohai 1992; Bullard 1990; Lee 1987; Martinez 1991; Miller 1993; Shepard 1994). The UCC substantiated its hypothesis through extensive research on the correlation between the location of uncontrolled and controlled hazardous sites and that of communities of color on a national scale. The study revealed that the preponderance of environmentally hazardous facilities in non-white communities is not a random phenomenon. Of the dozens of variables tested, race consistently proved to be the most statistically significant, even to the point of suggesting it as a determining variable in an intentional pattern of siting. With the exception of disproportionately nonaffluent white "host" communities, the study revealed that class was not a deterrent. While sitings within communities of color were found to cross all class boundaries, a surprisingly inverse correlation with economic status emerged. Higher-income communities of color were *more* likely than low-income white communities to house hazardous sites (Lee 1987). *Toxic Wastes and Race* has proven to be an invaluable source of support to residents of West Harlem in their struggle against the NRSTP, giving them a broader context in which to bring forth and defend their case to the New York City administration.

Backed by the evidence presented in *Toxic Wastes and Race*, West Harlem residents were able to validate their claim that the siting of the NRSTP in their neighborhood was a clear case of environmental racism. The North River Sewage Treatment Plant was originally slated for construction in the 1950s on the Upper West Side of Manhattan, an affluent white community (*Amicus Journal* 1994; Miller 1994; Schwartz 1966; Shepard 1994). Until recently, it was commonly believed in Harlem that residents of the Upper West Side had fought a successful battle to have the plant relocated. In actuality, the all-white, all-male New York City Planning Commission made the decision in 1962 behind closed doors, based on land use and siting projections for future development on the west side of Manhattan. In fact, in 1992 Donald Trump received approval to build a 1,500-unit apartment complex in an area which includes the original Upper West Side site[7] (Dunlap 1992; Kennedy 1995). The Planning Commission determined that West Harlem would be economically and aesthetically more appropriate as a location for the North River Sewage Treatment Plant since there were no major developments planned for the future. In reaching this decision, the Planning Commission seemed to consider neither the inhabitants of West Harlem nor the quality of their lives. Likewise, maintaining the aesthetic and cultural design of the community was not deemed important (*Amicus Journal* 1994; Miller 1994; Schwartz 1966; Shepard 1994). The decision to relocate the NRSTP was clearly influenced by the socioeconomic and political conditions of West Harlem which, in turn, are a direct result of the racial composition of that community. The judgment reflected not only the racial differences between the largely white Upper West Side and the primarily African-American West Harlem, but also their disparate power in the political process (*Amicus Journal* 1994; Bryant and Mohai 1992; Bullard 1990; Lee 1987; Shepard 1994).

Initially, the Community Coalition Against North River provided an effective forum in which to organize community opposition to the NRSTP. Strengths within the community surfaced and the coalition successfully forged links among the varied sectors of West Harlem. Using its alliance with borough president David Dinkins, CCANR entered into the city political discourse and was invaluable in the arduous process of reconciling the divergent agendas surrounding North River. At a certain point, however, CCANR began to lose its ability to negotiate and bargain effectively with city officials. Rather than continuing as an independent organization solely committed to mitigating the negative effects of the NRSTP, CCANR slowly became one more "political actor" in West Harlem and was ultimately divided by West Harlem factional politics. A schism formed between those who sought the support of white elected officials and those who believed that the only way to achieve their goals was as a separate entity outside the political influence of elected officials. Additionally, this latter group realized that the NRSTP was just one of several facilities producing disparate impacts on West Harlem. Having identified a host of environmentally hazardous sites, they wanted to focus on the whole spectrum of environmental issues, not specifically on the North River Sewage Treatment Plant, as had become the case with CCANR. As a new avenue for the pursuit of their goals, two of the more active women leaders in CCANR – one of whom had been with the struggle since the inception of the West Harlem Independent Democratic Club – went on to form West Harlem Environmental Action (WHE ACT) in March of 1988, including members of both WHID and CCANR. The Community Coalition Against North River continued to hold monthly meetings and eventually became the North River Community Environmental Review Board, the official community review board mandated under the NRSTP operating agreement.

WHE ACT: IN A LARGER CONTEXT

At 6:30 a.m. on Martin Luther King's birthday, January 1988, West Harlem residents brought traffic on both the West Side Highway and Riverside Drive to an eventual standstill when they occupied the two roads in front of the North River Sewage Treatment Plant. The cold, sleet, and ice of this winter day did not impede the protesters in their efforts to impress upon city residents, commuters, elected officials, and the media the seriousness of their struggle and their determination to acquire compensation for the hardships suffered by the West Harlem community. This emphatic group, primarily composed of senior-citizen women, physically blockaded the roads with their bodies until they were apprehended and summoned to court. Their action demonstrated unwavering commitment to reaching an acceptable mitigation of the impact of the NRSTP on West Harlem, even through acts of civil disobedience. It marked the new direct, activist approach that the soon-to-be formed West Harlem Environmental Action would bring to the struggle.

This initial act of civil disobedience was important in a number of ways: it was an invaluable experience for residents of West Harlem to realize that there were many avenues to pursue and that some involve direct action. This event also clearly reaffirmed that the West Harlem community was still very serious about redressing the situation created by the NRSTP. The Martin Luther King Day demonstration gave the unmistakable message that West Harlem was not going to settle for inaction on the part of the New York City government.

The construction of North River and the resultant efforts to mitigate the myriad adverse conditions it produced have galvanized the West Harlem community about what it means to fight for community control, land rights, preservation and, perhaps most importantly, self-determination of the quality of life by the people who live in a given community. WHE ACT has channeled this unprecedented self-awareness and willingness to engage the political machine into effective environmental activism (Miller 1994). While WHE ACT conducted work similar to previous efforts by CCANR, it proceeded in a very different manner. West Harlem Environmental Action became a cohesive, organized group with specific goals. Its members were persistent in their advocacy with elected officials and it became the first community organization in New York City to introduce the concept of environmental racism in a citywide dialogue.[8] WHE ACT formed important relationships with mainstream environmental organizations, especially the Natural Resources Defense Council (NRDC) and the Environmental Defense Fund (EDF).

In many ways, WHE ACT redefined the struggle against North River with the realization that the magnitude of the problem in West Harlem went far beyond the sewage treatment plant and that the discriminatory treatment of their community could not be adequately addressed without recognizing the interconnectedness of issues and the processes which created them. What was initially defined as an isolated assault on quality of life was reassessed as one which was not only environmental in nature but also social and political and could not be separated from the broader issue of quality of life in West Harlem. Prior to North River, residents of West Harlem had a "common sense" understanding of what was happening to their community, but environment was not part of the West Harlem vocabulary. It has since become recognized as intrinsically connected to people's lives and the quality of their life experiences.

People who live in places such as West Harlem and have to endure the daily assaults from noxious waste facilities have long been aware of the idea of "environmental racism" as a fact of life. To outsiders environmental racism may be a new concept and may take some time to comprehend, but to those who have lived it, this idea is nothing new. Deprived of the right to breathe clean air, drink clean water, and ultimately live without being physiologically assaulted, West Harlem residents have been facing these issues since the day the first highway was built in their community. Through the issues brought forth with North River, many of them realized that if existing practices did not change, eventually there would not be a community to struggle for. In response to changing perceptions, WHE ACT began to organize against multiple environmental assaults within the broader context of the environmental justice movement.

At the forefront of this struggle have been the senior citizens of West Harlem, who have been witness to the inevitable transformations that the ebb and flow of socioeconomic change have brought to their neighborhoods. They have lived there for decades, many of them since they were children, and, not surprisingly, they have a unique, firmly rooted investment in salvaging a decent quality of life in West Harlem. Because the seniors are the ones who tend to be "in the community" day in and day out, they have more of a direct connection to the ways in which the community is continually degraded. Their involvement has been invaluable to the struggle; without them there might not have been community opposition or West

Harlem Environmental Action. This coterie of senior citizens, overwhelmingly women, has been a consistent source of strength, commitment, and "local knowledge." Technological advances notwithstanding, the oral transmission and keeping of the records of what has happened in West Harlem is something that only the elders of the community can do. They have experienced the transformation of their beloved community and they are passionate about the need to restabilize and fight back.

Genevieve Eason is one such West Harlem resident whose impassioned participation and valued "local knowledge" have been fundamental to the success of WHE ACT. She is older than 70, but no one knows her true age. Mrs. Eason has lived in the community for at least forty or fifty years and she is enthusiastically committed to making West Harlem a special place to live again. Mrs. Eason, like numerous other senior citizens, would like to see West Harlem once again offer the safe and healthy environment that they once knew, free of the daily abuses of today's society, including physiological assaults from the plethora of noxious and hazardous facilities which abound in urban areas. From her involvement with WHE ACT, the Community Board, and the Baptist Church to helping build a Boys' Club and a Girls' Club, Mrs. Eason has been persistently engaged at every level with community revitalization and has a reputation as thorough, articulate, and focused on everything she does. Aside from her vigor and commitment, Mrs. Eason has brought to WHE ACT an invaluable historical account and personal record of what has happened in West Harlem over her lifetime. She is able to encapsulate three decades of struggle against the NRSTP, forgotten by most people, and she can support them with original documents. She continues the struggle with unflagging energy in every venue possible. Mrs. Eason is but one of many senior women involved in WHE ACT and other community renewal endeavors; each is unique and invaluable in what they have done for West Harlem.

The fact that it was predominantly women who organized against the NRSTP and continued to respond to issues over the years is in keeping with traditional expectations for women's roles, while at the same time it allowed women to explore new realms of activism. Particularly in African-American communities, women have long filled the unpaid, yet crucial ranks within community organizational structures. Throughout West Harlem, in almost every community organization, it is women who are responsible for day-to-day education and social welfare. It is women who somehow find a way to address these community issues in addition to their work, family, and home. For example, while every denomination and church in Harlem is headed by a man, they all depend on the labor and commitment of women to organize and carry out their activities. While there are men who are concerned and involved at the community level, women, as a group, seem to have a special connection to their homeplace and play a disproportionate role in voluntary community initiatives and leadership.

The deeply embedded gender divisions of community work, coupled with an impassioned group of senior citizens, has provided the context for a vibrant and committed movement working to preserve a decent quality of life in West Harlem. The nature of the community responses to the NRSTP was directly shaped by these socially constructed gender roles. While men initially reacted to employment issues, women later mobilized around health concerns for the community and their children. The issues that were raised, the women who voiced them, and the focus on dialogue with the city all broadened the traditionally "male" discourse of the Harlem community's struggles to reflect women's new found voice.

Women challenged their traditionally "apolitical" position within the West Harlem community when they acted on their opposition to the NRSTP in a highly visible and politicized arena. The creation of WHE ACT, as primarily a women's endeavor, empowered women to take a "visible" stand against a major contaminant in their lives.[9] Women who had traditionally been the linchpins of community and church organizations, and had subsequently been deprived of the opportunity to rise to outward leadership roles, were able to experience such roles for the first time through their involvement in WHE ACT. Many women found strength and leadership in the struggle against the NRSTP. They moved out of the "acceptable" realm of "behind the scenes" advocacy into an expanding dialogue. As they engaged the city administration and elected officials in a citywide dialogue and began networking with organizations outside their community, the old mold began to break. The efforts of WHE ACT provided an avenue in which change could begin and paved the way for future generations of women to follow.

These pre-established, socially defined gender roles made it possible for women in West Harlem to unite against a common threat. Yet their efforts to challenge these gender expectations elicited strong resistance, and even disapproval, from the predominantly male political machine. It was considered appropriate for women to take on activist roles in community affairs only insofar as they remained in unpaid, "secondary" positions. The paid, visible, political arena was seen as the male domain and women's efforts to enter it with their own agendas were met with scorn. It was in this context that Peggy Shepard, cofounder of WHE ACT and WHID, ran for political office in West Harlem and experienced firsthand the sexism embedded in New York City and Harlem politics.

WOMEN ENTER THE POLITICAL ARENA

Peggy Shepard had long been an active West Harlem resident and integral member of the West Harlem community. She had played an important, and at times critical, role in community level empowerment and politics. She was actively involved in a variety of community-based groups, both multicultural and primarily white, including the Rainbow Coalition, as well as in Democratic Party politics. Since 1985, Peggy Shepard had been a Democratic district leader, an elected unpaid party position on the lowest tier of the elected representative structure. Subsequently, in 1985 she cofounded WHID and served as one of the elected officials required in a recognized party club. In 1988, she again illustrated her conviction and concern for West Harlem as cofounder of WHE ACT.

In 1991, New York City passed a new charter to expand the number of City Council seats through redistricting. In an effort to achieve self-determination for West Harlem, WHE ACT and WHID fought for a separate West Harlem City Council district. Having successfully achieved that goal, WHE ACT members made a collective decision to nominate Peggy Shepard for the council seat. In keeping with a deeply ingrained pattern of contention between the Harlem communities, they received very little support from their Central Harlem neighbors. The Harlem-elected officials, including other African-Americans, feared an independent base which might one day challenge their political machine. In the end, the city gerrymandered[10] the districts in such a way that West Harlem became an appendage of a council district which extended north to include Washington Heights, a heavily Latino neighborhood, and Inwood, a predominantly white neighborhood, which was held by a well-established, twelve-year, white incumbent (Roberts 1991; Shipp 1991).

Peggy Shepard campaigned to represent Washington Heights and Inwood as well as West Harlem, since even with the newly drawn district she posed a significant threat to her white male opponent and there was no guarantee he would win. Her long progressive history of involvement in community issues and coalition building at every level was consonant with the neighborhood preservation initiatives and environmentally conscious nature of Washington Heights and Inwood. However, as soon as Peggy Shepard stepped out of the traditional community service arena and became an active political player, she lost the support of Harlem's male-dominated political structure. Her advocacy was acceptable, even expected, when she was in an "unthreatening," volunteer role. But she and other West Harlem women crossed the unspoken line when they determined that there was no reason why independent women like herself should not have access to equal power and political opportunity. Peggy Shepard had even been a "team player," working with men collectively and maintaining positive relationships, but when it came time for her to run for elections they refused to give her their endorsement. One of Harlem's most powerful national and local politicians vigorously campaigned for the white candidate rather than "create a monster in his own backyard" through supporting Peggy Shepard. At least he could "control" the well-established incumbent, but not an African-American woman and her allies. In a bitter battle, the African-American vote was split between Peggy Shepard and a popular, well-known African-American Nationalist attorney from Riverside Drive. The incumbent from Inwood took the solid white vote and won. The result of this election was that half of Harlem was to be represented by a white man, largely because the protective African-American male elected officials preferred a white male candidate over an African-America woman from their own community with an independent base in feminist, multicultural, and environmentalist politics.

Many women were also reluctant to support Peggy Shepard's candidacy. While there have been women elected to political office in Harlem, they have played a male-defined political game and have not raised gender issues and addressed differences between men and women as constituents. Peggy Shepard is chair of the Manhattan Women's Political Caucus, a board member of the National Women's Political Caucus, and has a well-established history as an activist around reproductive rights, abortion, and gay rights.

These issues were central to her campaign, especially in the context of her community work and that of WHE ACT. Her political agenda was clear to the electorate and in a community that has not fully recognized these issues, her views were threatening not only to men, but (in some sectors) to women as well. Harlem and the African-American community are still very much influenced by the Church with its conservative religious ideology and conservative politics. Peggy Shepard, and WHE ACT, had made such an independent base around feminist and environmental issues that the Harlem political establishment, including many women, felt threatened and would not support her candidacy.

In 1992, Peggy Shepard ran for political office a second time. She campaigned to represent the 70th district in the state assembly, which includes all of Harlem, and was again faced with opposition from much of her community and the political establishment. Even though she built her campaign around community quality of life issues, the conservative political machine could not embrace her as an outspoken, progressive feminist. There was never a sense from African-American women in elected positions that they could stand together against the "old boys" who have always controlled Harlem politics. They viewed Peggy Shepard as competition rather than an ally. Eventually, the race came down to Peggy Shepard and the son of a well-known, popular African-American judge, and she lost by less than eight hundred votes.

WHE ACT BROADENS THE SCOPE: LEGAL ACTION AGAINST NEW YORK CITY AND THE NORTH RIVER SEWAGE TREATMENT PLANT

While WHID was pursuing potential gains in the electoral arena, construction of the NRSTP, plans for River Bank State Park (RBSP), and WHE ACT opposition, continued. In 1992, the Sewage Treatment Plant was "completed" and, after a twenty-three-year fight to maintain the funding line for RBSP, West Harlem could begin to look forward to the imminent opening of this long awaited park. Activists in WHE ACT, however, were not elated with these plans. Their position had consistently been that not only was it absurd to build a park atop a sewage treatment plant, but that construction should at least be delayed until ambient emission tests could be completed. WHE ACT petitioned the state to allow the fund for RBSP to be used to refurbish Riverside Park and possibly build a facility in another location in West Harlem. They were informed that if the money were not used for RBSP, it would revert to the New York State treasury. West Harlem residents were not about to give up their decade-old fight for this necessary facility, even if its location posed a potential health threat.

On Memorial Day weekend 1993, River Bank State Park opened with great fanfare (Holloway 1993). There were marching bands, schools were let out early, school children were bussed in, and Governor Cuomo came to inaugurate the new park. More than 60,000 people used River Bank State Park each day of this first weekend. The park was a state-of-the-art, twenty-eight acre facility with a football field, soccer field, track, Olympic-size swimming pool, basketball, handball, and tennis courts, baseball fields, an ice-skating and a roller-skating rink, a carousel (under construction), a restaurant, five other food stations, a greenhouse, indoor gym, indoor cultural and performance center, and an amphitheater (Gold 1991a; Holloway 1992; Severo 1989). Under other circumstances it would have been a tremendous resource for the community, but it sat atop a sewage treatment plant that did not operate properly and had not had an adequate environmental impact study or ambient emissions tests (Gold 1991a and b; Miller 1994).

For more than half a century West Harlem had little to no open space, but now the community has perhaps the most magnificent facility, in terms of recreation, in New York City, if not the state. In the opinion of WHE ACT members, however, the $1.1 billion spent on the plant and the $128 million spent to build the park have been used to the detriment of West Harlem rather than for its benefit. Some of the community's most vulnerable inhabitants, children and young mothers with babies and toddlers, represent the park's most frequent users. Babies who cannot even walk yet and children who do not know better are unwittingly taken right to the source of pollution. Upon entering the park, fourteen towering stacks boldly present themselves, directly abutting the toddler playground and only fifty feet from the restaurant. Even if the plant operated safely, there is a certain inanity in building a substantial recreational facility on the roof of a noxious plant. There are days when the odor of raw sewage is like a brick wall which cuts though the park. River Bank State Park condemns innocent park users to unknown health risks and WHE ACT vehemently advocates that RBSP should be closed until substantial testing can be done.

From the onset of their formation into a new organization with a more comprehensive agenda, including a broader definition of environmental issues in West Harlem, WHE ACT members began to build relationships and networks with mainstream environmental organizations. They found that a variety of urban environmental problems were being addressed, but very few people had made the connection between social stratification (race and class) and environmental quality and protection. Wherever possible, WHE ACT worked at educating people about the concept of environmental racism and the environmental justice movement that has emerged to address these assaults on low-income communities and people of color. Eventually, through individual relationships with people in various organizations, WHE ACT was able to attract citywide attention to its conflict with the NRSTP and publicize the implications of its struggle for other marginalized communities.

In particular, WHE ACT was able to forge a crucial and lasting relationship with the Natural Resources Defense Council (NRDC). Peggy Shepard was instrumental in introducing Eric Goldstein, a senior attorney for NRDC and director of its New York urban program, to the problems facing WHE ACT and to convince him of the community's need for technical and legal assistance. Although Eric Goldstein was predisposed to collaborate on this issue, he and WHE ACT leaders engaged in a lengthy discussion process before they reached a point of clarity on the idea of environmental racism and began to work together on a variety of projects.[11] He had played a decisive role in the New York City gasoline deleading legislation, which served as a model for national policy. He had also been involved with clean water, water supply, and transportation policy for New York, so WHE ACT knew that he could potentially become an invaluable member of their team.

From 1990–1, the NRDC researched the issues surrounding NRSTP and arranged pro bono representation by an established and well-respected private law firm. In June of 1992, WHE ACT and NRDC, as coplaintiffs, filed a lawsuit against the New York City Department of Environmental Protection and the City of New York with David Dinkins as Mayor. They declared North River Sewage Treatment a public and private nuisance to the people who live and/or own property in West Harlem and had been prevented from enjoying the full fruits of their property because of the operations and malfunctions of the plant (Shepard 1994). The community was informed of the impending lawsuit and, through a screening process with the lawyers, seven West Harlem residents, including Mrs. Eason and the Hamilton-Granes Day Care Center – located right across the street from the NRSTP – were also chosen to serve as plaintiffs (Miller 1994; Officer 1993; WHE ACT 1993).

Throughout his long political career David Dinkins had been a consistent advocate for the West Harlem community, having lived there for thirty-six years before becoming mayor. In 1968, when the city's plans to construct the NRSTP along Riverside Drive were revealed, Dinkins was a Democratic district leader for West Harlem and has since been a staunch opponent of the plant. He was a member of a community group opposed to the siting of the plant in those early days, and has subsequently carried his opposition through to support the efforts of WHID and WHE ACT (King 1968; Miller 1994). His political career began in much the same way as Peggy Shepard's did in the 1980s with their comparable roles in the Democratic Party, commitment to similar issues, and involvement in community organizations. As mayor, he chose to take a more balanced position, but he never compromised his fundamental principles and relationship with West Harlem.

Although David Dinkins was a long-time family friend to Peggy Shepard, as well as to other members of WHE ACT, this unprecedented rise to the mayoral office by an African-American presented a fortuitous time in history for WHE ACT to take action against the city. Additionally, unlike his predecessor and potential successor, Dinkins was generally heralded by environmental groups for his pro-environmentalist agenda. While WHE ACT bore no personal grudge against Dinkins, as a lawyer and community activist he understood that his position as mayor offered them a rare opportunity. There was no assurance that he would win the next election in 1993, and WHE ACT wanted to guaranty that all of the good work and intentions of the Dinkins administration would be brought to fruition in the event that he lost the election. The reality was that policy-making is determined by the mayor and therefore, strategically, WHE ACT had to pursue the lawsuit while Dinkins was still in office.

Immediately after NRDC informed the New York City administration of the impending lawsuit, both city and state responded vehemently that the claim was ridiculous and that everything was already under control. Meanwhile, in an effort to build a case for dismissal, they began an extensive administrative process including a consent decree in which they outlined their plans to try to fix whatever problems there might be and budgeted an extra $55 million of capital funds toward work on the plant (*Amicus Journal* 1994; Miller 1994; Shepard 1994; Specter 1992; WHE ACT 1994). WHE ACT's case argued that the state was completely responsible for adequately monitoring the operations of North River and should never have allowed the plant to start operating in the first place. Amid dissension and disapproval from members of their community who argued that they were going too far, WHE ACT members continued to build their case against the city. They submitted numerous historical documents and briefs and the chair of the local Community Board filed an affidavit which included a history of the board's activities against the NRSTP since 1968. WHE ACT requested a motion to enjoin the City of New York from allowing North River to operate under the circumstances and require the city to bring in an independent consultant to undertake a review before continuing with discussions. The city responded by filing a motion for dismissal, pointing to the consent decree as evidence that it was in the process of rectifying the situation (Miller 1994; WHE ACT 1993).

The city administration also asserted that the state and the city represent the people and that, therefore, when the city and state create an administrative process the interests of the people have been amply represented. WHE ACT won a significant victory in May of 1993, when Alice Schlesinger – a progressive judge – ruled in their favor, stating that under no circumstance does a community have one sole voice of representation in terms of its interests, that is the state or municipality. A community, Judge Schlesinger wrote, always has the right to speak for itself.[12] With this unprecedented ruling, she wrote in her brief what has become a fundamental element in the concept of environmental justice: a community's right to seek redress of grievance. WHE ACT won against the city and the action to dismiss (Miller 1994; Shepard 1994; WHE ACT 1994).

Much of WHE ACT's legal success can be attributed to its representation by a large corporate law firm under a pro bono arrangement facilitated by NRDC. It was virtually unprecedented for a community-based group to acquire legal resources that match or exceed that of the entity it is opposing, particularly with the prestige and added corporate weight of the law firm which represented WHE ACT. The caliber of the case presented by WHE ACT left the City of New York with no other realistic choice but to consider settling out of court. The administration could not respond to the plethora of issues raised and documented by WHE ACT, NRDC, and their lawyers. It also soon became apparent that Dinkins would more than likely lose the impending election, and so WHE ACT had to weigh the possibility of an out-of-court settlement versus a potential loss to an unsympathetic, and perhaps even hostile, new administration.

Elections were held in the second week of November. Dinkins did not win and on December 30, 1993 he was to leave office. WHE ACT now had six weeks to settle with the city. On the final day of the Dinkins administration, the necessary signatures were collected. With only minutes to spare, the settlement agreement was brought from city hall to the courthouse for its final signature. After a brief overview of the forty-four page document, Judge Schlesinger signed the agreement, thereby making it law. The agreement called for a $1.1 million settlement fund for the West Harlem community to be administered by WHE ACT and the NRDC. WHE ACT was also made a party to the consent decree that the city and state had signed as an equivalent entity. In essence, WHE ACT was empowered to determine whether the actions of the New York State Department of Environmental Conservation and the New York City Department of Environmental Protection were sufficient to rectify the problems with the NRSTP. If not, WHE ACT could halt the plant's operations and bring in independent consultants to review the process and make recommendations to the judge who would then determine how and when the Department of Environmental Protection should continue. WHE ACT was also empowered to monitor flow and capacity in the plant, giving the organization a greater say in terms of economic development in West Harlem as well as lower Manhattan (*Amicus Journal* 1994; Miller 1994; Shepard 1994; WHE ACT 1994).

THE FUSING OF TWO WORLDS: NEW VOICES AND OLD RESOURCES

The process by which women in West Harlem challenged the racially based denigration of their community and the flagrant disregard for their quality of life involved a critical fusion of traditional gendered roles with a new activism and a willingness to confront the male-dominated political machine. The strength and wisdom of the senior women, combined with the conviction and drive of the younger women, empowered them to seek an effective means by which to continue their efforts and ultimately see their struggle to fruition. Women in West Harlem effectively asserted themselves in the political arena in an unprecedented manner and began the arduous, but necessary transformation of "old boy" politics.

The persistent efforts to ameliorate the negative impacts of NRSTP, as manifested by women uniting under a common concern for and investment in their community, was built upon a synthesis of two distinct yet overlapping cultures of women's activism in West Harlem. The traditional advocacy role of women within the community service arena, as exemplified by Mrs. Eason, coincided with issues of quality of life and brought a crucial cultural and historical foundation to the struggle. Peggy Shepard, and her generation of younger more activist-oriented women, provided a complementary and instrumental entrance to the corridors of power. The strengths that each group of women brought to the struggle were integral to its effectiveness.

As their opposition evolved and they moved together in a new direction, affecting the system, the two groups of women also influenced each other. Mrs. Eason and the other senior women found themselves stopping traffic in an act of civil disobedience. They learned that their experiences, concerns, and knowledge were a valid and invaluable resource for creating change in West Harlem. In the end, they discovered a new voice and were empowered to take a stand for their community. At the same time, Peggy Shepard was able to use her more visible and recognized position within the community political structure to engage "women's" issues and environmental concerns within the previously exclusionary male political discourse. While she drew on the commitment and passion of the senior women, these women gained inspiration from the respect which Peggy Shepard commanded within political circles. Together they opened the field, establishing and empowering a new cadre of feminist and environmentalist community activists and politicians for a new, more inclusive politics.

Because women define "environment," including "their" own environments, differently from men and perceive alterations and tolerate change on different levels, these newly expanded politics have come to revolve around a reconceptualization of environmental and quality of life issues. Environmental issues and feminist women who uphold them enter into an arena where they were not previously recognized as legitimate. The transformation of politics also involves the recognition of "local" knowledge as a valuable resource. This fusion of old with new, in this case traditional male politics and agendas with women's interests and concerns, is critical to disadvantaged communities effectively defending themselves against larger scale processes which disregard community integrity and interests.

This fusion is exemplified in the transformation of the local Community Board into an arena for the interjection of traditional "women's" issues in the political discourse. The same Community Board, which is responsible for all land use and zoning decisions in West Harlem and remained ineffectual throughout early opposition and struggle against the NRSTP, is now dominated by women's involvement and concerns. In an effort to have a greater impact on community development and avoid a repeat of North River, women in West Harlem have brought their agendas to this political forum and have made quality of life issues central to the Community Board dialogue. Women have begun to draw the connections between zoning, transportation, sanitation, and quality of life. They are now in the process of educating the larger community about these issues and about the relationship between them. Women community activists have committed themselves to fostering a healthy community with sound development initiatives.

This fusion of traditional values and women's community-based strengths with new energy and conviction is fundamental to the transformation of politics. As is demonstrated in the story of WHE ACT and the women of West Harlem, this process is critical to effective community struggles for environmental justice. Although women have entered the work force in significant numbers, they also continue in their community service roles and as the primary "caregivers." As in West Harlem, women throughout the world are often the first to notice and respond to changes in the quality of life in their neighborhoods and to draw connections between declining health conditions and the newest hazardous facility in their community. While environmental racism is the result of complex social and political processes, it manifests itself as an affront to the day-to-day life experiences within a given community. Perceiving these threats to their livelihoods, health and well-being, women have challenged traditional gender expectations and empowered themselves as the central activists for the preservation of their communities and a decent quality of life. Women in West Harlem engaged the political machine with an unprecedented conviction and showed that their concerns are valid and their activism central to combating environmental racism.

NOTES

1　This chapter represents a collaborative writing effort based on extensive interviews and discussions with as well as presentations by Vernice Miller recounting and analyzing the recent history of WHE ACT from her own position as one of the two founders.

2　The following abbreviations will be used frequently throughout the text:

CCANR	Community Coalition Against North River
NRDC	Natural Resources Defense Council
NRSTP	North River Sewage Treatment Plant
RBSP	River Bank State Park
UCC	United Church of Christ
WHE ACT	West Harlem Environmental Action
WHID	West Harlem Independent Democratic Club

3　When construction of the NRSTP began in 1972, the plant was designed to treat 220 million gallons per day. However, subsequent federal and city studies found less population growth throughout the 1970s than had been estimated and the Associated Engineers design plan for the North River Facility, completed in 1979, suggested that a smaller capacity plant would be adequate. The United States Environmental Protection Agency, which was funding 75 percent of the project, decided it would not pay for the larger plant. Thus, after construction had begun the capacity was reduced to 170 million gallons per day. As a result, since NRSTP began operating in 1986 it has been at or near full capacity (Gold 1991c; Miller 1994).

4　On July 1, 1992, the New York State Department of Environmental Conservation issued a Control Order in response to environmental violations at the North River Plant. The violations included exceeding the limitation for daily and annual dry weather flow; violations of parameter limits for fecal coliform bacteria; failure to notify the Department of Environmental Conservation of an interruption in chlorination; high nitrogin oxide emission rates; high hydrogen sulfide concentrations; failure to connect the odor control system when loading sludge onto a barge; excessive smoke emissions; and various odor problems (New York City 1992; Miller 1994).

5　"In 1990 the overall asthma hospitalization rate [in West Harlem] was 123 per 10,000, approximately three times the New York City rate within each age group. Hospital admissions were highest for the 0–4 age group" (Ford 1993).

6　In sufficient quantities hydrogen sulfide and sulfur dioxide deprive human cells of oxygen, further exacerbating sickle-cell conditions. North River repeatedly emitted both of these gases in excess of New York State air quality standards (New York City 1992; Severo 1989).

7　On February 3, 1995 – with disregard for the strenuous objections of environmentalists, Harlem residents, and some elected officials – New York City awarded Donald Trump permission to connect the first phase of his Riverside South commercial and residential development to the NRSTP, further increasing its overburdened capacity level (Dunlap 1992; Kennedy 1995).

8　Other groups in New York City concerned with issues of environmental racism, including the Toxic Avengers and the Magnolia Tree Earth Center, operate solely on the neighborhood level.

9　The founders of WHE ACT were Peggy Shepard, who is discussed below, Vernice Miller, one of the authors of this chapter, and Chuck Sutton. The two women have continued in leadership positions within WHE ACT and in liaison with the broader environmental justice movement.

10 "[To] divide (a territorial unit) into election districts to give one political party an electoral majority in a large number of districts while concentrating the voting strength of the opposition in as few districts as possible" (Merriam-Webster's Collegiate Dictionary, Tenth Edition, 1993). From Elbridge *Gerry* and sala*mander*, from the shape of an election district formed during Gerry's governorship of Massachusetts.

11 For example, working in partnership, WHE ACT and NRDC contacted West Harlem community leaders to convene and define their environmental agenda. The group known as The West Harlem Community Leadership Focus Group was asked to determine the primary, nonsocial conditions negatively impacting West Harlem's residents. They articulated the following twelve environmental concerns: land use, sanitation, toxins, pest control, parks and open space, infrastructure, consumer issues, air pollution, noise, vacant lots, transportation, and overall enforcement of environmental regulations (Shepard 1994: 742).

12 "[b]eyond these citizens' rights to be heard on an issue ... there is even a greater interest at stake. Individuals, while represented by elected and appointed officials have a basic right to seek redress of their grievances in a court of law even against those same elected and appointed officials" (WHE ACT 1993).

REFERENCES

Amicus Journal, The (1994) "In Defence of the Environment: West Harlem Justice," 16, 1: 53.

Anthony, C. (1990) "Why African Americans Should be Environmental," *Earth Island Journal*, Vol 5, 4: Winter: 43–44.

Ash, C. (1992) Remarks at Community Board Public Hearing on the Riverside South Development Project and the North River Sewage Treatment Plant at the Columbia University School of Law, by the Regional Director, Department of Environmental Conservation, February.

Browne, J. Z. (1994) "North River Sewage Plant Continues to Emit Odors," *Amsterdam News*, June 11: 6.

Bryant, B. and Mohai, P. (eds.) (1992) *Race and the Decline of Environmental Hazards*, Boulder, Colorado: Westview Press.

Bullard, R. (1990) *Dumping in Dixie: Race, Class and Environmental Quality*, Boulder, Colorado: Westview Press.

—— (1992) *People of Color Environmental Groups Directory 1992*, Riverdale, California: University of California, Department of Sociology.

Caro, R. (1974) *The Power Broker: Robert Moses and the Fall of New York*, New York: Knopf.

Commoner, B. (1987) Report on the North River Sewage Treatment Plant. Unpublished manuscript.

Dockery, D. *et al.* (1993) "An Association Between Air Pollution and Mortality in Six U.S. Cities," *New England Journal of Medicine* Vol. 329: 24, 1753–1759.

Dunlap, D. W. (1992) "Altered Riverside South Plan Wins Messinger Over," *New York Times*, August 27: Section B: 2.

Durning, A. (1989) *Action at the Grassroots: Fighting Poverty and Environmental Decline*, Washington, D.C.: Worldwatch Institute.

Engineering News Record (1984) "Sewage Plant Rises Above River; Manhattan Plant Will Treat Wastes While Supporting Park," June 21.

Ford, J. (1993) Presentation to the American Thoracic Society International Meeting, May 16.

Gold, A. R. (1991a) "Sewage Plant Near Capacity After 5 Years," *New York Times*, April 12: Section B: 1.

—— (1991b) "Turning Sewage Plants into Friendly Neighbors," *New York Times*, August 16: Section B: 3.

—— (1991c) "Despite Decades of Spending, Sewage Plants are Full Up," *New York Times*, August 18: Section 4: 16.

Goldstein, E. A. and Izeman, M. A. (1990) *The New York Environment Book 172*, Natural Resources Defense Council, N.Y., New York.

Grossman, K. (1992) "From Toxic Racism to Environmental Justice," *E: The Environmental Magazine* 3, 3: 28–35.

Holloway, L. (1992) "28 Acres of Roof and a Place to Play in West Harlem," *New York Times*, September 1: Section B: 1.

—— (1993) "Park, However it Smells, Blossoms on the River," *New York Times*, May 28: Section B: 3.

Kennedy, S. G. (1995) "Trump Development Clears Hurdle Despite Objections," *New York Times*, February 4: Section A: 23.

King, S. S. (1968) "Sewage Plan Voted Over Harlem Protest," *New York Times*, April 26: Section A: 1.

Lee, C. (1987) *Toxic Wastes and Race in the United States*, Washington, D.C.: United Church of Christ Commission for Racial Justice.

—— (1992) *Proceedings: The First National People of Color Environmental Leadership Summit*, Washington, D.C.: United Church of Christ Commission for Racial Justice.

Martinez, E. (1991) "When People of Color are an Endangered Species," *Z Magazine*, 5, 4: 33–38.

Miller, V. D. (1993) "Building on Our Past, Planning for Our Future: Communities of Color and the Quest for Environmental Justice," in R. Hofrichter (ed.) *Toxic Struggles – The Theory and Practice of Environmental Justice*, Philadelphia: New Society Publishers.

—— (1994) "Planning, Power and Politics: A Case Study of the Land Use and Siting History of the North River Water Pollution Control Plant," *Fordham Urban Law Journal* XXI: 707–722.

Muzynski, W. J. (1993) Public Announcement, by the Acting Regional Administrator, U.S. Environmental Protection Agency, May 3.

National Law Journal, The (1992) "Unequal Protection: the Racial Divide in Environmental Law," 15, 3: S1–S12.

New York City (1984) Department of Environmental Protection, "Report on the North River Water Pollution Treatment Plant 3," New York City: DEP.

—— (1992) Department of Environmental Protection, No. R2-3669-91-05 "North River Sewage Treatment Plant – Odor, Flow, and Air Emissions Control Order," New York State Department of Environmental Conservation, July 1.

Officer, D. (1993) "1.1 Million to Harlem Environmental Group," *Amsterdam News*, September 18: 4.

Public Hearing (1968) Capital Project No. PW-164, "North River Pollution Control Project," New York City Records, April 25.

Roberts, S. (1991) "Redistricting Oddities Reflect Racial and Ethnic Politics," *New York Times*, May 7: Section 1: 1.

Ryan, W. F., Congressman (1968) Memorandum to the Hon. Stewart L. Udall, Secretary of the Interior 2, April 18.

Schultz, H. (1989) "North River Plant in Bad Odor Because of Sewer," *New York Times*, December 29: Section A: 34.

Schwartz, L. (1966) Memorandum to S. W. Steffensen, Director, Bureau of Water Pollution Control 1 from L. Schwartz, Chief, Division of Plant Design, Bureau of Water Pollution Control, New York City Department of Public Works, May 9.

Severo, R. (1989) "Odors from Plant Anger Many in Harlem," *New York Times*, November 30: Section B: 6.

Shepard, P. M. (1994) "Issues of Community Empowerment," *Fordham Urban Law Journal* XXI: 739–55.

Shipp, E.R. (1991) "Just Where is Harlem?" *New York Times*, May 8: Section B: 2.

Specter, M. (1992) "Stench at Sewage Plant is Traced; Millions Pledge for Repair Work," *New York Times*, April 17: Section 1.

WHE ACT (1993) West Harlem Environmental Action vs New York City Department of Environmental Protection, No. 92-45133 (Sup. Ct. N.Y. County May 17).

—— (1994) West Harlem Environmental Action, Stipulation for Settlement, No. 92–45133 (filed January 4).

4

PROTECTING THE ENVIRONMENT AGAINST STATE POLICY IN AUSTRIA

From women's participation in protest to new voices in parliament

Doris Wastl-Walter

(Translation from German by Susanne C. Moser)

CRISIS IN AUSTRIA: UNORGANIZED ENVIRONMENTALISTS DEFEND THE LAST DANUBE RIVERINE FOREST AGAINST CORPORATE AND STATE INTERESTS

A local conflict as the occasion for a change of values and political structures

In Austria, as in many Western European countries,[1] a healthy environment and unlimited natural resources have usually been taken for granted. During the immediate postwar era the desire for a higher standard of living, more comfort, and more material goods was dominant. Hardly anyone cared about the abuse of the natural environment and the damage to the ecosystem as long as it served human needs. Nature was not regarded as something valuable and the preservation of the ecological heritage was neglected.

But the gendered experience of the environment and an emerging consciousness of the responsibility for future generations, finally led to a movement of women. Previously, women had been politically marginalized and were not used to expressing in public their discontent and disagreement with the prevailing economic aims and political practices. The gendered awareness of problems and the difficulties of gaining access to the political process encouraged women to create their own political organizations in order to address their feminist ecological concerns.

Women drew attention to and supported environmental agendas by means of nonviolent protest and by giving voice to the powerless. Environmental and green politics propelled women and feminist issues into the national political mainstream process.

THE CONFLICT

After World War II, Western Europe had to cope with extensive destruction and severe damage caused by the war. Infrastructure was deficient and industry was directed toward the requirements of war. In addition, the Nazi regime had brought about the loss of political and democratic culture. In 1945, economic recovery and the reconstruction of social and political structures were initiated. Austria's main objective was the re-establishment of an independent state with a workable parliamentary system as well as the development of a modern economy.

At that time "development" was regarded as economic growth, with little attention to any damage to the environment. Austrian politics were characterized by the so-called "social partnership" between the two sides of industry, that is employers and employees, and the coalition of the two major political parties. Political development was dominated by the cooperation of the leading, mostly male, political functionaries.

Only after 1980, at a time when the Western European economic rise became obvious and Austria had become one of the twelve richest countries in the world, did some people begin to see the burden being put on the environment. They realized the price that had to be paid for such enormous economic growth and considered it too high.

As usual, it was not the (male) political elite that demanded a debate about the value system of the modern consumer society. A political movement initiated mostly by women, who had hardly been active in politics before and who were neither experienced nor organized, questioned the basic consensus on the system. Their spontaneous, amateurish actions at various Western European locations led to the most important changes in democratic and economic development policy since 1945 and a shift in social values. Today environmental and feminist politics are spreading in Western Europe and economic growth, propagated predominantly by the older (male) generation, is being called into question by the younger generation, and by women of all age groups. All over Western Europe the consequences are reflected in election results: green parties especially, but also other political parties supporting the principle of equal rights, have become very successful.

In the early 1980s, planners in Austria debated the possibility of building a hydroelectric power plant in the last extensive, contiguously forested, riverine wetland area of Central Europe. Except for these 50 kilometers (about 30 miles) of riverine landscape, the Danube River in Austria was already harnessed to a chain of hydroelectric power plants. For the Donaukraftwerks AG (DOKW), the Danube Power Company Inc., it was only a logical extension of its previous activities to build another plant in this last remaining "unused" stretch of the river.

According to the facility planners, the ideal site for the new plant was Hainburg, a town of 5,731 inhabitants about 40 kilometers (25 miles) east of Vienna, located in a particularly scenic natural landscape. Untouched riverine forests, wetlands and regularly flooded areas form a unique ecosystem and habitat for many rare and some endangered species. For good reason, therefore, the area was put under landscape protection in 1982. This meant that no industrial or other facilities could be built in the area if they significantly impaired the character of the landscape, its beauty and unique character, and its recreational value for the population and tourism

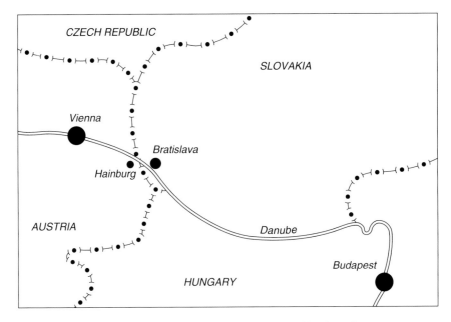

Figure 4.1 Regional and national map showing Hainburg forest

(Niederösterreichisches Naturschutzgesetz [Lower Austrian Nature Conservation Law] 1982, Section 6, Paragraph 2).

Approximately 500 hectares (about 1,240 acres) of riverine forest or 1.2 million trees would have had to be cleared for the planned construction of the hydroelectric power plant. Structures up to 18 meters (about 60 feet) high and 22 kilometers (14 miles) of 8-meter (about 25-foot) high levees would have had to be built, and the riverbed would have been altered over a 5-kilometer (3-mile) stretch. Together these changes would have amounted to an irreversible encroachment on this unique natural landscape with its irreplaceable recreational value for the Vienna metropolitan area. No national agency demonstrated any concern over this potential loss. In response to this inaction, various citizens' environmental initiatives formed to prevent the power plant's construction.

The interest groups and their agendas

In 1947, after World War II, companies were established in Austria to build and maintain hydroelectric power plants to ensure the necessary energy for the economic development of the country. To a large extent they are publicly owned and legally mandated to guaranty Austria's electricity supply. One of these is the Danube Power Company which constructs and manages the hydroelectric power plants along the Danube River and intends to erect more dams based on their managers' interpretation of the company's responsibility defined as "we need more energy for future development."

The construction industry and labor unions supported the Danube Power Company at a time of economic austerity, especially since worker teams and

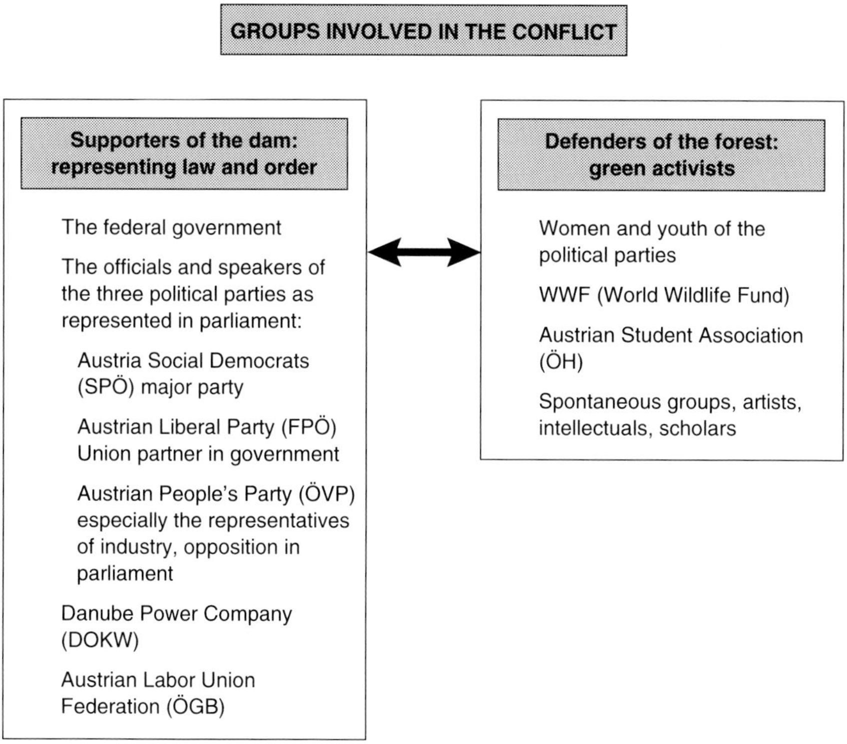

GROUPS INVOLVED IN THE CONFLICT

Supporters of the dam: representing law and order

The federal government

The officials and speakers of the three political parties as represented in parliament:

Austria Social Democrats (SPÖ) major party

Austrian Liberal Party (FPÖ) Union partner in government

Austrian People's Party (ÖVP) especially the representatives of industry, opposition in parliament

Danube Power Company (DOKW)

Austrian Labor Union Federation (ÖGB)

⟷

Defenders of the forest: green activists

Women and youth of the political parties

WWF (World Wildlife Fund)

Austrian Student Association (ÖH)

Spontaneous groups, artists, intellectuals, scholars

Figure 4.2 Organigram of groups involved in the conflict

machinery were available from previous power plant construction and they were in danger of losing their jobs. The Austrian Labor Union Federation (Österreichischer Gewerkschaftsbund – ÖGB), which enjoys substantial political clout in Austria, brought all its weight to bear to force the construction of the Hainburg power plant.

In 1984, the Austrian federal government was formed by the so-called small coalition between the largest party, the SPÖ (Sozialdemokratische Partei Österreichs – Austrian Social Democrats), and the smallest party, the FPÖ (Freie Partei Österreichs – Austrian Liberals). The SPÖ – which had previously governed alone for over a decade (from 1970 to 1983) and then declined to a coalition party – stood squarely behind the construction of the power plant and seemed strongly committed to the demonstration of political strength. As the traditional workers' party it was under massive pressure from the unions and it urgently needed to make up for its decline over the previous decade. Only a few years earlier, the first countrywide citizens' initiative, "Mothers Against Nuclear Power," had prevented the first Austrian nuclear power plant in Zwentendorf from going into operation. This led to a referendum that put an end to all plans for the production and use of nuclear power in Austria. This was probably another reason why the SPÖ-dominated government put all its efforts into pushing the construction of this comparatively "clean" hydropower plant.

The parliamentary opposition party ÖVP (Österreichische Volkspartei – Austrian People's Party) was divided on the issue. Traditionally, the ÖVP is very close to industry. Some of its members, especially the construction industry lobby, approved of the project. On the other hand, the voices of several very prominent members were raised against the project, in particular those of the subsequent Secretary of the Environment, Marilies Flemming, representatives of the Women's Organization of the party, and young members of the ÖVP.

The supporters of the power plant were mostly men of the war-and-reconstruction generation whose attitude toward life and livelihood was marked by striving for economic growth and increase of their individual material well-being. Growing energy consumption was regarded as an indicator of progress and ensuring this "progress" with growing numbers of power plants became a political goal. Unlimited resource use with all available technological means was viewed as desirable. Natural landscapes were of use only insofar as their potential could be utilized for material ends. Economic development meant putting to work natural resources which until then had not been economically utilized. "The preservation of toads, frogs, reptiles and mollusks stands no reasonable comparison to the economic value of a power plant" (W. Fremuth, Executive Director, Consortium of the Electric Industry, as quoted in Profil 1984c: 16).

The anti-Hainburg activists came from all political and social walks of life in Austria. Few of these men and women were politicians, managers or public servants. Initially, they were hardly organized or coordinated, and consisted of various grassroots movements that were hard to place in the conventional political fabric. What brought them together was a different perspective on reality. For them, the quality of life could not be reduced to increasing energy consumption and material wealth, but had to include nonmaterial values. Their approach was about a different interaction with each other and nature; their protest was based on nonviolent resistance.

The protesters criticized the unsparing abuse of nature in the name of a growth- and consumption-oriented society. They were not willing to sacrifice the last forested wetland landscape of Central Europe to the electric industry. For these women and men, the riverine forest had inherent value which they sought to preserve for future generations.

> The remaining unclaimed natural landscapes are gaining such significance in people's value system that we have to take this attitude change into account. Let us not make assumptions about the aesthetic values of coming generations. They will certainly look different from those of our Coca-Cola-society.
> (B. Lötsch, Ecologist, as quoted in Profil 1984c: 16).

The mighty agree and a grassroots movement forms

From the beginning of the debate, the voices of the environmentalists were not taken very seriously. The Danube Power Company had no intention of changing its plans to build the plant because of the protest movement. All the company needed was the official permit to begin construction. In November 1984, the state politician responsible granted the construction permit despite the fact that the civil servants in his department and many legal experts were convinced that the construction of the power plant was incompatible with the protection of the riverine forest mandated by the Landscape Conservation Law (Landschaftsschutzgesetz). The construction permit, therefore, was contrary to the spirit of the law but legally binding.

The construction permit unleashed great indignation among large parts of the population and was perceived as an extreme perversion of justice, if not as a breach of the law. However, the permit gave a green light to the construction companies and it was obvious that a clarification of the case in court would take months. The ruling political and economic powers of the country – the government, the unions, industry, and parts of the opposition party – were in favor of construction of the power plant. Federal Chancellor Fred Sinowatz announced that land clearing would be carried out quickly, and the Secretary of Agriculture, Günther Haiden, signed the permit. The construction companies were determined to begin clearing the trees immediately and to confront the protesters with a *fait accompli*. If the Danube forest was to be protected, something had to happen quickly because the environmentalists had no legal recourse other than the possibility of a protracted court case.

At this stage it was foreseeable that the perversion of justice through those in power could no longer be prevented simply by the usual procedures of the constitutional state and parliamentary democracy. A later affirmative judgment would not be able to undo deforestation already carried out, and the resulting loss of the riverine landscape. In this situation and in view of the danger ahead, the women who had thus far expressed their views on the subject only within the frame of parliamentary and party structures took an unprecedented initiative.

On November 29, 1984 women of all political parties and various citizens' initiatives called a press conference. The speaker, Freda Meissner-Blau, stated in her preliminary remarks:

> The coming together of women across all political boundaries has created confusion in all parties. They are not grasping yet that environmental matters – survival matters – cannot be solved by the most powerful party alone, nor by all parties together, but only through the input and cooperation of the entire public. What party can prevent the death of our forests, climate change, or the destruction of our groundwater resources? . . . We are no experts. But if you consider the dead end into which the experts have led us, this appears to be a strength rather than a weakness . . . . Fighting vehemently against the destruction of the very basis of the lives of our children, is to us women as much a matter of common sense as a matter of the heart.

Freda Meissner-Blau (born in 1927) comes from a wealthy, upper middle-class and cosmopolitan background, which, however, did not include political roles for women. Compared to other Austrian women, Meissner-Blau has an atypical history: she is competent in several languages and has lived abroad (England, Zaire, France) for many years. The revolts in Zaire in the 1950s and in Paris in 1968 sharpened her political consciousness. She took the political path rather late when she was already over 40 and had three children. Since the late 1970s she has been a symbolic figure of the "green movement" and has strongly influenced women's consciousness in Austria and its neighboring countries.

Media response to this first extensive action of "negligible women" was more astonished about the fact that women had come together across party boundaries than about their substantive criticism of environmental policy. This first joint action of established party women with other women was based on the motto expressed by Maria Hampel-Fuchs: "To solve environmental issues, the creation of new coalitions is necessary. We as women in particular have the damned duty to do so" (M. Hampel-Fuchs, Austrian People's Party, as quoted in Profil 1984a: 21). This was disquieting to the parties, and the women were accused of lacking party discipline. The common action of "green" women left the impression that their respective parties were not really listening to them. Even so however, the women did not succeed immediately in creating environmental awareness within their organizations and in gaining greater influence.

At the same time, some of the same people who had organized the press conference formed a party-independent activist coalition, an alliance of local citizens' initiatives, the Austrian World Wildlife Fund, committed individuals, and the alternative section of the Austrian Students Association (Österreichische Hochschülerschaft – ÖH). This activist coalition organized the infrastructure necessary – offices, contact addresses, public education material to promote awareness, and fund-raising – to support a broadly based movement and large-scale mobilization.

Since those in power remained largely unimpressed by the press conference and other activities, and the legal means of protest were largely exhausted, a small group of activists decided to take the struggle to the Danube forest itself.

Resistance against the state: The occupation of the Hainburg forest

In the first days of December 1984, many young people went to the Hainburg forest to see what was happening on the spot while others organized a protest march for December 8 which brought people from all over the country together in Vienna to go to Hainburg. What no one had expected was that 8,000 to 10,000 people would march to the forest. Freda Meissner-Blau summarized the mood of the movement: "Thousands of people began to care about their own business, not only to protect the most extensive forested wetland in Central Europe against destruction, but also to save the constitutional state and democracy" (Meissner-Blau 1989: 56).

In the next few days, in spite of the cold, thousands of people decided not only to visit the Hainburg forest but also to stay in order to prevent its logging with their presence. It was their goal to confront the state's power through nonviolent protest: "Nonviolence is the most radical form of resistance because it robs power of its aims" (Meissner-Blau 1989: 59). Yet, the Hainburg forest was declared a construction site and the activists' presence became an act of trespass, that is a criminal offense. Thus, the constitutional state transformed protectors of the environment into violators of the law. They violated the law even if it was only a minor infringement that could be regarded as civil disobedience. This infringement is the most extreme form of nonviolent protest, of appealing to the state against unjust measures, outdated norms, and inhumane actions.

This type of resistance is often used by women (see Weizsäcker and Bücking 1992; Mies and Shiva 1993) who do not want to respond with violence to individual or state violence, and who do not have any other effective means of protest within the existing law. This civil disobedience is not directed against the principles of public order per se, and so does not constitute anarchism, as some officials suggested. Rather this mobilization conformed with the characteristics described as "civil" disobedience by Christian Bay in the *International Encyclopedia of the Social Sciences* (1968). Thus "civil" disobedience is observing one's civic duty, in a nonviolent, civilized, public fashion, oriented toward public welfare. The Danube forest occupants understood their commitment in this manner; they were not extremist left- or right-wing anarchists, as the government alleged them to be. Their actions corresponded much more to the idea of civil disobedience put forth by Henry David Thoreau in the late 1840s: "I think that we should be men [*sic*] first, and subjects afterward. It is not desirable to cultivate a respect for the law, so much as for the right" (Henry David Thoreau in Canby 1937: 790).

Paul Blau captured the spirit of the struggle in the Hainburg forest in this account:

> The nonviolent resistance began. With sleeping bags and tents the first five hundred moved into the forest and waited for the attack. At dawn loggers, bulldozers, and the police came. The wetland protectors sat down in front of the machines, hugged the trees and covered them with their bodies, just as they had learned from the Chipko women in India. The logger teams withdrew.[2]
>
> (Blau 1992: 180)

In spite of the icy temperatures of December, 3,500 to 4,000 men, women, and children continued to camp out in the Hainburg forest. They lived in six tent camps as well as hastily built mud huts, and the regional population and the action committee supplied them with food, equipment, and information. Men and women were equally involved in all types of work; women made up about 50 percent of the Hainburg forest occupants in this "biggest operation of civil disobedience in the Second Republic" (Kolba 1985: 119).[3] They built protection walls; the highest was called the "Mahatma Gandhi Barricade" as an expression of their nonviolent resistance. In response to this spontaneous campaign an unexpected wave of sympathy rolled over Austria and was taken up by many mass media, which provided the activists with moral support. From every corner of the country donations and gifts-in-kind arrived to support the public information campaign and the forest occupation. The environmentalists badly needed this support since – in contrast to the institutions promoting the dam – they had no other resources.

State and industry representatives stressed both the formal correctness of the decision regarding the permit, and the need to maintain law and order. "Law has to remain law ... hence one has to accept that legally made decisions will be implemented" (F. Verzetnitsch, Secretary of the Austrian

Labor Union Federation, as quoted in Profil 1984: 35). Hardly anyone in the government or the parties – except for a few women with little say – grasped that the mood of the general public had already turned. Officials held on to their power and their understanding of legality. "One just has to accept this point of view of the parties, the government, and the Federal President" (F. Verzetnitsch, as quoted in Profil 1984c: 34).

Righteous workers versus green extremists

The powerful state and corporate interests were not initially swayed by the protest. On the contrary, they fought back – physically and verbally. Police forces were sent to the Hainburg forest, but initially exercised restraint. There were few injuries and few arrests. All the more aggressive were the verbal attacks: the unions played a leading role in the confrontation and emphasized that they would not allow an environmentalist minority to terrorize them. Union officials and other power plant supporters used rough language meant to provoke the construction workers into acting against the environmentalists. The environmentalists were slandered as anarchists, extremists, provocateurs, and parasitic students who prevented honest workers from doing their jobs. As one of the speakers expressed it:

> It is those we call coffeehouse ecologists, . . . that are acting as opponents to the Hainburg power plant, as job destroyers so to speak . . . those who turn night into day, who go to bed when other people are rushing to their work places at six o'clock in the morning . . . those who deny people who are willing to work their right to do so.
> (J. Hesoun, Chairman of the Construction Workers and Lumbermen's Union, now Federal Secretary for Labor and Social Affairs, as quoted in Profil 1984b: 14)

Despite thirteen hours of negotiations in parliament and meetings between the protesters and the Federal President and several ministers, the government also heavily criticized the protesters. They were labelled as "forces that want to destabilize democracy" (K. Blecha, Federal Secretary of the Interior, as quoted in Kolba 1985: 119), and the Chancellor suggested that if the government gave in, "one might as well shut down the republic" (F. Sinowatz, Federal Chancellor, as quoted in Kolba 1985: 119).

These remarks further increased the tensions. Large parts of the population sympathized with the wetland protectors, and even the construction workers were more understanding than many expected. As husbands and fathers, many of them were open to the arguments of the environmentalists. The union leadership, however, continued to try to mobilize the workers. They issued a call "to the silent majority of workers" to come to the Hainburg forest as a "demonstration of determination." The union promised transportation and provisions for up to 30,000 workers.

To understand the reasons for the mental shock this proposal generated in the Austrian population, one has to remember the time before Austria's "Anschluss" to Hitler's Germany. The still young democracy of the 1920s had been unable to overcome peacefully the fundamental ideological differences between the Christian Conservatives and the Labor Party by legal, national means. During the time of the First Republic street battles and even a civil war took place because the major parties had organized and mobilized their supporters militantly, so that they existed in opposition to the powerless state army. The lack of respective democratic traditions and adequate strategies for the solution of the conflict finally led to the dissolution of the parliamentary system and to the call for a strong man, which made Hitler's success in Austria possible. This historical consciousness and the resulting fear that old enmities could be reopened is still alive in the population. Consequently, the threat of the mobilization of the workers forced the government into action. It also demonstrated the perplexity of the powerful corporations in the face of this new, green and nonviolent movement, and their inability to deal with it.

When a further attempt to log the Hainburg site failed, the Secretary of the Interior decided to have the police clear the area. Hundreds of policemen – equipped with shields, helmets, riot sticks and guns and accompanied by attack dogs – proceeded against the entirely unarmed protesters. In the course of the "clearing" demonstrators were beaten, kicked, dragged away, and some of them arrested. One of the first seriously injured persons was a 70-year-old woman.

All of Austria was in turmoil over this brutal police intervention against peaceful, unarmed women and men. During the afternoon and evening of the same day, about 40,000 people participated in a spontaneous demonstration in Vienna in solidarity with the environmentalists. No one remained untouched by the events; many private initiatives sprang up. The spontaneously formed committee "Scientists and Artists for Democracy" issued an admonition to the politically responsible of Austria. Union and party members protested strongly and distanced themselves from top-ranking officials. There were numerous public demonstrations and protests, but all of them were peaceful since everyone was conscious of the danger of street fights. People were speaking of the need for a Christmas peace and the need to find a new and peaceful way to solve this type of conflict in a democracy.

By the end of December 1984, the mobilization against the forest clearing and the dam had grown into a much larger and broader movement. There was a storm of protests from inside and outside the country, and at this point, the construction of the power plant was no longer of primary importance. The manner in which the constitutional state had proceeded in this conflict with its citizens, and the interpretation of civil law, had become the central subject of debate.

The point of contention shifts from power plant to democracy

The protectors of the riverine forest had, in fact, broken the law by occupying the Hainburg site and by hindering the officially permitted logging. However, it was the letter of the law versus the environmentalists' sense of justice: they saw their occupation legitimized by the authorities' failure to comply with the Landscape Conservation Law. Eventually, a Supreme Court ruling, in fact, upheld the environmentalists' view. These women and men felt responsible for the stewardship of nature and the preservation of a beautiful and biologically rich forest ecosystem for future generations, even if carrying out their intent had been rendered illegal by the legal decision of the politician responsible.

For the second time since World War II, there was a public confrontation between Austrian women and men and the plans of government and industry. But while women had taken the initiative in the protest against nuclear power ("Mothers Against Nuclear Power"), both men and women were actively involved in the Danube forest. Many men, among them many artists, scientists, and intellectuals, adopted the women's perspective favoring democratic processes and preservation of nature over economic growth. From this standpoint, they fought together against the entrenched notions of economic development and "progress." "While those in power still believe in a final victory of technology over nature, a new Austrian nation has been constituted in the Hainburg forest" (F. Hundertwasser, artist, as quoted in Profil 1984c: 11).[4]

In the course of the conflict, the spontaneous protests against an irreversible destruction of nature turned into a sustained debate over democratic rights, power and violence, responsibility, and social values. The issue at stake was how minorities (and silent majorities) can participate in a political system – a question that particularly concerned women who were and still are barely represented in political parties and institutions. In Austria, as in many other parts of the world, "women's visions of their rights, roles and responsibilities are changing" (see Rocheleau, Thomas-Slayter, and Wangari, Chapter 1, this volume). Feminist and environmentalist agendas converged within a broad wave of democratic concerns over the nature of public participation in government and the values guiding national political life. Protesters and movement sympathizers demanded more opportunities for participation of the public – as in referenda and general polls – and sought greater influence over important decisions with perceived long-term impacts. Repeatedly the occupants of the forest stressed the ecological, economic, and democratic aspects of their concerns. Representatives of corporate and state interests looked at these concerns from an entirely different perspective, and failed to understand any other approach.

In fact, "very few people think political participation is a bad thing" (Birch 1993: 80), but there are very different opinions about what are acceptable forms of participation and opportunities for various groups to represent their interests (see Birch 1993: 80–94). Even among women, there are very different opinions on this issue. What many women hold in common is their rejection of patriarchal domination and state power, backed up by so-called experts.

In contrast, Secretary of State Erich Schmidt stated in an interview: "The introduction of a direct and participatory democracy in Austria would bring with it considerable disadvantages. On many subjects it is impossible to provide every voter . . . with the comprehensive information . . . necessary to be able to weigh sufficiently the advantages and disadvantages of the various alternatives of a political project" (E. Schmidt, Secretary of State, as quoted in Profil 1984b: 21).

Yet it was precisely expert knowledge that was rejected by the wetland occupants. They criticized government reliance on reductionist science in the

Plate 4.1 Demonstrators marching at Hainburg forest, 1 January, 1985

Source: Doris Wastl-Walter

service of a growth-oriented consumer society and advanced against it the idea of responsible use of resources and respect for nature in its capacity to support life on earth.

In this spirit, about 4,000 people spent the days before Christmas in the Hainburg forest. December 21, two days after another harsh confrontation, they expected a further attempt to clear the Hainburg site by 2,400 policemen from Vienna. But at one o'clock in the morning the order was withdrawn. Chancellor Fred Sinowatz proclaimed peace until January 3 to allow reflection on the issues before a political decision was made. This decision never transpired, however, because the Austrian Supreme Court prohibited any logging until the resolution of the court case. On Christmas eve, 10,000 people from all over the country came to celebrate the Christmas mass in the Hainburg forest.

In January 1985, the federal government appointed an Ecology Commission, a forum of renowned scientists who were to determine whether the hydroelectric power plant at Hainburg was acceptable from an ecological point of view. In the fall of the same year, the Commission submitted its final report and the Hainburg site was rejected for ecological reasons.

The confrontation of the year before had impacts far beyond preventing the construction of the power plant: Austrian women and men had woken up from their blissful submissiveness and began to develop the political self-confidence to question and resist powerful state and corporate interests. Since then, more voters have become mobilized; absolute majorities have become harder to obtain in elections; and the spectrum of parties has become more variable. Currently (1995) there are five parties represented in parliament; the two most recent – the "Liberal Forum" and the "Green Party" – are chaired by women. Today ecological and women-friendly remarks and arguments are in vogue. There are discussions about possibly changing (as one of the first Western European countries to do so) the tax system in favor of ecological concerns. The Chancellor was right when he said that much would be different after Hainburg – in a different sense, however, than he had expected.

THE CONSEQUENCES OF HAINBURG

"Freda for President!"

The victory of the ecological movement around the Danube forest was to turn my life upside down. We had gotten to cherish the taste of democracy from down below and we wanted more of it.

(Freda Meissner-Blau 1990: 6).

During the last days in the Danube forest, people discussed future political perspectives around the campfires.

"When are the next elections?" somebody asked.
"Presidential elections, next year, May 1986," was the reply.
"Freda for President," rang the voice of a woman.
"Nonsense, I never would. . . . "
"Yes you will, it's about time there was a woman-candidate in Austria. Elections have since long deteriorated into male-beauty-contests. Having a green woman candidate, would get us at last media attention and would help to get our message across. . . . "
I laughed.

(Freda Meissner-Blau 1990: 7)

Over the course of the year 1985, however, Freda Meissner-Blau's opinion changed: there were three candidates on the ballot for the elections to the Federal Presidency: one each from the two major parties – the Social Democrats (SPÖ) and the People's Party (ÖVP) – and one from the Liberal Party (FPÖ), a medical doctor who had been prominent during Nazi times. When Freda Meissner-Blau heard this and realized that he was the only choice for the protest voters, she changed her mind. If she did not run, the usual protest potential of 5 percent would go to this FPÖ candidate, and would internationally be interpreted as an Austrian Nazi vote. To offer a woman and green alternative she decided to run for the presidency.

Freda Meissner-Blau explained her motives during a press conference:

One can resign or one can commit oneself to a cause. I chose to commit myself in my personal life and to the big public issues that move me: the environment . . . civil rights, the misery in the Third World, the armament madness that is part of the reason for this misery, the income disparities in the face of growing unemployment and poverty. . . . And last but not least I am running for the citizens' access to their rights. . . . In that respect also my candidacy will be a lesson, and may even set a hopeful example.

(Freda Meissner-Blau, as quoted in Paul Blau 1992: 184)

After a strenuous election campaign – primarily by public transportation, without any money or party apparatus to support her, and mostly ignored by the mass media – Freda Meissner-Blau won 5.5 percent of the votes, while the FPÖ candidate received less than 1 percent. A runoff election became necessary, something that had not happened since 1951.

The candidacy of a woman was a significant demonstration of courage. Only once before had a woman run for the Federal Presidency (1951), and had received only 2,132 votes (0.05 percent). Traditionally, this election is a competition between the candidates of the two major parties who also have access to the appropriate means for a campaign. For Austria, the candidacy of Freda Meissner-Blau was a sensation, and her 5.5 percent of the votes constituted a political landslide that rendered the old political patterns questionable. Throughout her election campaign, Freda Meissner-Blau stressed a gender-specific perspective of justice and responsibility, resource use and preservation, and power and democracy, resulting in an increasing discussion of these perspectives by the public. She contributed to a consciousness-raising process among the population that showed its first impacts during the next elections for the National Council (the Austrian parliament) six months later.[5]

The Greens in federal and state parliaments

The mega-dam symbolized the final victory of the technosphere over the biosphere, the sealing of our children's future with concrete. Against this nightmare we went out to protect trees and democracy, got frostbite, beatings, and bruises, experienced the power of our own nonviolence, the joy and relentlessness of community. We developed a taste for more democracy.
(Meissner-Blau 1989: 53).

This taste for more democracy and the increased self-confidence brought by success in the political arena led to the candidacy of the "Green Alternative List" organized by Freda Meissner-Blau in the parliamentary elections during the fall of 1986. Laborious negotiations among all the spontaneous groups before the elections led to the formation of a group that was represented in all Austrian states and whose basic principles included gender-parity. All this had been previously unimaginable in Austria: a woman as head of a party, a democratic and discursive procedure for choosing candidates, ecological themes, and a consciously women-friendly basic principle.

But the time was ripe for such a party, and so the Greens made their entrance into the National Council, just as the German Greens had entered the German Federal Parliament two years earlier. In the fall of 1986, this new Austrian party obtained eight out of 183 seats (and 4.82 percent of the vote). That was a big, and largely unexpected, success because for the first time there was a fourth party in parliament. It was surprising even to this young movement, which brought new standards and themes into parliament, but could not always live up to its own expectations, especially with reference to basic democratic issues and gender equality. Freda Meissner-Blau recalls a typical experience:

One thing that, at the time, weighed heavily on my mind was the question to what extent the situation as such was our own fault, the fault of us women. Of the eight seats, I had expected four to be held by women, but then all of a sudden I found myself standing there with seven men.
(Freda Meissner-Blau, personal communication, March 1994)

Before the elections, the members of the green alternative groups in all states had agreed to enter women and men alternately on their candidate lists.

All groups had complied with this decision, but all (except for Vienna with Freda Meissner-Blau at the top) had listed a man first. Since the number of votes sufficed only for one candidate each in eight of the nine states, the men were in the majority. But the Greens learned quickly: in the next state parliamentary elections they listed female candidates first, and since the National Council elections in 1990, half of the Green seats have been held by women. The success of this previously ridiculed group began to stir up even the established parties.

Societal changes: friendliness toward women and a bit of green in all parties

> Hainburg has had significant impacts on the political system in Austria. That holds true for the contents, methods and consequences of the conflict. ... Hainburg made all established forces, all established parties "green." Since then at least lip service has become commonplace – lip service paid to ecological goals. This also means an inflation of ecological rhetoric, a tendency towards noncommittal and indistinguishable political parties.
> (A. Pelinka, Professor of Political Sciences, University of Innsbruck, as quoted in Monjencs and Rainer 1989: 61)

There was a sudden awareness in the population and in the media – men could no longer say the things they matter-of-factly used to say before. Open discrimination against women became history. But altered patterns of speech did not yet mean a change in consciousness and major actions. Environmental laws and directives to treat women and men equally have been issued, and are as such welcome progress; but they are phrased in such general terms to often remain without impact in practical terms. They are easily circumvented, and violations are hardly ever punished.

The same is true in the environmental arena. The facade has become "green," but behind it is still the exploitation of women and nature. The Ministry of the Environment, for many years now run largely by women, has hardly any executive power, yet it serves as an alibi in the more heated debates.

> So a federal ministry was created with environmental protection as part of its name. Unfortunately, this ministry was and is not in charge of environmental protection, and that is not only an embarrassing absurdity, but the documentation of an unforgivable inconsistency. The establishment of this ministry provides undeniable evidence that the government recognizes the inevitability of environmental politics. Omission of efforts to protect the environment can no longer be excused under the pretense of ignorance.
> (J. H. Pindur, Deputy Secretary, Federal Department of Health and Environmental Protection, as quoted in Meissner-Blau 1989: 54)

On the surface the improvements for women are enormous: of the five parties in parliament, the two small opposition parties are led by women. During the last presidential election in 1992 there was again a woman candidate, and after a reorganization of the federal government at the end of March 1994, a quarter of the members of government are now women. Yet a real positive change in women's matters has yet to happen. Freda Meissner-Blau criticized the political situation:

> One thing that has become increasingly obvious to me is that women who have gotten into higher positions through men, say within parties, are the ones I would call "male-adapted women" who are quite adapted to the goals and thinking of men. ... Women who are consciously women, and who consciously want to put through the perspective of women, and who exercise solidarity among women, those women still have a hard time, because they give offense to conventional thinking.
>
> (Freda Meissner-Blau, personal communication, March 1994)

But even this conventional thinking is changing at least in parts of the population, and there are important and encouraging steps toward addressing the basic concerns of a feminist political ecology. For example, in Tirol in April 1994, a young woman, a political scientist, became a member of the state government as the top candidate for the Green Party.[6]

Generally, however, women in Austria, as elsewhere, do not tend to make their careers in politics. Freda Meissner-Blau suggested that this situation derives in part from their values, rooted in daily experience:

> I would say that women have an entirely different relation to power than men do, simply because they have never had power, and therefore are not power-hungry, but tend to see their strength in conflict resolution. I know that myself. I have three children. For me it was important that when we had a quarrel we could somehow resolve it. We tried to show the children ways to not be mad at each other for days or not to hit each other over the head in order to solve problems. Women learn that with their children . . . . This not insisting on power but being oriented toward problem resolution is part of the reason why women can be found more often in grassroots movements, ecological movements and green movements than in political parties.
>
> (Freda Meissner-Blau, personal communication, March 1994)

This, however, still leaves Austria with a long way to go until it becomes a matter of course for politics-as-usual to mirror feminist environmental values. Maria Mies and Vandana Shiva call for "the practical and theoretical insistence of the interconnectedness of all life, in a concept of politics that puts everyday practice and experiential ethics, the consistency of means and ends in the forefront" (Mies and Shiva 1993: 321). This outlook may be true of some grassroots movements; it has not yet permeated parties and politics in corridors of national power, however.

The debate over the Danube forest continues

At the national level, fundamental changes, such as a decreased influence of the energy producing industry, have not taken place. Although the Ecology Commission made its recommendation against the power plant in Hainburg in 1985, the debate over the power plant siting in that section of the Danube River has not ceased. Efforts since 1984 to establish a national park in this area have to date been unsuccessful. The concerned parties remain at a political impasse over the Hainburg riverine forest, because the conflict of interests persists and all involved groups fear a loss of face.

The Danube Power Company continues to plan new power plant alternatives in this section of the Danube River. The Austrian World Wildlife Fund (WWF), in its "Bailing Out Nature" campaign, obtained 85 million

Austrian schillings (about US$8.5 million) from 120,000 donors and was able to buy a core region of the riverine forest to protect it from any future grab of the area. In a representative poll in June 1993, 88 percent of Austrians were in favor of a "Danube Forest" national park. But how the conflict will be put to rest in the end depends primarily on the economic and social situation at the time. The Danube Power Company has learned a lot: it invests much money in polishing its image and argues that its investments in the environment and water quality demonstrate its concern for nature.

In spite of all attempts to restore the landscape, as done after the construction of other power plants, no naturally grown, highly complex, dynamically balanced, and interconnected ecosystem can simply be replaced. It is difficult to assess how in times of growing unemployment the argument for economic development will be evaluated. For many, the process of rethinking has not yet begun. It has not become clear yet that conventional solutions to economic and social problems no longer work, and that it is necessary to develop a new understanding of what constitutes responsible interaction with nature and our fellow human beings.

Prospects for the future

The prospects are rather dreary. Austria became a member of the European Union (EU) in January 1995. In terms of ecological consciousness and women's politics the EU does not set a glowing example. Many Austrians thought (and the government suggested) that Austria could keep some of its comparatively high ecological standards and even push the EU to improve its regulations. In contrast, one of the first laws Austria has to change is the animal transportation law, because in the EU animals (cattle) can be transported for a longer time and longer distances. Another issue concerns the question of transit in the alpine valleys (Tirol). Despite many declarations it seems not to be possible to improve the international railroads and to transport goods by rail instead of by the trucks now polluting the small valleys.

Another major international problem for Austria concerns the (unsafe) nuclear power plants in the former socialist countries beyond Austria's borders. Austria's population (again women in grassroots movements!) is very anxious about the dangers these nuclear power facilities represent and tried to find support in the European Union. But Western European politics to date do not allow very much scope for a feminist political ecology, because economic growth has absolute priority, and a strong demarcation against the outside is intended. A critical discussion of worldwide economic interests and the ecological implications of economic development are not issues of debate in the political and bureaucratic centers of power. Throughout Western Europe the clash of gendered views of development and gendered responsibilities can be seen: as quantitative versus qualitative employment, and economic growth versus environmental and resource protection. A different theory of the role of the state, based on a feminist political approach, emerged when Freda Meissner-Blau asserted: "Mr. Chancellor, the Austrian Republic is not a construction firm" (Meissner-Blau 1986: 8).

With regard to democratic political developments and participatory aspirations, high hopes for the future are hardly appropriate either. Even if women are more visible now in Western European politics, gendered power relations still exist in terms of official functions and lobbies versus small-scale organizations in parliamentary opposition. Ways and means must be found to empower women to express the gendered view of development, political rights, and environmental sustainability in democratic processes. Women's political participation in local, national, and global politics is required to implement a feminist environmentalist perspective.

NOTES

1 The term Western Europe is used to differentiate certain countries from those previously part of the Soviet Union. It includes the countries of Central Europe, such as Austria. The term Central Europe, however, will also be used to separate Austria and its immediate neighbors from the other countries of Western Europe.

2 The success of the nonviolent resistance of the Chipko women in India against the deforestation of their home country is well known in Austria and an encouraging example for green movements (Shiva 1989).

3 After hundreds of years in the Habsburg Empire, Austria was proclaimed a republic in 1920. The time between the two World Wars was the so-called "First Republic;" after World War II Austria became the "Second Republic."

4 Of course we know that the notion of nature is a social construct and that it has a different meaning in different historical periods, different cultures, and different geographical surroundings. In this chapter it is used as it was understood within the movement, in the popular sense: forests, rivers, wetlands, and the ecosystems that support life on earth with very little human influence.

5 Austria is a federal state, consisting of nine states. Each has a state council and a government. The national council is the legislative for the whole federal state. The federal government, headed by the federal chancellor, is the executive for the whole state. The federal president has less political power than the French or American president, but rather a more representative function like the British Queen.

6 Tirol is a very conservative, Catholic Austrian state; but the inhabitants of the small Tirolian alpine valleys also suffer from the international traffic between Germany and Italy. Consequently, the elections meant a decision between green local initiatives and traditional international political networks and lobbies.

REFERENCES

Argawal, B. (1988) *Structures of Patriarchy: The State, the Community and the Household*, London: Zed Books.

Bay, C. (1968) "Civil Disobedience," *International Encyclopedia of the Social Sciences*, 2: 473–87.

Birch, A. (1993) *The Concepts and Theories of Modern Democracy*, London: Routledge.

Blau, P. (1992) "Wehren bewährt sich," in C. Weizsäcker and E. Bücking (eds.) *Mit Wissen, Widerstand und Witz: Frauen für die Umwelt*, Freiburg (Germany) and Vienna: Herder.

Canby, H. S. (ed.) (1937) *The Works of Thoreau*, Boston: Houghton Mifflin.

Gerlich, P. (1995) "Freda Meissner-Blau," in H. Dachs, P. Gerlich and W. C. Müller (eds.) *Die Politiker: Karrieren und Wirken bedeutender Repräsentanten der zweiten Republik*, Vienna: Manz.

Klose, A. (1987) *Machtstrukturen in Österreich*, Vienna: Signum.

Kolba, P. (1985) "Ziviler Ungehorsam und Demokratie," in G. Nenning and A. Huber (eds.) *Die Schlacht der Bäume*, Vienna: Hannibal.

Kontos, S. (1991) "Zum Verhältnis von Autonomie und Partizipation in der Politik der neuen Frauenbewegung," in B. Schaeffer-Hegel and H. Kopp-Degetthoff (eds.) *Vater Staat und seine Frauen: Studien zur politischen Kultur*, Vol. 2, Pfaffenweiler: Centaurus.

Meissner-Blau, F. (1986) "Let the Danube Flow," *Resurgence* 115, March/April: 8-9.

—— (1989) "Hat Hainburg die Republik verändert?" in I. Monjencs and H. Rainer (eds.) *Hainburg – 5 Jahre danach*, Vienna: Verlag für Wissenswertes.

—— (1990) "A Thread in the Web of Her Story," unpublished manuscript.

Mellor, M. (1992) *Breaking the Boundaries: Towards a Feminist Green Socialism*, London: Virago Press.

Mies, M. and Shiva, V. (eds.) (1993) *Ecofeminism*, London: Zed Books.

Monjencs, I. and Rainer, H. (eds.) (1989) *Hainburg – 5 Jahre danach*, Vienna: Verlag für Wissenswertes.

Niederösterreichisches Naturschutzgesetz (1982) Landesgesetzblatt (LGBL) Nr. 5, Paragraph 6, Absatz 2.

Profil (1984a) Nr. 46: "Irrlichter," 21.

—— (1984b) Nr. 51: "Die Welt des Josef Hesoun 'Jene die die Nacht zum Tag machen'," Auszüge aus einer Rede in der Bauten-Budgetdebatte, 14; "Wählerfang," 21.

—— (1984c) Nr. 53: "Aufhören! Anfangen," 10–11; "Au – Du mein Österreich," 14–17; "Es gibt keinen Krieg," 34–5.

Rinehart, S. T. (1992) *Gender Consciousness and Politics*, London: Routledge.

Schaeffer-Hegel, B. (ed.) (1990) *Vater Staat und seine Frauen: Studien zur politischen Theorie*, Vol. 1, Pfaffenweiler: Centaurus.

Schaeffer-Hegel, B. and Kopp-Degetthoff, H. (eds.) (1990) *Vater Staat und seine Frauen: Studien zur politischen Kultur*, Vol. 2, Pfaffenweiler: Centaurus.

Shiva, V. (1989) *Staying Alive*, London: Zed Books.

Weizsäcker, C. and Bücking, E. (eds.) (1992) *Mit Wissen, Widers tand und Witz: Frauen für die Umwelt*, Freiburg (Germany) and Vienna: Herder.

5

SPANISH WOMEN AGAINST INDUSTRIAL WASTE

A gender perspective on environmental grassroots movements

Josepa Brú-Bistuer
(Translation from Spanish by Moya Hallstein and Annette Ramos)

This chapter explores the gendered perception and experience of the impacts of industrial waste disposal in minority regions of Spain. Three case studies demonstrate women's distinct experiences with an environmental problem that affects industrialized countries as a whole and which constitutes a significant form of social and territorial marginalization. In each case, the threat posed by industrial waste disposal led to large-scale community movements in which women played a pivotal role that has been systematically ignored by the mass media and the "official chronicle of events."

Gibraleon, a small town in the Province of Huelva on the southwest coast of Spain, is the site of the first case study. With a population approaching 10,000 inhabitants,[1] Gibraleon is located 15 kilometers (about 9 miles) from the provincial capital of Huelva where one of the most important chemical-industrial complexes on the Iberian Peninsula is located. This facility poses very serious regional contamination problems, especially for the surrounding communities.[2]

Late in 1987 the Environmental Agency of the Autonomous Regional Government of Andalusia proposed a hazardous waste disposal site on a Gibraleon farm. This marked the onset of community activism against the discriminatory treatment of the inhabitants of Gibraleon. The farm site was chosen because an extension of the railroad line passes through it, thus providing adequate access to the area for unloading industrial by-products. The community response was initiated in 1988 and lasted until May 1989 when they succeeded in preventing the waste site from being located in their community and the project was canceled. There were almost two years of conflict, including a whole year of continuous community resistance with several very difficult situations the affected population will find hard to forget.

The second case study comes from Bilbao, the capital of the Basque region on the northern coast of Spain.[3] The study area includes the neighborhoods of Old Bilbao and San Francisco, both located in the old city center on the slope of Mount Mirivilla, with a combined population of approximately 9,700 inhabitants. These two communities are typical of many historical, yet decaying, inner-city neighborhoods which have an increasing tendency toward high rates of unemployment, racial segregation, drug abuse, and a disproportionately large senior-citizen population.

The women in these two communities mobilized a local social movement after industrial waste – accumulated illegally in an abandoned mine in Mount Mirivilla – spontaneously caught fire during the fall and winter of 1989. Due to the quantity of accumulated by-products and their high flammability, it was extremely difficult to contain and completely extinguish the fire. This prolonged combustion of noxious chemicals resulted in an entire month of toxic emissions produced by the burning of PCBs, cyanide, and pesticides, among other substances.[4]

The third case study examines the community opposition efforts against a government waste disposal plan in four small towns of Catalonia, on the northeast coast of Spain. The neighboring towns of Solivella, Sarral, Pla de Santa Maria, and Rocafort de Queralt are situated within a radius of just a few kilometers of each other and have a combined population of almost 4,000. Although they are clearly rural communities, these four towns are closely connected to the coastal urban-industrial complex through a network of highways which extends from the metropolitan area of Barcelona to Tarragona, the center of the Catalan petrochemical industry.

When the regional government – the *Generalitat* of Catalonia[5] – made public its proposal for a "Waste Management Plan" in Catalonia, which included an incinerator and a large-scale industrial by-product disposal site primarily for chemical waste from the Tarragona area, members of these communities, especially women, reacted with strong unified opposition. Their efforts lasted from January through September of 1990, at which time the government announced its decision to abandon the proposed Waste Management Plan.

HYPOTHESIS AND METHODOLOGY

Through the comparison of three gendered community opposition movements experiencing a similar threat to local integrity and quality of life under different conditions of setting and circumstance, two hypotheses can be developed about the relationship between women and the environment.[6] The first is that due to their gender-specific roles and distinct investment in "their" community, women have a unique perception of the environment, what constitutes their environment, how it is constructed, and how it affects quality of life, especially in terms of healthcare and family well-being. Second, because of these gendered experiences of community and environment, women participate in community activism and social movements in ways that are gender-specific and directly correlate with the roles and functions of gender in society.

This study involved a combination of field research and information gathering techniques, the most substantive of which were the comprehensive, in-depth interviews with sixty-four women who had been active participants in these community movements. The combination of open dialogue with a standard questionnaire allowed for the comparison of results without relying entirely on standard survey techniques, which would not have been adequate for the nature of this study. Additional information was collected through consulting technical documentation pertaining to each case, analyzing information published in newspapers, magazines, bulletins, posters, and through key informant interviews with people who have had technical, political, or social responsibilities in each case.

The women interviewed were contacted through the leaders of each social movement. This was the only way to conduct the interviews with the necessary degree of trust while also avoiding potential jealousies. Despite the lack of randomness of this selection process, the participating groups of women from the different locations were a representative cross section of the activist women in each community.

THE ROLE OF WOMEN IN LOCAL SOCIAL MOVEMENTS

Through the analysis of these three case studies it is apparent that the role and involvement of women and their influence in social movements of this nature should be examined from at least three angles: investigation of these women as significant leaders, recognition of their active membership in the community, and exploration of the energy, commitment, and spirit they bring to their community struggles for change and their daily strategies of resistance.

In Catalonia, the majority of the women involved were over 40 years old and had completed a primary level of education. In Bilbao, the women's level of education was the same, but their age ranged evenly from about 25 to older than 40. In Gibraleon, most women were between 25 and 40 and had completed middle school. Catalonia was the only case in which a small percentage of the women had a university education. In all three case studies, a high percentage of the women stated that they were housewives, married with children – more than two thirds in Gibraleon, almost half in Bilbao, and more than one third in Catalonia.

Gibraleon

From the beginning, the women of Gibraleon incorporated unique strategies of resistance and struggle into their daily lives. Perhaps the most significant was their unprecedented, constant presence on the street, clearly indicating the exceptional nature of the situation during almost two years of conflict. The existing gendered networks, relations, and uses of spaces became the venues through which women diffused new information and devised new strategies.

At noon each day, the women met at the gates of city hall to share ideas and discuss the progress of their opposition. They would come alone or with their youngest children and stay until they had to go prepare the evening meal. The women would often bring their domestic work to the streets, frequently sewing, knitting or cleaning vegetables while taking care of their

Plate 5.1 Women of Gibraleon bang pots in the streets in protest against the planned
hazardous waste disposal site on a nearby farm

Source: M. J. Florencio (1991) *Gibraleon: punto a punto*, Seville: Imprenta Sanda

children. In the afternoon, the whole family – women, men, and children
of all ages – marched through the streets to the central plaza. At night, from
ten to midnight, the women went to the windows or roofs of their homes
to strike pots and pans in protest at and rejection of the waste disposal site.

Women also played a decisive role in the coordination of a central office,
which was created to facilitate and organize community opposition. The
responsibilities the women undertook multiplied and became increasingly
diverse as the movement evolved. These included daily contact with the
local radio stations and newspapers to inform them of new developments
in their efforts, writing press releases, and organizing public events to dissem-
inate information.

Women also brought invaluable strength and support to members of the
movement during the most difficult periods of opposition which culminated
in several confrontations with the police. It was the women who visited those
who had been arrested and informed their families, attended the trials, and
kept in touch with attorneys. Under such precarious, yet critical, circumstances the women kept the community united and maintained peaceful
relations with local officials.[7]

Three women formed a musical group to motivate and educate the
community with their music. Through the creation and performance of their
popular songs, they shared stories of community resistance to the waste
disposal site and inspired the community in their efforts. Their songs reflect
the popular perception of the struggle and also provide an innovative history
of the movement. Through this record of important events, the creation and

Plate 5.2 Neighbors of Old Bilbao demonstrate in front of city hall and display a Basque language placard announcing the dangers of flammable waste dumps in a nearby abandoned mine

Source: Correo Espanol del Pueblo Vasco newspaper, November 11, 1989

destruction of characters in the struggle, and the portrayal of popular perceptions, these songs have become an authentic locally based mythology of the conflict.

This use of popular song is particularly interesting in that it reflects a gendered strategy of resistance and provocation through the expression of values, images and relations directly linked with women's culture and gender roles in Gibraleon. The women varied their musical forms throughout the year according to the season, using Christmas carols, the brass band sounds of the carnivals, and Easter songs. In this way they were able to link the development of the conflict and its popular perception directly to the life of the community.

Bilbao

The community response and the role of civil disobedience in Bilbao was completely different from that in the other case studies. The spontaneous blaze of the industrial waste was perceived as one more affront in a long line of difficult and conflictive incidents, rather than an isolated occurrence. Aside from the health and environmental risks imposed on the community, people understood the event not as an accident, but as one more consequence of the process of marginalization in their neighborhoods.

In a situation like this, women in the community are already engaged in daily struggles for the preservation of their quality of life in the face of poverty, marginality, unemployment, and the drugs trade. They know what it means to attempt to defend their families and to bring dignity to their cherished neighborhood under difficult and oppressive conditions. In this context, the involvement of women in the community movement was inherent in the nature of the struggle, women's roles and investment in the community, and their gendered sense of responsibility for their community. They felt they were fulfilling their responsibilities as women, mothers, and housewives.

Much of the strength, direction, and organization of this social movement came from women in the community, most of whom were members of the Old Bilbao Neighborhood Association; nine of its fourteen board members and the president were women. Most of the women had joined the organization before the conflict and successfully rescued it from a state of neglect. Through their use of the neighborhood association as the vehicle for community opposition, it has become the stronghold of Bilbao's struggle against the growing degradation of their neighborhoods. The women were effective in contacting people and inspiring their involvement through persistent telephone contact and an effective information network in the local bars and stores. Today the neighborhood association maintains this leadership role. The women involved in the organization manage it as they would their homes, with responsibility, affection, and absolute dedication. Without the constraints of a regular work schedule, they are able and willing to do anything at a moment's notice, even forgoing weekends and vacations, for the benefit of their community.

Catalonia

Unlike the situation in Bilbao, the communities in Catalonia are all rural and enjoy a decent quality of life. This is clearly evident in the commitment to preservation and continuity of the rural landscape and environment demonstrated by the women interviewed. Although the distinction between rural and urban is increasingly difficult to establish in this highly cohesive region, the dichotomy between country and city remains fundamental to both individual and community identification with place and their emotional attachments. As that distinction is increasingly blurred with the national restructuring of regions to serve specific functions, many inhabitants of rural areas experience a sense of insecurity and threat from urban encroachment. Regardless of how much the authorities and certain intellectual quarters promote the necessity of territorial solidarity, rural communities consider the imposition of a waste disposal site to be a genuine act of aggression.[8] They are unwilling to bear the potentially negative impact on their quality of life from the by-products generated in urban-industrial areas.

The women in Catalonia reacted to this threat to their quality of life by assuming a role as guardians of the continuity of the landscape and the rural milieu for future generations. They became involved both for personal reasons and because as women they felt it was their responsibility to preserve a decent quality of life for their communities. The decision on the part of the women involved to act in defense of their families' well-being and the

Plate 5.3 A group of Catalan women protest the proposed industrial waste disposal site at Castellbisbal in the Barcelona metropolitan area

Source: Aqui newspaper, March 1, 1990

quality of their home environment was difficult. Their commitment to the social movement would imply some initial instability and adjustment for their families which could be particularly hard on their children.

Perhaps one of the strongest threads of solidarity among the Catalonian women came from their shared conflictive and even contradictory experience of fulfilling a responsibility to their community while at the same time feeling that they were neglecting their children who felt insecure and abandoned as a result of their mothers' new activism. If the gender division of involvement in and commitment to this social movement, given all it entailed in terms of community activities and constant availability, had been different, the children might not have been so aware of the situation. They became conscious of it because of their mothers' increased involvement compared with their fathers'. As a result of their involvement, women also ceased to fulfill their traditional roles of filtering problems and cushioning family

Table 5.1 Women's average ranking of impact/damage from waste dumping

Site	Bilbao	Catalonia	Gibraleon
Environment	9	9	9
Health	10	9	10
Economy	6	8	8

Note: Importance ranked 1–10

conflict. Subsequently, through the nonconformist attitudes of their mothers, children experienced a sense of being unsheltered. They also saw themselves indirectly implicated in a situation of civil disobedience they were incapable of comprehending. This predicament disrupted their sense of order, which until then, they had considered as the only possible one.

Despite the difficulties their children had to face due to the involvement of their mothers in the community opposition, not one of the women interviewed had any doubt about joining the social movement, nor did they have any regrets later. Their involvement was simply something that had to be done and no one else was going to do it for them.

THE INTERVIEW RESULTS

Women's perceptions of the impact on local environments, health, and economies

On a scale of 1 to 10 the women from each community were asked to rank the magnitude of local environmental, health, and economic problems (Table 5.1). In all three case studies, health was considered the highest priority followed by the environment and the economy.

Agreement on this gradation shows not only in the middle values, but also in the number of abstentions: there are practically none when health or the environment are being considered, but when referring to the economy they make up two thirds of those interviewed in Bilbao, and about a third of those interviewed in Gibraleon and Catalonia did not even rank the economy. It is clear, therefore, that health risks constitute the most important and clearest threat when it comes to the perception of the problem.

The results in Gibraleon reflect the feeling among women that the waste disposal site introduced a destructive element in an area that until then had been protected from degradation, unlike other areas of the province. This is a region that has traditionally had a relatively prosperous agricultural economy which has continued to flourish and maintain its integrity in the face of urban and industrial encroachment. The proposed waste disposal site threatened to end a positive situation perceived as exceptional, at least in the context of the metropolitan area of Huelva.

Although the interview samples are small, age plays a significant role in women's perception of the threat to their community's health, economy, and environment. While the concern for health is an absolute priority for women over forty, the younger women often emphasized the environment. For example, in Catalonia – the largest sample group – the older women gave

Table 5.2 Catalonia. Women's reasons for participating in protests

Reasons	First reason	%	Second reason	%
Family health	11	35	5	16
Community health	10	32	8	26
Environmental protection	1	3	7	23

Table 5.3 Gibraleon. Women's reasons for participating in protests

Reasons	First reason	%	Second reason	%
Family health	5	23	5	23
Community health	6	27	8	36
Environmental protection	2	9	6	27

median scores of 10 to health and 9 to the environment whereas the younger women weighed the two the same at 9. The youngest group also equally valued the environment and health.

Perceptions of the extent of the problem

The women were asked about the first image that came to their mind when they thought of their perception of the problem. The most prevalent response in Bilbao was that the problem had occurred there because of Bilbao's location in "the middle of nowhere." This seems to be indicative of feelings and experiences of isolation characteristic of marginalized communities, such as this neighborhood in Bilbao. In Catalonia – even though almost one third of the women did not know how to respond – a recurring answer was that "the problem was an example of the inadequate environmental policies of the Regional Autonomous Government (*Generalitat de Catalunya*)." Contrary to what occurred in Bilbao, this answer illustrates a globalized perception of the problem and its essentially political nature. The women from Gibraleon – although again almost one third abstained from answering – overwhelmingly responded to "the need to halt the project to avoid the risk of similar initiatives or problems occurring in the entire area." This attitude seems to confirm a sense of the importance of maintaining high environmental quality, especially in an already seriously threatened province and metropolitan area.

Motivations for women's involvement in community activism

The women were asked to prioritize the motivating factors for their involvement in the community opposition movements in terms of the protection of family and community health, and protection of the environment; and

Table 5.4 Women's attitudes and feelings during the conflict

Attitudes	Bilbao	%	Catalonia	%	Gibraleon	%
Personal autonomy	0	0	1	3	5	23
Enthusiasm	1	9	10	32	7	32
Useful to community	7	64	21	68	18	82

the reaction to an externally imposed threat, or a sense of endangerment of their community and quality of life.

Concerns about family and community health and environmental quality were unanimous, rendering the remaining categories insignificant. In Catalonia (Table 5.2) the preservation of family and community health was the primary motivation for two thirds of the women. Similarly, health was most important to 50 percent of the women in Gibraleon (Table 5.3). Community health, however, took precedence over family health, which speaks for the community character of the problem experienced.

In all three case studies environmental protection was given little importance as a first motivating factor. It was most significant for only 9 percent of the women in Gibraleon and 3 percent of the women in Catalonia. As the second motivation, however, it was important for 27 percent of the women in Gibraleon and 23 percent of the women in Catalonia.[9]

Feelings and attitudes during and after the conflict

Aside from the issues that motivated the women to take part in the community movements, this study explored their feelings about this participation and whether these feelings had provoked permanent attitudinal changes. To determine this the interviews pursued issues such as personal autonomy, enthusiasm, and belonging to the community (Table 5.4).

In the three case studies, the most commonly felt sentiment – for more than two thirds of the women – was that of feeling useful to the community. In Catalonia and Gibraleon, up to one third of the women interviewed felt enthusiastic about the struggle. Twenty-three percent of the women in Gibraleon experienced a sense of personal empowerment. These results suggest that many women had a distinct public experience with the conflict as well as a more personal one. Both the nature and magnitude of these community social movements led to a forum in which women could project and validate their gender roles through publicly asserting themselves, their values, and their concerns. At the same time, they were also able to experience a growing sense of personal empowerment.

In terms of new perceptions and attitudinal changes, most women felt they had acquired a broader awareness of environmental issues and problems. In each case study women reported having acquired a wider sensitivity about environmental problems: 84 percent in Catalonia, 64 percent in Bilbao, and 55 percent in Gibraleon. After environmental awareness, women in

Gibraleon felt they had gained greater confidence in themselves. One third of the women in Bilbao shared these feelings, with another third responding that they felt they had attained more significant roles in the community. A slightly smaller group in Gibraleon gave the same response.

Participation in these community movements has generated wider consciousness of environmental problems, but it has also begun an important process of personal empowerment for many of the women involved through their increased range of activities, self-assertion, and voice within the community.

Perceptions regarding gender differences

The women were asked to compare differences in motivational factors and forms of action employed by men and women. Interestingly, in all three case studies, the most frequent response were that there was none. The women from Catalonia were the most emphatic on this point with 87 percent responding negatively, followed by the women from Gibraleon with 71 percent, and the women from Bilbao with 64 percent. When asked for a justification to their answer, they replied almost unanimously that "the problem affected us all equally."

They were also asked whether they had spoken more frequently with men or women about the situations facing their communities and whether it had been easier to communicate and understand each other with one group or the other. The responses were similar to those of the previous question, with 81 percent of the Catalan women and 68 percent of the women from Gibraleon affirming that there were no differences. Interestingly, in Bilbao, with one exception who had spoken with more women than men, they all answered that there had been no prevalence either way, yet they also provided independent information that indicated that 46 percent had spoken with more men.

To the question about who stayed home the most frequently if necessary, most women responded that "it was arranged in such a way that nobody's role or involvement in the movement would be compromised." The women from Catalonia said that in the event someone did need to stay home, they would do it, whereas the women from Bilbao said, on the contrary, that the men would stay home if necessary.

In analyzing the interview results, it is apparent that there were some difficulties with the question for two reasons. First, the self-image of the women interviewed, especially the younger women, as modern and autonomous can often result in an unconscious denial of gender inequalities. Second, many women felt defensive about the feminist slant of the study and responded that the study seemed to be "searching for differences rather than seeking equality." Stereotyped conceptions of feminism, femininity and gender relations, combined with a desire to fit into an ideal model of "modern" woman, also precluded most of the women interviewed from engaging in a detailed appraisal of gendered participation in the struggle, in the terms in which it was articulated in the question.

Table 5.5 Women's changes in feelings toward the community

Type of Change	Bilbao	%	Catalonia	%	Gibraleon	%
No change	0	0	4	13	5	23
Greater loyalty	5	45	10	32	11	50
Fear of new occurrence	5	45	19	61	9	41
Wish to leave	1	9	2	6	2	9

Changes in the perception and evaluation of the environment and region

The objective of the initial question in this section was to uncover the extent to which the women's involvement in community activism served to outline, amplify, or change the perception of the nature and dimensions of women's living space. For more than three fourths of the women from Gibraleon the experience seems to have expanded their knowledge of interpersonal and institutional relations. It also provided the vehicle for a better understanding of their environment, their geographic and political place within the country, and its significance. In Catalonia, the responses were similar, although more emphasis was placed on the latter. In Bilbao, most of the women cited both of these changes as equally important.

Interestingly, in all three case studies, the question which provided the widest scope of change – "knowledge of the role of their place in the industrial and urban economy of the area" – was given the least importance. Nevertheless, about two thirds of the women from Bilbao and almost a third of the women from Catalonia and Gibraleon affirmed that they are now more conscious of this issue. Direct contact with the matter of hazardous waste gave the women from Bilbao greater consciousness of the nature and breadth of the problem.

In terms of women's changed feelings about place (Table 5.5), in Catalonia fear of the situation occurring again was the most prevalent response – as stated by almost two thirds of the women. In Gibraleon, increased commitment to their community was most important, followed closely by a fear of the situation occurring again. In Bilbao, both feelings are equally present. Despite their fears of a recurrence, very few women expressed a desire to move away.

This attachment to place is consistent with the reasons women gave regarding their feelings when the conflict began. The women all felt deeply rooted and committed to their homes, communities, and towns. Concerns about having to relocate were high: 77 percent in Gibraleon and more than half in Catalonia. Of the women in Gibraleon, only 5 percent communicated a fear of risk; in Catalonia it was only 7 percent. In Bilbao, on the other hand, fear of risk surpasses the feelings associated with not being able to stay in their neighborhood. This difference is most likely a reaction to the waste dump fire constituting a present, not a future danger, as in the other two cases.

Table 5.6 Gibraleon. Prioritization of strategies for environmental protection

Indicators	Consumption	Protests	Education of children	Self-realization	Public education
Average	9	6	9	7	7
Maximum	10	10	10	9	10
Minimum	7	2	7	5	5
Stv	1	3	1	2	2
NS/NC	9	10	7	11	11

Note: Importance ranked 1–10

Table 5.7 Bilbao. Prioritization of strategies for environmental protection

Indicators	Consumption	Protests	Education of children	Self-realization	Public education
Average	10	9	10	6	9
Maximum	10	10	10	8	10
Minimum	9	6	9	1	7
Stv	1	2	0	3	2
NS/NC	5	1	4	7	5

Note: Importance ranked 1–10

The environment and the daily lives of women

On a more objective level, the women were asked about their daily activities and responsibilities in relation to the environment. When asked about the possible changes in their shopping habits or household management as a result of the events in their communities, more than half the women in Old Bilbao and up to one third of the women from Gibraleon had not made any changes, while in Catalonia fewer than 10 percent attested to not having made changes. The women from Bilbao, despite their small number, cite changes almost in equal proportion in all three spheres presented to them: shopping habits, conservation of resources, and recycling. The women from Gibraleon emphasize the first two points equally, granting little importance to the recycling and the Catalan women responded with recycling first, followed by shopping habits and conservation of resources.[10]

Regarding the magnitude of the changes, the impression that their involvement in these social movements has left on the daily activities of the women is most profound in Catalonia, notable in Gibraleon, and less important in Bilbao. These results are consistent with the way the problem was posed and developed in each place. In Catalonia, the opposition to the *Pla de Residus* generated a debate regarding waste management that involved the whole autonomous community and led to a high degree of knowledge and consciousness raising within the community. In Gibraleon, however, this only occurred at the provincial level, largely getting caught up in political disputes and rivalries. In Bilbao, this process barely spread beyond the old neighborhood limits, remaining just one more conflict in a neighborhood besieged by problems.

Table 5.8 Catalonia. Prioritization of strategies for environmental protection

Indicators	Consumption	Protests	Education of children	Self-realization	Public education
Average	10	8	10	8	8
Maximum	10	10	10	10	10
Minimum	6	5	8	5	6
Stv	1	1	1	1	1
NS/NC	8	14	10	13	16

Note: Importance ranked 1–10

Assessment of women's roles in protecting the environment

The women were asked to evaluate, on a scale of 1 to 10, in what activities they thought they could best collaborate in the protection and preservation of the environment: control over consumption, activism, education of children, personal empowerment, and public education.[11] The women from Gibraleon and Bilbao (Tables 5.6 and 5.7) responded that their main contributions were in the education of their children, followed by control of consumption.

In Catalonia (Table 5.8) the women prioritized control of consumption and the education of their children. Important differences occurred in the second and third place rankings. The women from Catalonia considered personal empowerment next in importance while the women from Bilbao felt that their participation and public education in the movement were more important than their personal empowerment. The women from Gibraleon ranked both personal empowerment and public education in third place, and placed involvement in the movement last.

It is interesting, that even though all of the women interviewed were directly involved in community activism, they value their roles in it so differently. They all attributed the success of their opposition movements ultimately to "the strength of the people."

The fact that the Catalan and Gibraleon women value control over family consumption and education of children as their most important contribution to the environment may reflect an established principle in Catalonia of "normalization" of the environmental problem and its assimilation by the community. This allows them to begin to think about collective strategizing. In contrast, in Gibraleon, except for those directly affected by the proposed project and those involved in the opposition movement, the community has not yet engaged in a collective process of environmental management.

The women from Bilbao felt most strongly about the family strategy of education and consumption control as well, but also place great importance on their participation in the movement and community activism. Given that the waste site has not yet been completely emptied and it continues to smoke, it is not clear that their problems are over. The women remain vigilant and ready to mobilize again at any moment. The perception of these women is that the only way they will accomplish goals is "to take to the streets." This is a phenomenon that does not occur in Gibraleon or Catalonia, where the women perceive their presence in the movements as an exception to the norm and they hope not to have to repeat the process.

NEW CONCEPTS OF ENVIRONMENT, RESPONSIBILITY, SPACE, AND TIME

Health, well-being, and affection for place: a different concept of the environment

During the interviews, it was rare that women framed their feelings of threat in relation to the environment. They spoke about their concerns for health, especially that of the children, and about a feeling of loss of a place and a milieu they could consider their own. They expressed frustration over thwarted expectations with respect to quality of life, as families and as individuals. There is a need to revise the "stereotyped" nature of the concept of "environment" so that it is possible to capture, comprehend and explain the distinct perceptions and experiences of women. A new concept must be found which also embodies the manifestations of environmental impacts on household life and on the desires and aspirations of people and communities. Since the unit of analysis would be individuals and families, contrary to current definitions, analysis would focus on the microscale. This lack of an adequate definition of environment in the academic as well as the waste management spheres, indicates a need to explore the androcentric character of current, formal environmental knowledge, which is considered scientific and universally valid.[12]

Women as activists, contrary to the myth of passivity

In all the cases analyzed, the women have demonstrated a degree of radicalism and boldness that openly contrasts with the stereotype of the woman who is passive, indecisive, and fearful when faced with anything political and technical that she does not fully comprehend or when confronted by physically risky situations. In this regard, their distinct commitment and connection as women to their families and their communities, both sources of strength for their existence, seem to have been the fundamental impetus behind the participation of women in these very gendered social movements. Under such circumstances, the town or neighborhood becomes a community with social meaning. A crucial network of women's sociohistoric, personal, and emotional relations emerges, which women defend and preserve and seem intricately committed to.

When involvement in community activism is derived from a gendered sense of responsibility for the quality of life in the community and the family, women tend passionately to safeguard a healthy environment for present and for future generations.[13] This gendered objective of attaining a better future for the family and the community explains why the women did not waver before unconditionally participating in a wide array of necessary activities for the success of the social movements. They fulfilled this longer term obligation despite any suffering that they might have felt as mothers as a consequence of any unavoidable inattention to or momentary abandonment of material or emotional tasks and responsibilities in their homes.

Struggle, perseverance and resistance: in "defense" of domestic life

Women handled these situations in an instinctive, stoic, and even good-humored manner. This is remarkable given that these events were in every respect exceptional in their lives and that of the community. These are situations which push one to the limit, that entail enormous personal sacrifice, require continuous readaptation to changing circumstances day by day, and versatile, multidirectional action. In trying to explain why women seem to have this capacity to remain on an even keel in these situations, it becomes clear that the attitudes required to mobilize an effective social movement, although different in intensity, are the same as those that women have had to develop as their socially defined gender roles and responsibilities in domestic and community life have evolved.[14] To the extent that the patriarchal model has situated women in the private sphere, making them responsible for family life, it has fostered very gender-specific and specialized coping mechanisms and strategies for addressing everyday situations. This model has conditioned both the nature of women's work and their methods of fulfilling their responsibilities as well as their value and role in society.

Here, it is pertinent and indispensable to comment about the integrative nature of home life, and its relation to socially constructed femininity. If we consider the abilities women typically develop in their domestic roles, integration emerges as a defining characteristic. This generalized role – which is projected clearly in the activities of the "housewife" – is valued negatively, undervalued or, at best, ignored when juxtaposed with the specialization of the public sphere. This dominant specificity is used to define, classify, and grant value to people in a society run by "specialists."

In this context, an event such as a social movement, which primarily affects the private sphere, disrupting domestic patterns of everyday life, values, and work, is difficult for the dominant, androcentric paradigm to incorporate and can be socially and personally disruptive. In these three communities, women demonstrated a facility for understanding these types of situations and readily integrated them into their lives, almost automatically, as a part of everyday life.

The "spilling" of private space into public space

In each of the case studies, the social movements led to a type of "spilling" of the domestic sphere outside the home gates to the point of a genuine invasion of the public sphere. This subsequently engaged the public arena in a foreign process which was not considered part of its nature or "essence." The participation of women in civil disobedience dissolved many of the barriers that exist between functionally differentiated spaces. The strategy employed by the women in each case study, of taking to the streets and making themselves constantly present, transformed the public space into an excellent stage for women's self-affirmation. In all the cases, women took private activities and attitudes to the street: they remained there preparing food, sewing, taking care of young children, conversing about domestic problems and about news of the mobilization, discussing strategies, constructing banners, organizing raffles, and even setting up improvised candy or prepared food stands to raise funds.

This process seems to be a direct consequence of the previously mentioned transformation of local space – town or neighborhood – into community. Through their visible and active involvement in the social movements, women's historically subordinate roles in the towns and neighborhoods became public. Their gendered identities and use of space temporarily became dominant due to their active agency in the organization, development, and success of the movements.[15]

From public times to private times, or the feminization of periods of civil disobedience

The overlap of public and private spaces and times affects the use and significance of time in two ways: first in terms of the rhythm and intensity of the movements, which is a function of domestic time; second, personal time comes to articulate itself based on the simultaneity and compatibility of a new set of tasks of different natures. A "typical" day of community activism is organized in terms of domestic time. In the morning, news is circulated, community actions are planned, and consciousness raising events are developed to "fire up" and keep the movement alive. Despite the objective importance of the morning and the fact that those involved in the movement recognize its importance, the public image of the problem usually renders this time and the activities of the women invisible.

The afternoon/evening and holidays provide the stage for public acts. Assemblies, rallies, and other public education activities or civil disobedience can be most effective at these times because the men, who would otherwise be at their paid jobs, can participate. Often, given their public, external nature, it is only these most "visible" acts that public opinion identifies, erroneously, as community action. In the most "public" of these events – meetings and assemblies – the participation of men is frequent. This potentially misleads those not directly involved and may promote the erroneous impression that the men are at the helm of the social movement, subsequently devaluing the tremendous amount of work, although less "visible," done by the women.

Regarding the organization of individual time, the androcentric model of compartmentalization and specialization of tasks is broken and substituted by the feminine model of simultaneity and versatility, characteristic of the domestic sphere. This gender-differentiated model for time allocation is imposed over the more typical male approach because it adapts best to spontaneous emergency situations. As discussed earlier in the chapter, this process can cause women to regard the social movements as an overload, but not as a qualitative change. Men, on the contrary, may find themselves, to a large degree, outside of themselves and their usual routine, which provokes a sense of disruption and stress.

In terms of community time, the coordination of public community activism with the annual cycle of popular festivities proved critical to the effectiveness of the social movements. Invariably, in all the cases, popular events that mark annual festivities served as a platform to make known or to diffuse information about the problems confronting the community. Women, as the main instigators of these movements, took the initiative to organize public events and feasts, making them their own instruments and means of activism and struggle.

CONCLUSION: TOWARD A FEMINIZED MODEL OF ECOLOGICAL CONSCIOUSNESS AND ACTION

This final section proposes a new political formula – in the broad sense of the word – and a social agenda with regard to the issues raised in this chapter. It has been illustrated repeatedly, from different perspectives, that the involvement of women in social movements in defense of the environment brings distinct gender-specific perspectives, experiences and values to the forefront. Because of their role in society, women have their own perspectives and conceptualization of the environment that differs from that which is considered the norm. Women's gendered perceptions and experiences give them the framework and perspective to develop alternative definitions of the environment. The traditional roles of women in the domestic sphere result in a distinct scale and milieu – health/well-being/affection for place – in which environment is defined in a way that seems more adequate to address the everyday problems of the vast majority of people. Women's gender-specific influences also play a significant role in the organizing strategies or ways in which space is used and time is structured. This confirms the importance of women's contributions to the formation and success of community activism and social movements.

The role, significance, and critical involvement of women in the social movements in Bilbao, Catalonia, and Gibraleon suggests the existence of a "feminine model of environmental consciousness and action"[16] which, far from being limited to women, can serve as a model for the design of alternative, non-androcentric forms of addressing environmental problems. These alternative approaches, while based on women's experiences, values, and contributions, must also undertake the necessary redefinition of the private sphere. They are inscribed in a global re-evaluation of the personal such that it serves as an arena for individual, social and political participation.[17]

The involvement of women in the denunciation and resolution of ecological problems not only possesses specific gendered characteristics, but it creates the possibility for a feminist political ecology. Moreover, it contains the seeds for social change of great scope. Women's environmental activism can contribute to a society in which men and women, liberated from the tyranny of gender ascriptions, can live fuller lives, and dissolve the limits between the personal and the public, immersed in an environment in which economic profit and technical rationality are at the service of the health and well-being of all of humanity.

NOTES

1   Population data for the three case studies from 1990 census.

2   Chemical contamination constitutes a serious problem that affects the Huelva urban area as a whole and the coastline in particular, extending to wetland zones of high ecological value. AMA (1987) and Adaro (1990a and b) present successive plans for decontamination.

3   For a complete description of the structure and development of industry in the Basque region, see García Merino (1987). A sociological study of the city is included in Leonardo (1989).

4   The city government commissioned a chemical analysis after the blaze to determine the composition of the by-products. For the results, see IMPOLUSA (1989) and EPYPSA (1989).

5   For a complete description of industrial waste problems in Catalonia and the contents of the proposals for the "Waste Management Plan," see Junta de Residus (1989a and b).

6   For more in-depth theoretical background material, see for example Bellucci (1992); Biehl (1991); Brú (1993); Garcia-Ramon (1989); Heller (1977); Heller and Feher (1985); Norwood and Monk (1987); Offe (1988); Yeatman (1990).

7   The attitude of the mayor and the regional government, who interpreted the movement almost as a political revolution, exacerbated the situation to unbearable limits. From November 1988 to May 1989, when the conflict ended, armed forces established themselves permanently in Gibraleon, applying strong repressive tactics.

8   For an interesting analysis of the conflict from a population perspective and that of ecological movements, see Borras and Perales (1990).

9   In the case of Bilbao, there were some difficulties in the implementation of the question. Only three women answered, but their responses were consistent with the results from the other two

10   It seems, then, that when it comes to the type of changes, daily consumption is given much importance – the most in all cases except in Catalonia, where it is surpassed by recycling.

11   The rate of abstention was high in all case studies, at times reaching 50 percent, perhaps indicating a lack of familiarity with questions of this nature.

12   The theoretical implications of these conclusions have been developed more broadly in Brú (1993).

13   In the analysis of the perception and attitudes of women regarding gender roles, I locate my work in the lines of "domestic feminism." In this sense, and in the Spanish-speaking world, Bellucci (1992) is of great interest.

14   In referring to sociological change and the political implications of domestic life, it is essential to consult the already classic works of Heller (1977); Heller and Feher (1985); and Offe (1988).

15   For a panoramic vision regarding the contribution of geography to gender studies, see Norwood and Monk (1987) and Garcia-Ramon (1989).

16   For a complete development of this concept, as well as that of women's ecology, see Brú (1993).

17   For an interesting analysis of the new ecological perspectives at the heart of the re-evaluation of private life in postmodernist society, from a gender perspective, see Biehl (1991) and Yeatman (1990).

REFERENCES

Adaro, E. N. (1990a) "Estudio Sobre las Alternativas de Reubicación de las Industrias de la Avenida Francisco Montenegro," Avance (Estado del Proyecto), encargado por la Agencia del Medio Ambiente de la Junta de Andalucía.

—— (1990b) "Estudio Sobre las Alternativas de Reubicación de las Industrias de la Avenida Francisco Montenegro-Huelva," encargado por la Agencia del Medio Ambiente de la Junta de Andalucía.

AMA (1987) Agencia del Medio Ambiente de Andalucía, "Plan de Corrección de los Vertidos Industriales Contaminantes en el Litoral de Huelva," Sevilla: AMA, Junta de Andalucía.

Bellucci, M. (1992) "De la Participación al Protagonismo de las Mujeres: Estrategias de Supervivencia de las Mujeres Pobres Urbanas, Son Formas de Participación Cotidiana o de Protagonismo Social," in VVAA *Mujeres Hoy*, Buenos Aires: Fundación TIDO.

Biehl, J. (1991) *Rethinking Ecofeminist Politics*, Boston: South End Press.

Borras, M. and Perales, E. (1990) *La Merda a Catalunya, Qui la fa i qui se la menja*, Barcelona: Llibres de l'Index.

Brú, J. (1993) "Medi Ambient i Equitat: La Perspectiva del Génere," *Documents d'Analisi Geografica* 22: 117–30.

EPYPSA (1989) "Estudio de Localización de un Depósito de Seguridad de Residuos Inertes," Madrid.

García Merino, L. V. (1987) *La Formación de Una Ciudad Industrial, El Despegue Urbano de Bilbao*, Bilbao: HAEE/IVAP.

Garcia-Ramon, M. D. (1989) "Para no Excluir del Estudio a la Mitad del Género Humano: Un Desafio Pendiente en Geografia Humana," *Bol. Ass. de Geógrafos Espanoles* 9: 27–48.

Heller, A. (1977) *Sociología de la Vida Cotidiana*, Barcelona: Península.

Heller, A. and Feher, H. (1985) *Anatomía de la Izquierda Occidental*, Barcelona: Península.

IMPOLUSA (1989) "Estudio de los Residuos Vertidos en un Area de la Mina San Luis," encargo de la Caja de Ahorros de Vizaya, Madrid.

Junta de Residus (1989a) *Pla Director per a la Gestió dels Residus Industrials a Catalunya*, Barcelona: Departament de Política Territorial i Obres Públiques, Generalitat de Catalunya.

—— (1989b) *Projecte de Llei de Mesures Urgents per a la Gestió de Residus Industrials a Catalunya*, Barcelona: Departament de Política Territorial i Obres Públiques, Generalitat de Catalunya.

Leonardo, J. (1989) *Estructura Urbana y Diferenciación Residencial: El Caso de Bilbao*, Madrid: Siglo XXI/CIS.

Norwood, V. and Monk, J. (1987) *The Desert Is No Lady: Southwestern Landscapes in Women's Writing and Art*, New Haven, Connecticut: Yale University Press.

Offe, C. (1988) *Partidos Políticos y Nuevos Movimientos Sociales*, Madrid: Sistema.

Yeatman, A. (1990) "A Feminist Theory of Social Differentiation," in L. Nicholson (ed.) *Feminism/Postmodernism*, London: Routledge.

ACKNOWLEDGMENTS

This chapter is the summarized version of a study financed by the Spanish Department of Social Issues, Institute of Women. The contributions of María Angeles Alió and Nuria Ferrer, who collaborated as technical advisors, and Juana Bernabé and Maribel Cervantes, who participated in the gathering of general information, are gratefully acknowledged.

Part III

GENDERED RESOURCE RIGHTS

6

GENDERED VISIONS FOR SURVIVAL

Semi-arid regions in Kenya

Esther Wangari, Barbara Thomas-Slayter,
and Dianne Rocheleau

INTRODUCTION

Gendered land and resource tenure and structural changes in the property regimes within agrarian systems have been instrumental in determining gendered roles in environmental management and economic development in the semi-arid lands of Eastern Africa. Of particular interest are gendered rights of resource use, access, and control; gendered use of resources; the gendered basis of knowledge and responsibility; and gendered organizations and political activity in semi-arid areas. We argue that the above factors have been shaped in part by specific attributes of capitalist, colonial, and neocolonial economic and political structures.

This chapter presents three case studies which provide insight into the ways that social, political, cultural and ecological factors interact with gender and how these are expressed in gendered relations of production and strategies for survival. Empirical data are drawn from the semi-arid settlement and agricultural frontiers of Embu and Machakos districts, Eastern Province of Kenya. The two districts, while similar in many ways, especially in the implementation of land tenure reform programs, exhibit variations in ecosystems, climate, cultural, and farming systems. Recurrent famines and the conditions which create them have occurred amid dramatic changes in land use, spawning in turn radical changes in gendered rights, responsibilities, and knowledge within local livelihood systems.

The first case examines the impact of agricultural development policies – in particular the land registration process that was introduced in Mbeere, Embu District, in the 1970s – on gender relations at the household and community levels and on the environment. It also explores the actions taken by local women to overcome their difficulties, many of which can be linked to a land reform process which institutionalized and further exaggerated women's subordinate position to men within the social system.

The second case examines the involvement of *mwethya* women's groups in collective action for environmental change in the two communities of Mbusyani and Katheka in northern Machakos district. It focuses on the risks and uncertainties faced by households and how the *mwethya* women's groups are addressing them collectively at the community level. This case considers the impact of the groups upon the resource base, landscape, and economy in the context of underlying patterns of gender relations.

The third case deals with the gendered knowledge of plants in relation to tenure and organizations. It illustrates the positive results that can be achieved when development agents call upon both women and men in rural communities to participate in development programs that affect their lives. Women in the market center of Kathama in Machakos district, equipped with their local skills and knowledge of their own environment, particularly the trees, are turning many plots of bare land into green, productive land through their involvement in agroforestry, water management and soil conservation. They are also struggling to maintain ecological and economic diversity in landscapes increasingly dedicated to annual crops for both cash and home use.

In spite of reports affirming widespread economic and environmental recovery in Machakos (Tiffen *et al.* 1994), and a variety of agricultural developments in Embu, the rural poor of the districts, and women in particular, continue to experience considerable distress during droughts and famines. They suffer from hunger, malnutrition and related diseases, and the depletion of livestock herds affected by disease and starvation or the need to sell at low prices in times of distress. Droughts and the consumption of trees for fuel have also led to deforestation. Some families have little money or none at all and their farms are not only small, but the land is marginal. When crops fail, most poor farmers in Embu and Machakos districts resort either to casual labor on plantations or to charcoal production and sales in order to earn the necessary cash to purchase food. Since many men have migrated to urban areas and plantations in search of wage labor, it is the women who have become increasingly responsible for rural life. These responsibilities include home maintenance, farm management, soil and water conservation, as well as drought response, famine prevention, famine response, and recovery.

Local and national contexts as well as international development programs, as filtered through national programs, also shape the circumstances under which people in Embu and Machakos districts operate. The privatization of land, the introduction of cash crops, low commodity prices, and the lack of social services due to structural adjustment programs have left many people landless, hungry, and poor. These misguided programs have developed into a source of social disintegration and environmental degradation.

Land registration, instead of redistributing land and encouraging agricultural development, has increased stratification in these communities along the lines of both class and gender. The privatization of land has been based on a trickle-down policy in which the wealthier have been able to acquire more land and the poor less, generating a landed and a landless class, as Swynnerton (1954) predicted. In addition, land reform, by recognizing only males as "heads of households," has effectively institutionalized women's customary subordination to men. The rich and powerful have easily acquired land, while the poor and women have been left out (Wangari 1991).

In response to national agricultural programs, farmers have shifted land previously used for food production to cash crop production. In Machakos, many households struggle to grow coffee or cotton; in Embu they try to grow cotton and tobacco. Families hope these crops will cover school fees and meet other requirements of modern life. But the soils are relatively poor for these crops, and the yields are a fraction of those found in more fertile parts of Kenya. Moreover, the prices for these commercial crops are not determined by the regional markets or even in Nairobi, but by the General Agreement on Trade and Tariffs (GATT) in Geneva. The welfare of these households is shaped by the occurrence of drought in Brazil, strife in Colombia, or national policy toward tobacco farmers in the United States. Yet the price of cigarettes in the United States may go up while the price of tobacco in Mbeere remains constant. Returns on Kenyan cash crops vary based on events taking place far away and well beyond the control of farmers. Many women farmers have responded by shifting to alternative crops such as citrus, papaya and other fruits, and green vegetables for the national market.

Structural adjustment programs advocated by the International Monetary Fund (IMF) have led many families to operate under marginal circumstances and have pushed them to the brink of poverty. Structural adjustment programs call for national austerity measures such as the devaluation of currency. The altered exchange rates are expected to increase exports by making them more competitive in the world market. That means exports become cheaper internationally. However, the policy does not account for countries that do not have adequate industries for manufacturing capital goods. For example, a farmer in Kenya has to import capital goods, fertilizer and other inputs at a devalued exchange rate. For the poor, and especially women in these districts, the results of these new policy measures have been particularly devastating, reducing income and increasing the price of inputs for cash crops. The IMF has also put pressure on the government of Kenya to reduce social services and keep expenditures down. These types of austerity measures have affected the adequacy of educational opportunities, health services, and other benefits for rural communities.

These national and international factors are found to affect gender relations at the local level, the formation of local organizations, as well as the rates at which commodity production, fragmentation of land, and the processes of privatization and subdivision of holdings are proceeding within a community. Inexorably, people in these communities of Kenya – and other arid and semi-arid lands of Africa – are being drawn into broader ecological, political, economic, and social systems propelled by an increasing need for and reliance on cash.

Women in Embu and Machakos districts have embarked on their own strategies of survival. Frequent droughts have made them more alert to alternative strategies to sustain their communities. The 1984–5 drought, for example, brought severe hardship to millions of people throughout Kenya. The rural people in the arid and semi-arid farming communities, however, responded to the drought and subsequent famine with a combination of long-established local knowledge and innovation. Many of the rural poor mobilized and pooled their knowledge and skills – ecological, economic, and political – at individual, household, and community levels. Women in these regions of Kenya are reclaiming their knowledge and skills to make a differ- ence in their communities. They are looking for alternative visions for survival. These visions include strategies of coping with droughts, environ- mental crises, and the high prices for food, water, fuelwood, and healthcare. Collective actions exemplified by participation in women's groups have become essential. In the following we explore the conditions leading to these strategies for survival and consequent collective action.

CASE I: LAND REGISTRATION AND WOMEN'S INITIATIVES IN MBEERE, EMBU DISTRICT

Embu district in eastern Kenya covers an area of 2,714 square kilometers and has three divisions: Siakago, Gachoka, and Runyenjes. The arid/semi- arid divisions of Siakago and Gachoka, with a total area of 2,073 square kilometers, represent more than 75 percent of the district. Coffee, cotton and tobacco are the main cash crops grown on Embu's 244,000 hectares of agricultural land.

The current status of the land as well as women's well-being are tied to the history of land tenure. In 1954, under the so-called Swynnerton Plan, the colonial government of Kenya implemented a land reform program charac- terized by adjudication, consolidation, and registration of land (Swynnerton 1954). The program came to be viewed by policy-makers as instrumental for agricultural development in rural areas regardless of variations in ecosystems. The rationale for land registration was that security of land tenure – through the provision of title deeds securing the private ownership of the land – would give farmers in rural areas the incentive to improve their farms. Policy- makers and international financial institutions also pointed out that farmers with title deeds could acquire credit to invest in their farms. This credit would result in increased land productivity, income, and employment.

Land reform was started in the higher rainfall areas of the country, and it was not until the 1970s that it was implemented in the arid and semi- arid areas of Embu district. This case study examines the impact of land reform on the people living in Mbeere, an ethnically distinct area in the Siakago and Gachoka divisions of Embu. We focus particularly on how land reform has negatively affected women, the poor, and the environment and how the women of Mbeere have responded to their substantial difficulties.

Conditions in Mbeere

A sample of 180 Mbeere households were interviewed in 1989; half of the households had gone through the process of land registration, the other half had not participated in the process. In the total sample of registered and unregistered farm households in Mbeere, about two thirds of the farmers owned less than twenty acres (8 hectares) of land. The arid/semi-arid area of Mbeere is not suitable for growing most crops found in high potential areas due to its sandy, infertile soils, and unreliable rains. Farmers grow two crops of corn as well as the cash crops cotton (6,224 hectares) and tobacco (642 hectares). On many farms, cotton and tobacco have replaced such local crops as sorghum, bulrush millet, beans, sweet potatoes, green gram (lentils), cowpeas, and pigeon peas.

Women spend more time than men on farms in Mbeere. In the survey, 73 percent of households were headed by males, while 27 percent were headed by females. Women in both types of households participated more in agricultural production than men: 72 percent of the women, but only 53 percent of the men, were engaged in agricultural production full-time. Only 13 percent of the women were involved part-time, whereas 23 percent of the men were only part-time farmers.

Households in Mbeere cannot depend on incomes from farming activities to meet their daily needs, not to mention school fees, school uniforms, and school building funds (Wangari 1991). Given the conditions in the region, nonfarm activities are as significant for the welfare of households as farm activities. Primarily men, but also some women, migrate to the cities in search of wage labor with which to supplement the family income. Young married men often migrate leaving their mothers, sisters, or wives to take care of the farms. Many households in Mbeere remain female-headed because husbands or sons do not return home from their jobs in the cities or the women were divorced or widowed. Other households were headed by unmarried women with children.

Problems with land reform in Mbeere

Land registration in Mbeere, and elsewhere in Kenya, institutionalized a new form of private ownership of land which was different from customary concepts of land allocation and use rights within the region. The new form conceptualized the male in a household as the sole owner of the land and the legal "head of the household" and failed to recognize women's rights of use and responsibilities for management. Previously, land was under lineage control and was managed with the oversight of clan elders and members. With land registration, men began to regard their state-granted status as "heads of households" as more important and therefore more powerful than their connection with their clan. Men would no longer accept the restrictions from elders or clan members about what they could or could not do with their land.

In addition, the new form of private land ownership institutionalized women's subordinate position in relation to men within the local social system. Although on paper the Registered Land Act (Republic of Kenya 1977) does not legally exclude women from land ownership, the way it has been implemented has done just that. Implementors of the land registration have followed local customary laws and religious beliefs which limit a woman's right to "own" land. Instead of offering land ownership to women, implementors combined Western and local social practices and distributed title deeds primarily to male "heads of households." Land allocation officials ignored women's roles in household decision-making, their contribution to the production, reproduction, and maintenance of families.

In Kenya, this practice is reflected in the fact that only about 5 percent of women are landowners. In Mbeere, only 2 percent of the women in the sample of 180 farm households had title deeds (Wangari 1991). Over 40 percent of those interviewed in Mbeere indicated that landlessness among women was due to the fact that the clans did not allow them to have land. Over 51 percent of the respondents pointed out that women were not allocated land (that is were not recognized as legitimate independent users) through the land registration program (Wangari 1991).

Thus, in Mbeere, as in most of Kenya, the majority of women are "legally landless." They may live and work on their husbands' farms, but they have no legal rights to the land or decision-making power over land use changes. There is no explicit legal barrier to women's ownership of land, and some women do own their own plots. However, the majority of women live with the double legacy of:

1 patrilineal customary practices that grant use rights to women and control over resources to men, and
2 modern land tenure reform under colonial and national governments which exaggerated the previous inequalities by legalizing men's ownership and failing to recognize women's use rights.

The few women who obtained title deeds in Mbeere did so through market mechanisms after the initial land tenure reform and land survey process. Many women with no husbands had not been allocated land in the original land tenure reform process, and purchased land using earnings from off-farm employment or businesses. Those not able to purchase land often became dependent on sons or were forced to search for employment and lodging in cities or on plantations.

Land adjudication and registration in Mbeere has been very slow due in part to inevitable disagreements among clan members over land claims and boundaries (Land Adjudication Office, Embu 1987). The transformation of lineage claims of land ownership to private ownership has been at times cumbersome and further complicated by claims from outside lineages. Furthermore, because of the complexities of land ownership in polygamous households, divisions of land among the sons of these households has been contentious, and decisions that involve selling land have often been conflicted.

The land reform program in Mbeere has undermined what little sense of control over or input into household decision-making with regard to land women may have had under customary law in relation to clan elders. Women now find it even more difficult to defend their rights when a "head of household," for instance, wants to sell the family's land. At the time of the survey, for example, a farmer with two wives wanted to sell five acres (2 hectares) of his land to a doctor. The second wife, his favorite, and her children supported the move, while the first wife with her children were against it. The case was taken to the Land Board and, after much deliberation, the farmer was not allowed to sell the land, based on customary law. However, as in most cases where there are no clear statutory laws to protect women and children, the farmer sold the parcel of land anyway. It was later revealed that he had colluded with the officials of the Land Board.

This example illustrates how a combination of local customary laws and colonial policies can leave women and children with no reliable recourse in defending their rights to land (Wangari 1991). Depending on the "mood, whim or politics of the situation, African culture may be invoked to oppress women or modern society cited in efforts to eliminate the very cultural practices which safeguard the rights of women" (Ooki and Mbeo 1989). As implemented, land reform has served to reinforce the continuing subordination of women in the household, particularly with respect to property ownership and development programs.

Class, gender, credit, and crops

The land tenure reform process has also increased stratification by class, with special repercussions for women. In the words of the architect of the colonial land tenure reform, the privatization of land was based on a trickledown policy: "energetic or rich Africans will be able to acquire more land and bad or poor farmers less, creating a landed and a landless class" (Swynnerton 1954: 10). Swynnerton's prediction has, in fact, come true in Mbeere and elsewhere in Kenya, with women in particular suffering from this development. Many women from near landless families have been left alone to farm what little land remains available to them while their husbands work as wage laborers in cities and on plantations.

The privatization of land was predicated upon the use of land as collateral for bank loans. As title deeds were held by very few women in Mbeere, few women had successfully received credit from financial institutions at the time of the study. Farmers in Mbeere commonly sought credit from financial institutions such as commercial banks and particularly the Agricultural Finance Corporation (AFC). In 1985–6, 210 households out of a total of 50,241 in Embu district were qualified to receive credit from the AFC, based in large part on their title deeds. This included both small and large loans. Out of the 210 households, only nine women received loans, using other forms of collateral to obtain credit (Agricultural Finance Corporation Office, Embu 1986).

Farmers in Mbeere, as in the district as a whole, also had difficulty obtaining credit. Of farmers growing cotton, only 15.9 percent obtained credit from financial institutions (Wangari 1991). Instead, informal credit was widely used for agricultural purchases such as buying hoes, and food in times of drought. Farmers reported being discouraged and even intimidated about applying for loans because of stringent credit requirements, lack of knowledge about filling out forms, lack of transportation, the long application process, and lack of confidence in the system. In Mbeere, 92.2 percent of the sample's registered farmers felt that financial institutions did not favor farmers with small farms or without other sources of revenue (Wangari 1991). Over 40 percent of the registered farmers were reluctant to use their title deeds as collateral in case they would not be able to repay the loans (Wangari 1991).

For the poor and for most women, applying was a moot point. Poor farmers often did not have the necessary collateral (off-farm income) and most women had neither a title deed to land nor off-farm income. None of the women in the sample group in Mbeere had acquired credit. The low percentage of credit from financial institutions in Mbeere discredits the argument that land registration was a fundamental tool for encouraging agricultural development. Much of the reason for that failure lies in the fact that absentee men maintain exclusive legal rights over lands cultivated and managed by women.

Analysis of cotton production in Mbeere revealed that land registration was not a fundamental tool for encouraging agricultural development. In the sample of 180 farmers, 45 farmers grew cotton. In Siakago, farmers harvested an average of 7.5 bags each per year while farmers in Gachoka harvested about 3 bags (Wangari 1991). It was anticipated by the creators of the land tenure reform and agricultural development policies for the region that farmers would have intensified their farming systems through the use of fertilizers, new plowing techniques and improved seed varieties and that this would have resulted in a higher output per acre. In our survey, however, the mean output per acre in Siakago indicated that despite land registration, productivity has, in fact, not increased and deliveries of inputs have been unsatisfactory. Corruption within all levels of management of the Cotton Lint Marketing Board was also prevalent at the time of the survey, resulting in often unreliable and delayed payments for cotton deliveries.[1]

In searching for explanations for this phenomenon, we found that farmers were concerned about declining productivity of their cotton land, prices for cotton, availability of family labor, and lack of credit, all of which interact with gendered access to and control over land. Farmers noted how monocropping cotton led to the degradation of their soils and left little land to grow food. They could no longer grow traditional food crops such as legumes, sorghum, or millet. In addition, the average annual income from cotton at the time of the survey was 1,202 Kenyan shillings (about US $45) after deductions for inputs. Farmers could not justify growing cotton with the devalued exchange rate, the rising costs of many inputs and basic foods as well as education and social services. Incomes could not feed a family or send a child to school. Almost two thirds of the cotton farmers (about 64 percent) noted that the availability of family labor was not sufficient to manage cotton cultivation.

Given the out-migration of men and division of labor in Mbeere, which puts women in charge of food production, it was women who most keenly felt the brunt of these difficulties. Less land meant less food for their families. Since they could not afford to buy food, they had to find other alternatives, one of which was to intercrop beans with the cotton where the agricultural extension officers would not catch them.

Just as they were responsible for food production, women in Mbeere were in charge of the health of their children. Since food was not adequate, poor families and especially female-headed households suffered nutritionally and their health declined. Searching for food added stress on women and reduced time for them to be with their children. When family members were sick, women trekked for miles to health facilities which often did not have medicines because of reduced expenditures on social services.

Overall, cash crop production and land tenure reform have resulted in poor economic returns and degradation of farmlands, as well as heavy work-loads and insecure benefits for women. Land consolidation and the development of more fixed boundaries have also reduced the flexibility and diversity of farming systems, and have contributed to the overexploitation of resources in the larger landscape, such as wood and grazing land. This has had a disproportionate effect on women, since they are more directly dependent on these resources than most men for meeting their daily needs and responsibilities.

Women and their strategies for survival in Mbeere

While some women migrate like their male counterparts to urban areas or plantations in search of wage labor, it is middle-aged women who are often left behind. In Mbeere, it was this group of women who were left to cope with the fallout from the land registration program. In response to the environmental degradation, increased social stratification, lack of livelihood options, and lack of support from the government, their husbands, and clans, some women in Siakago Division formed a self-help group. Unlike some women's groups that have financial support from nongovernmental organizations, this group had no such support. Women in the group had one common goal, to help each other.

One of the main activities pursued by the group was tree planting and the gathering of knowledge on trees and other local plants that could be used for medicinal applications. The closest health center was more than a few miles away and more often than not was not staffed with a doctor and had no medication. In addition, the lack of proper sanitation and clean water, as well as increased runoff due to soil erosion, had led to an increased incidence of diseases such as malaria, bilharzia, and typhoid in Mbeere. Women were searching for roots, trees and leaves for the prevention and cure of various diseases in the area. The women in the group recognized that they must reclaim their knowledge about the plants in and around their communities in order to overcome these health and environmental problems. Since colonial times, the political and economic climate in Kenya has been hostile to this knowledge and has devalued it.

Other group efforts included weeding parties, rotating credits and soil and water conservation activities. The idea of weeding parties, weeding as a group for pay, was developed as an income-generating activity to fund some of the other projects. Using these funds, the women's first project was to build each member a house. Once they had shelter, the women started more long-term income-generating projects such as buying a goat for each member. Each goat produced offspring which could be raised or sold and provided meat and milk for sale and to feed the family. Income was then used to cover educational expenses or to meet other household needs.

With their earnings and their own labor the women made simple drip irrigation systems which enabled them to grow fruits and legumes for home use and sale. Initially, they sold their produce to middlemen who bought them at cheap prices and resold them at high prices in the cities. Given poor rural roads and lack of transportation to the district market, the women thought that they would have to be dependent on middlemen to sell their produce. Eventually, however, they formed a cooperative and built a store in Siakago to sell their produce. The process involved extensive negotiation and paperwork with the chief and the town counselors, all of whom were men. Most women did not know how to read or write and were at the mercy of the politicians, but ultimately they succeeded.

So in practice, land reform and the distribution of title deeds in Mbeere has not given all farmers access to credit, nor has it necessarily led to increased land productivity, income, and employment. Overall, land reform has had a negative effect on rural women in the semi-arid lands of Siakago and Gachoka divisions. Until ingrained patriarchal attitudes toward women change, there will be little progress toward an equitable and sustainable society. Women's role in sustaining the human environment as well as managing the physical environment starts with planting a shade tree, a pumpkin, or sweet potatoes around a homestead and is important for providing food and fuel as well as preserving the environment. Women's access to resources of all kinds is crucial for sustainable development in this and other regions. In Mbeere and similar areas women can facilitate effective management of resources since they are responsible for their immediate environments. This implies that the development programs have to integrate time, nutrition, health facilities, energy, water education among others as components that address women's needs in rural development. Attention to gender relations with respect to resource use, access and management could improve development policy design and formulation for both economic and environmental ends.

CASE II: MERGING GRASSROOTS ACTIVISM AMONG WOMEN IN THE SEMI-ARID LANDS OF MACHAKOS

As the preceding discussion of Mbeere reveals, most communities in the semi-arid regions of Kenya face persistent problems related to poor development planning, a patriarchal social system, a harsh environment, and environmental degradation. In many instances, local people are actively addressing the decline of the resource base and the changing pressures upon the environment and the economy. These responses to the environmental crisis are largely initiated, collectively, by women who see the very basis of their livelihood system eroding. Their efforts are organized through self-help groups working on shared community problems or household matters. In Machakos women have long been active in *mwethya* groups, a form of labor exchange and collective work among the Akamba people.[2]

Today, many groups are reformulating themselves to address the issues arising from pressures of environmental degradation combined with the region's increasing incorporation into the larger polity and economy (Asamba and Thomas-Slayter 1991; Mwaniki 1986; Ondiege 1992). They build and repair bench terraces and dig cut-off drains on individual family farms or "shambas." They cooperate, building check dams in gullies that run between

Plate 6.1 Women's groups work on soil conservation and land reclamation throughout Machakos district

Source: Barbara Thomas-Slayter

farms and may border on as many as twenty or thirty separate land holdings. They help one another with individual tree planting or farm weeding. In addition, the group members cooperate on a variety of income-generating activities. These include maintenance of tree nurseries, tree planting, horticulture – with emphasis on beans, cabbages, tomatoes, and onions – brickmaking, poultry, beekeeping, rabbits, and making baskets.

We explore women's involvement in collective action for environmental change in the two communities of Mbusyani and Katheka in northern Machakos. This case draws on research carried out in these communities in 1987, 1989–91, and 1994 through household interviews, focus group discussions, surveys, and related participatory data-gathering techniques.[3]

Mbusyani is located approximately 90 kilometers east of Nairobi and Katheka lies just to the northeast in the next location. Each has a population of approximately 7,000 people. The terrain is stony, the climate dry, and the land gently sloping to hilly. Both communities are located in agroecological Zone 3 characterized as a marginal coffee zone suitable also for corn, beans, and pigeon peas. Fragile soils, eroded hillsides, and land only marginally productive are characteristic of many parts of these sublocations. Rainfall is low and unreliable, with drought occurring in approximately one out of four years (Thomas-Slayter and Ford 1989; Munro 1975; Silberfein 1984).

We ask specifically what are the forms of risk, uncertainty, and insecurity that these households in Mbusyani and Katheka face? How are they being addressed collectively at the community level? What is the impact of women's collective efforts, as illustrated by these sublocations within Machakos district?

Are *mwethya* groups managing to secure the conditions needed to assure survival? Are the women members diminishing risk, especially environmental risk, and creating new opportunities for themselves, their families, and their households?

In exploring these issues we consider:

1 the origins of *mwethya* groups;
2 the circumstances which shape an individual's decision to join a *mwethya* group;
3 some of the instrumental factors determining the effectiveness of the group; and
4 the impact of these groups upon the resource base, landscape, and economy.

Throughout we assert that the underlying patterns of gender-based roles, responsibilities, and opportunities within the household and within the public domain shape participation in "gendered" organizations for environmental action.

The origins of *mwethya* groups

In the past, women have worked together in small groups of two, three or four, sharing tasks according to the agricultural season. This practice has been commonplace for many communities in East Africa (Thomas 1988). Such a mutual aid system provides firm underpinnings for the activities of women's groups. Among the Akamba the system of *mwethya* groups has been both systematic and widely practiced. For many generations, the Akamba people have used *mwethya* groups consisting of men or women organized along clan or family lines to provide emergency assistance, to perform special needed functions such as house building or clearing new fields, and among women, to share regular tasks such as weeding. The custom of *mwethya* had slipped away during the colonial era and was replaced with a more formal system of work groups and conscripted labor units such as those coerced into soil conservation programs (Thomas-Slayter and Ford 1989; Collier 1989). These and other coercive practices, such as land alienation and destocking, left a legacy of resentment, not only against the coercion and confiscation, but against organized resource management practices (Munro 1975).

In the decades following independence, people began to recognize the problems of soil erosion and deforestation and to consider organized forms of resource management. Today, there is a qualitative and quantitative jump in the kinds of endeavors in which women's groups are involved. The women's organizations operating in Mbusyani and Katheka are quite different phenomena from local agricultural support groups or age-group collectivities or the conscripted labor of the colonial era. Many originated in the 1970s and early 1980s with the encouragement and support of the national government and political leaders. Since this time, the *mwethya* group practice has been reinvigorated and transformed under the broader national support for new or customary groups engaging in environmental management and rehabilitation.

Plate 6.2 A *mwethya* group leader leads a soil conservation work party in Machakos district

Source: Barbara Thomas-Slayter

In some communities, such as Mbusyani, the groups have become the backbone of resource management initiatives. Each group consists of approximately 20 to 40 people (mostly women), usually from a neighboring cluster of farms within a village. Today, they are not necessarily organized along clan lines but instead among households with common interests. For example, in Mbusyani the groups join together in what is called the Mwangano Group for public works such as digging at a dam site, or terracing a hillside where erosion is particularly bad. In so organizing, the members are acting with a new community-based spirit, strengthened in some ways by clan interests, but not limited to immediate family concerns.

For a variety of reasons, national government policy continues to support the formation of women's groups. This policy derives in part from a growing international interest in women's issues, from the United Nations impetus behind the International Year of Women in 1975, the 1985 Women's Conference in Nairobi, and a worldwide strengthening of women's organizations. Most importantly, the groups do not challenge the existing cultural and structural underpinnings of the national government. In fact, there is a "convenient" joining of government policy concerning self-help and environmental conservation and the capabilities, needs, and interests of rural women.

Indeed, evidence suggests that a patriarchal ideology and the presence of women's organizations have sometimes provided officials with the opportunity to control women's collective labor for public purposes. Labor-intensive, community-based tasks, such as school maintenance or feeder road repairs, are frequently undertaken by women's groups at the request of local officials.

Motivation for group participation

Data reveal that participation in *mwethya* groups comes from all socio-economic categories, but disproportionately from poorer households. In Katheka, for example, in the late 1980s, 64 percent of all households had a family member in a *mwethya* group, whereas among the households with less than 5 acres (2 hectares), 85 percent of the respondents belonged to *mwethya* groups. Of those with more than five acres, 53 percent belonged.[4]

Membership in a *mwethya* group is often one element in a household's repertoire of livelihood strategies. For the majority of households access to capital is negligible, as is access to land. Women may participate in rural women's organizations to

1 improve access to productive assets, that is land, labor, and capital;
2 generate exchange opportunities – both market and nonmarket – for cash, goods, services, information, and/or influence; and
3 obtain access to common property, including resources such as water and communal grazing, or institutions and services such as schools and health clinics (Thomas 1988).

In Mbusyani and Katheka the most important reasons for joining a women's group are to preserve the resource base, both public and private, and to ameliorate the problems of labor scarcity. Given the current land scarcity in Machakos district, access to land for most households is virtually fixed. The continuing processes of privatization and fragmentation greatly affect women's rights to land. Households may have more than one parcel of land scattered within the sublocation. This is an important strategy for diversifying risk, yet the time spent to reach these parcels, which are often tiny, has a negative impact on labor allocation for food production. As the chief food producers, women are under increasing pressure to devise ways to improve yields under circumstances in which methods of agricultural production are changing from extensive to intensive.

Women in female-headed households find *mwethya* groups useful as a source of labor. Both Mbusyani and Katheka have a high percentage of de facto female heads of household because men go to Nairobi, Machakos, and other urban centers to seek work in order to supplement the household's income (Asamba and Thomas-Slayter 1991). These realities put enormous pressure on the women of the community – given women's routine responsibilities – to assure the family's food supply. Access to labor through a *mwethya* group can be a great advantage to a woman managing household food production under such circumstances.

For all the women involved, group activities and group ownership of new assets constitute a way to increase the household's access to new technologies, to new forms of capital investment, and to new services. The one certain resource is the household's labor, and in many instances that is primarily adult female labor. Groups help women to optimize labor at crucial times. For many women who are managing farms while their husbands are elsewhere, this need recurs regularly. These women have learned how to increase their access to a critical factor of production by sharing the only factor over which they have control – their labor.

Groups enable women to share labor not only at critical times in terms of the agricultural cycle, but also on a regular basis to prevent the deterioration of the resource base on which agriculture depends. Their immediate objective is to deal with soil erosion and water retention in their fields by working as a group to construct bench terraces, cut-off drains, and check dams to prevent soil and water loss and to increase crop yields. Much of their effort focuses on bench terracing, which is critical for agriculture in this hilly and arid region. In joining *mwethya* groups, the women are responding not only to the harsh environment and evidence of soil loss, which is often alarming, but also to their pressing poverty. The *mwethya* group efforts in soil management are legion.

In general, the groups work two mornings a week on the farms of members, rotating systematically among their membership. The owner of the farm where work is being undertaken on a given day may decide on the tasks to be performed. The groups maintain records and assure each participant's turn in the rotation cycle. Customarily, the groups work ten months a year, two mornings a week. Each household has the benefit of the agricultural labor force three to four times per year. Specific measures are used to ensure accountability in each member's work effort. If new bench terraces are to be constructed, each member is responsible for digging a specified amount, usually 1.5 – 2 meters per person per session. New terraces dug and old ones repaired range from 1,500 to 4,000 meters per group per year. Other tasks customarily performed include weeding, cultivation, mulching, or carrying manure.[5]

In sum, women perceive group activities and group ownership as a way to improve their resource base and to increase their access to labor, services, new technologies, and new forms of capital investment. Under present circumstances acquiring new land is virtually impossible, and there is little or no transfer of earnings from the commercial crops into improving food crop production or into other investments which provide the household with income. Groups offer some new opportunities.

The effectiveness of self-help groups

Group activity in Mbusyani and Katheka has deep philosophical and social roots found elsewhere in Machakos and more broadly in Eastern Province. While the basis for mutual cooperation and mutual enterprise is found in a social tradition, it is also supported by the firm dichotomy between male and female roles and the tradition of both male and female solidarity. Women's roles involve management of agricultural resources in order to provision the household, as well as home and childcare. Women's groups are building upon the customary responsibilities and duties of women.

The groups offer women an opportunity to expand their successes within a traditional domain: improving food security, the nurture of children, and the welfare of their homes. Where the groups are edging into business, it is through the link with women's major roles as cultivators and providers of food. Only as women's associations move into cultivation of fields for cash crops or ownership of shops or equipment and technology are they diverging from traditional roles.

So far, community support for these new roles exists. An indication of this support lies in the actions of the county council. In Machakos, the local governing authorities have given women's associations permits to open shops or maize mills, and the councils have leased or occasionally allocated land for the use of women's groups. Some assistant chiefs work closely with women's groups providing support, counsel, and "official" approval for their activities.

Education is also an important factor contributing to group effectiveness. Parts of Eastern Province have relatively high levels of education for women and men. Mbusyani has had a secondary school for nearly twenty years. Increasingly, women who have completed upper primary or lower secondary schools are willing to assume leadership positions in women's groups. In some communities women are emerging as active participants in formal organizational and political structures, no longer confined, as in the past, to the role of "volunteering" their labor. They are willing and able to take on management responsibilities, planning activities, or liaison work with the government or private agencies. Moreover, these women usually have ideas, energy, and determination, enabling the group to make up, in part, for what it may lack in capital assets.

In addition, institutional linkages are emerging in some communities. There is potential for collaboration based on a growing institutional infrastructure, involving churches, women's groups, NGOs, youth groups, and other organizations. The *mwethya* groups are beginning to take advantage of these opportunities.

Mwethya groups serve a variety of functions vis-à-vis the households and community in which they exist and consequently they have diverse impacts on the resource base, landscape, and economy. Their environmental impacts are particularly important. In Katheka and Mbusyani, as in many rural communities in semi-arid lands, poor women are particularly dependent upon access to shared natural resources for fuelwood and other forest products (Hoskins 1983; Agarwal 1986; Shiva 1989; Hayes 1986; Chavangi 1992; Davison 1988). With a decline in public, clan, or other shared lands, and with increasing privatization, they have been among the first to suffer losses. In the face of this decline, networks and associations are proving valuable instruments for providing households with increased access to productive and exchange resources. They become key elements in individual and household strategies for survival, accumulation, and mobility.

At the community level, responsibilities are clearly delineated along gender lines. When there are community-based, labor-intensive tasks to be done, it is the women, not the men in the community, who are mobilized to undertake them. If a road needs repair, if gulley erosion must be stabilized, if a school requires maintenance, it is the women who are asked to provide these services. The request will come from the assistant chief, in collaboration with the sublocation branch chairman of the national political party, the Kenya African National Union (KANU), to the leaders of the *mwethya* groups. They, in turn, notify their membership of the task at hand and the schedule proposed. On the designated day the women will turn up with their tools or supplies to carry out the assignment.

Why are only women asked to supply voluntary labor for the benefit of the community? Several reasons emerge from observations in Mbusyani and Katheka. These are confirmed by the community development officer at Machakos district headquarters, a woman with long experience in observing women's roles and organizations around Kenya.[6] First, there is a history of women's involvement in public works which emerged in the colonial period in the form of conscription of both women and men for various resource management activities. Second, positions of authority within the community are held by men, with the exception of officers of the *mwethya* groups. In Katheka, for example, the assistant chief, the KANU chairman and secretary, the head of the KANU Youth Wingers, church leaders, and school principals are all male. Third, women's traditional roles have related to family economic survival. For the most part this has involved them in agricultural pursuits and resource management on their own farms. By extension of their responsibility for family needs, women undertake community effort on behalf of public concerns. Fourth, the presence of the women's groups makes it easier to mobilize this membership than to mobilize individuals, whether male or female. There are officers elected to positions of authority within the group, and thus, there is someone, the chairwoman or secretary of the group, who can be held accountable if the members fail to appear for the specified task. The organization of women into *mwethya* groups for labor on their respective farms clearly facilitates their mobilization for other kinds of responsibilities. Moreover, the intertwining of traditional and contemporary public authority with accepted patterns of gender relations is evident in the relationship between the assistant chief and the *mwethya* groups.

Why is it not customary for the men in the community to join in the "voluntary" mobilization for public works? The answers are complex. They lie in perceptions of status and prestige, of appropriate gender roles in manual labor related to economic pursuits, in the traditional hierarchy of male dominance and female subordination. Whereas men in Katheka will shovel sand needed for construction purposes for a minimal wage, they will not, in general, shovel dirt needed for road repair, or build a check dam on a volunteer basis. Women, who would not seek the employment shoveling sand, can be mobilized to shovel dirt for the roads and construct the dams. It is clear that the process of utilizing female labor on public works is deeply enmeshed in historic gender relations.

This exploration of Mbusyani and Katheka suggests that women's involvement in *mwethya* groups addresses critical problems of environmental deterioration. It improves access to land, labor, capital, and information, including access to common property, and it increases the productivity of their land. Through their organizations, women are beginning to gain access to the political system. They do not yet have the institutional capacity to demand accountability from local leaders. Observations based on Mbusyani and Katheka suggest, however, that the organizational capacities are emerging which will enable these groups to sustain an agenda not only for addressing practical problems but also for seeking strategic, long-term interests within their immediate communities and beyond.

CASE III: GENDERED KNOWLEDGE AND FAMINE RESPONSE IN KATHAMA

Gendered knowledge plays a major role in farming, resource management, and survival of droughts and famines in the semi-arid farming communities of Eastern and Central Kenya. Yet half of local ecological science has been obscured by the prevailing "invisibility" of women, their work, their interests and especially their knowledge to the national and international scientific communities. The failure to recognize and learn from gendered approaches to rural environments and resource management has consequences for rural women's futures and those of their communities. The third case study explores the gendered nature of rural people's own sciences of production and survival within the context of a formal research project and the community response to drought and famine in one cluster of villages within Machakos district.

Kathama: the place and the people

The experience of the people in the Kathama market area illustrates the interaction of gendered local knowledge and technology innovation with external technical interventions and experiments in a semi-arid farming community before, during and after the drought and famine of 1984–5. Kathama is a market center that currently serves two sublocations, Ulaani and Katitu, in Mbiuni location, Machakos district. It is an Akamba farming community in a dry forest and savanna zone, nested between the ridge of the Kanzalu Range, the Athi River, which traverses the valley 300 meters below, and the Yatta Plateau. The area encompasses just under 50 square kilometers and embodies many of the problems and opportunities of the transition from agropastoral to mixed farming land use systems, a process which has occurred over the past seventy-five years (Rocheleau 1984).

The Kathama Agroforestry Project

During the course of a long collaboration on agroforestry technologies and land use change, social and ecological researchers worked with the people of Kathama during the 1980s to understand connections between gendered work, knowledge, rights, and responsibilities in rural landscapes and in rural people's lives. The Kathama studies helped to clarify the relation of gendered work and knowledge to the restructuring of ecologies and economies in rural life and explored approaches to support both women's and men's knowledge and vision in shaping their own futures. The farmers and community groups involved in the initial agroforestry research project pursued both customary and experimental agroforestry practices and allowed researchers to document their own independent efforts long after the formal experiments had ended. The community responses to drought and famine in 1984, and the use and expansion of gendered knowledge in that difficult time, illustrate the resourcefulness of rural women's organizations and the complexity of rural people's sciences of survival – both ecological and political.

In 1983, a team of social and ecological scientists began to work with self-help groups to follow up prior on-farm trials of trees to produce mulch, nitrogen, and fuelwood in croplands. They recognized the importance of women's self-help groups and their role in community-level soil and water management work. The groups gradually revealed themselves as associations of individuals and households, formed as reciprocal work groups and mutual aid networks rather than public works organizations. For example, five self-help groups (officially registered *mwethya* groups, predominantly women) that had worked at a common site chosen by local officials and the agroforestry project, noted that they preferred to work separately in the next season's tree planting activities.

The project's entry into this work coincided with the onset of the drought and subsequent famine of 1984, which began early in the drier parts of Machakos. The self-help groups acquired greater importance due to the drought, as access to public relief food and/or "food for work" projects was expected to depend on individual wealth and influence or on group membership and participation. As the drought wore on, many women faced the overwhelming reality of hunger, as well as water and fodder shortages, which threatened the well-being of their families and the survival of their draught animals and small stock. The famine and fodder shortage spurred a resurgence of interest in a wide variety of indigenous trees as reserve fodder sources, in tree crops rather than livestock as assets, and in a diversity of wild fruits and vegetables. Wild foods were said to provide nutritious snacks, combat the effects of malnutrition and to serve as substitutes for other foods (Rocheleau *et al.* 1985; Wachira 1987).

Whole families took to the woodlands, bushlands, and in-between places (fencerows, roadsides, and streambanks) in search of possible fodder sources. Some people tested leaf samples of several tree species on their cattle. The elders in the community made a concerted effort to recall which tree leaves had served in the past as drought reserve fodder. However, unlike the last drought and fodder shortage of similar magnitude in 1946, there was less grazing land, less flexibility of livestock movement within the region, and in over 60 percent of the households women, not men, were responsible for livestock management as supervisors or directly as herders. The men were working as wage laborers in cities and on plantations or serving in the army and police force. Women relied heavily on their prior knowledge of wild foods and acquired new knowledge about fodder plants through hearsay and widespread experimentation with trees and shrubs in range and woodlands.

Given the role of wild foods and medicinal plants in general as poor people's drought and famine reserves, researchers approached women's groups with the possibility of protecting and managing some of these plants *in situ* or domesticating them on-farm within agroforestry systems. Group members entered into a collaborative survey effort on edible plants. Moreover, they insisted on including medicinal plants, an unexpected turn for two reasons. While the project team had associated herbal medical practice with a few specialized men, there was a well-developed practice by specialized women herbalists, who were midwives and general practitioners. There was also widespread knowledge and practice of basic herbal medicine among women over thirty years of age, who described themselves as "unschooled" and taught by mothers and elder women. The women of the community also expressed a widely shared concern over the local disappearance or scarcity of particular medicinal herbs as well as specific indigenous fruits and vegetables.

A survey of men's and women's botanical knowledge in the five villages started with the "general public," proceeded to the specialists and went back again to the women's groups and their children (Rocheleau 1984, 1989; Wanjohil 1987; Munyao 1987; Wachira 1987). Together they identified 118 indigenous or naturalized wild plant species used for medicine and forty-five for food (Rocheleau *et al.* 1985; Wachira 1987; Rocheleau *et al.* 1989). Of these, participants selected five fruit trees, three vegetables and three medicinal plants for potential domestication in agroforestry systems or small gardens. They also named several fruits, vegetables, and medicinal plants as candidates for special protection in place, although women were quite cynical about their ability to enforce management rules in public and shared lands.

While men's and women's priorities varied, they knew many of the same places, classes of ecosystems, and plant associations. They tended to know and use different species and different products from the same species. Whereas men's widely shared knowledge of local plants had been developed in rangeland food and fodder, their out-migration, sedentarization and formal schooling had militated against the transmission of this gendered science and practice to the young. Some men knew a great deal about specific classes of wild plants for specialized purposes, such as charcoal, brickmaking fuel, carving, local timber, bee fodder, and medicine, but the knowledge was unevenly dispersed and decreased markedly among the younger men in the community. Among women there was a widely shared, high level of general knowledge about wild food, craft, and medicinal plants, but an overall reduction in scope and depth of proficiency among younger women. However, the knowledge gap between generations of women was not nearly as pronounced as that for men (Rocheleau 1991).

Some members of the community attributed the persistent decline in local botanical and medicinal knowledge to formal schooling and rejection of non-Western knowledge and practices by the young. Moreover, men's out-migration had removed adult men as tutors and created a labor shortage and double workload for women, leaving little time for traditional education in multigenerational groups of either sex. Women now had different rights and responsibilities than in the past and had to acquire and maintain an ever broader range of new knowledge and skill.

The differential erosion of local ecological science among men and women may also reflect their respective rights, responsibilities and opportunities in

Plate 6.3 Women's groups in Kathama mix social occasions and soil conservation on land rehabilitation site

Source: Dianne Rocheleau

farm versus wage labor sectors. While women maintain livelihoods and retain their rights of access to land through residence and agricultural production, young men can aspire to leave home and to succeed as wage laborers in nearby cities and towns, without fear of sacrificing their long-term access to land. The feminization of famine and drought response and the requisite science of survival reflects the new spatial division of labor between men and women into rural and urban domains. This experience also demonstrates that the boundaries of gendered knowledge are neither fixed nor independent. Content and distribution of gendered knowledge influences and is influenced by the gender division of rights and responsibilities in national, regional, and local context.

The importance of wild plants during the drought was obvious, as was the increasing responsibility of women to maintain the knowledge about them for the community at large. The botanical survey also confirmed that most women normally drew upon fodder, fuelwood, and sometimes wild food sources beyond the boundaries of household land, as did their children. However, those most reliant on resources outside their own land stated that their children were unlikely to enjoy the same facility of access to shared lands in the future.[7] They noted that community-level land tenure, land use, and vegetation changes proceeded on their own momentum, outside the control of individuals and small groups.

Since smallholders and most women[8] relied on the shared use of private lands to make use of wild plants, they needed to focus on social strategies to secure and maintain access to wild plants or on alternative ecological and economic strategies for meeting contingencies (Rocheleau 1989; Rocheleau and Fortmann 1988). Their future access to these resources on shared lands would depend on careful cultivation of social and political networks, as would their influence on soil-, plant- and water-management decisions taken by largeholders and male owners of family plots. Poor women's experience during the drought exemplified the careful interweaving of social and ecological knowledge to survive in the crosscurrents of erratic environmental conditions with uncertain terms of resource use, access and control.

When drought gave way to famine, the women's groups emerged as a critical link to shared use of private lands. As the community tacitly declared a state of emergency, they called upon those with greater resource endowments to share an increasing proportion of those resources with others. However, this social pressure applied to the act of sharing, not to the naming of the beneficiaries; in fact, participants would be a more apt term, since those in need were recognized by largeholders according to longstanding relationships of reciprocity, most often and most predictably in the context of kinship or women's self-help groups (Rocheleau 1991, 1990).

The fact that men actually owned most of the private plots and formally controlled the public lands set the stage for a gendered struggle for access to resources which was waged with finesse and skillful manipulation by individual women and self-help groups. Poor women's knowledge of botany was necessary but not sufficient for coping with drought and famine. They also required access to resources controlled by men at household and community level. Women's groups and individuals mobilized substantial political skill to legitimize and tap their "social credit" at household, group, and community levels.[9]

All subsequent discussions of species preference, tree management, and land use for the next two years were influenced by the experience of the drought. Both women and men had rediscovered trees and wild plants as sources of food, fodder, and medicine, and they expressed interest in planting fodder trees and in the introduction of fruit trees for *both* home use and sale. Many people (particularly women) had acquired a healthy skepticism about overreliance on cash income to offset the effects of famine.

This last point illustrates the learning, storage, and transmission of knowledge about social, political, economic, and environmental change in the form of oral history, particularly in the naming of events. In Kathama in the 1980s most men related historical events to years, by number, or to the major wars of the twentieth century: World Wars I and II and Kenya's own independence struggle. Women, by contrast, said: "We are poor and unschooled. . . . We reckon time in famines, and remember them by name."[10] While all people in the area formerly used famines and similar events to mark historical periods, men – who had access to schooling and outside employment – had increasingly adopted numbers and modern categories,

Plate 6.4 Ana Ndungwa and her grandchildren watering seedlings at her *mwethya* group nursery in Kathama

Source: Dianne Rocheleau

while women – who had very few educational and employment opportunities – retained the more descriptive, local words and images for recounting the past. The name of the 1984–5 famine, as reported by an elder leader of a women's group, captures the painful irony of the changing times: "I shall die with the money in my hand."[11]

When cash from distress sales of livestock and other major assets failed to secure food for their families, many people lost faith in market mechanisms to resolve food shortages. Poor rural people learned that the terms of the exchange with the national market could fluctuate markedly and unpredictably. The famine name suggests that the people of Machakos district had reconsidered the terms of their integration into national markets and had come away with renewed resolve to maintain a greater degree of food self-sufficiency.

The codification of knowledge in the form of famine names records the central surprise of the last famine, makes sense of the experience, pre-empts the surprise of similar incidents in subsequent years and informs practical, popular planning measures to prevent future famines altogether. As exemplified by the name of the last famine, rural women's local knowledge extends well beyond the confines of botany and agriculture, and into the domains of environmental history and practical political economy. Moreover, it is not a knowledge guarded exclusively by women, but rather is increasingly carried and nurtured by them on behalf of the community at large.[12]

The experience of women and men in Kathama clearly indicates the need to recognize and document gendered local science and practice within existing research and development programs as well as in special efforts

Plate 6.5 In Katangi Location, Machakos district, women are both at home and at work in the landscape. They weave these baskets on the way to work, providing cash income for the household

Source: Dianne Rocheleau

focused on local environmental knowledge. For example, an action research program might facilitate the discussion and transfer of knowledge between men, women, and children, as their roles, responsibilities, and interests change.[13] This process could result in renegotiation of the division of rights and responsibilities as well as domains of knowledge and skill, though not without substantial struggle over conflicting interests at household, community, and state levels. The mere recognition and documentation of survival as a gendered science in harsh and unpredictable environments – political, economic, and ecological – could also effect change at local and national level. Such a process could serve to legitimize and strengthen rural women's *and* men's separate, shared and interlocking knowledge as tools to shape their own futures.

LINKING GENDERED ENVIRONMENTAL AND LIVELI-HOOD REALITIES IN THREE COMMUNITIES

Women in sub-Saharan Africa are the major food producers in their developing countries. Yet political and economic structures continue to deny women rights to resources. Harding (1991: 3) argues that Western sciences "continue to be in complicity with racist, colonial, and imperial projects. . . . Western sciences fail to situate their understandings of both nature and sciences within maximally realistic and objective world histories." Left with no alternatives in terms of power and resource base, developing countries echo the mechanisms used to devalue their knowledge and culture.

The three case studies have illustrated that gendered resource tenure – land, water, trees – as well as access to agricultural credit and related inputs are instrumental in the gender relations of production, development, and environmental management. The studies have also shown alternatives to coping with the negative effects of prevailing resource tenure policies on people and ecosystems in semi-arid farming communities. The coping strategies are often related to grassroots activism on environmental and related matters, usually in the form of participation in women's self-help groups. The case studies suggest that women's access to and control over land, water, and plant resources, as well as their own labor, is crucial to resource management and economic development in semi-arid areas in Kenya in particular and more broadly in other areas.

Thus, women's involvement in collective action for environmental change is rooted in existing gender-based interpretations of appropriate behavior within the household and outside its confines (Thomas-Slayter 1992; Thomas-Slayter and Rocheleau 1995a and b; Cubbins 1991; Collier 1989; Muthiani 1973; Stichter and Parpart 1988). It rests upon perceptions of status and prestige, of appropriate gender roles in manual labor related to economic pursuits, and the traditional hierarchy of male dominance and female subordination. Women's self-help groups, collectivities of women working on shared community problems or household matters, have emerged from a history of gender relations in which they are enmeshed.

Increasingly, women are seeking ways to redefine expectations at the intersection of gender-based responsibilities and the requirements of the state and environmental sustainability (Thomas-Slayter and Rocheleau 1995a and b). They seek ways to involve men in the work of managing and improving the community resource base, and to involve women in the decision-making which shapes the community's future. Women have long been creative at using whatever resources are available for meeting the livelihood needs of their families. Through their organizations and through a growing network of supporting institutions, women are beginning to "find voice" in the public arena, to shift the boundaries between public and private, and to raise as common concerns to all – whatever class or gender – the issues of environmental deterioration, resource management, and economic security. These new voices emerged in the three communities discussed in this chapter and are illustrative of changes occurring in many other communities in the semi-arid lands of Kenya and elsewhere in sub-Saharan Africa (Ondiege 1992; Thomas-Slayter and Rocheleau 1995a and b).

NOTES

1 According to the *Nairobi Standard* newspaper (May 11, 1989), several members of parliament urged the Ministry of Agriculture to pay farmers promptly for their cotton deliveries.

2 The Akamba people constitute one of the major ethnic groups in Kenya with traditional lands largely in the present-day districts of Machakos and Kitui.

3 Team members included Kenyan associates from the Government of Kenya's National Environment Secretariat (NES), as well as researchers from the Ecology, Community Organization and Gender Project (ECOGEN) of Clark University, Worcester, Massachusetts.

4 From a survey conducted by the Kenyan National Environment Secretariat (NES) and Clark University in 1987 and further cited in Thomas-Slayter and Ford (1989) and Thomas-Slayter (1992).

5 National Environment Secretariat (NES) and Clark University survey data, 1987; cited in Thomas-Slayter and Ford (1989) and Thomas-Slayter (1992).

6 Interview, July 15, 1987, with Ms. Mbari, Community Development Officer, Machakos.

7 The term "shared lands" is used as an alternative to common land since the formal definition of the latter (see Bromley 1989) excludes the complex pattern of use, access, and control described here.

8 See first part of this chapter on Mbeere, Embu district, for an explanation of women's customary and legal rights to land throughout the areas treated in each of the three examples.

9 See Shields *et al.*, this volume, for a more elaborate discussion of "social capital."

10 Alice Mwau, personal communication (1989).

11 Alice Mwau, personal communication (1989).

12 The fact that women have had less access to formal education and employment has also reinforced their role as the primary culture-bearers with respect to local ecological knowledge about farming, forestry, and medicinal herbs.

13 For a separate discussion of the methods used in the field work and those appropriate for further action research, see: Chambers, Pacey, and Thrupp (1989); Lightfoot and Ocado (1988); Jiggins (1988); Rocheleau and Fortmann (1988).

REFERENCES

Agarwal, B. (1986) *Cold Hearths and Barren Slopes: The Woodfuel Crisis in the Third World*, New Delhi: Allied Publishers.

Agricultural Finance Corporation (AFC) Files, Government of Kenya District Offices, Embu (1986).

Asamba, I. and Thomas-Slayter, B. P. (1991) *From Cattle to Coffee: Transformation in Rural Machakos*, Worcester, Massachusetts: Clark University, ECOGEN Case Study Series.

Bromley, D. (1989) "Property Relations and Economic Relations: The Other Land Reform," *World Development* 17, 6: 867–77.

Chambers, R., Pacey, A. and Thrupp, L. (eds.) (1989) *Farmer First: Farmer Innovation and Agricultural Research*, London: Intermediate Technology Publications.

Chavangi, N. A. (1992) "Household Based Tree Planting Activities for Fuelwood Supply in Rural Kenya, The Role of Kenya Woodfuel Development Programme," in D. R. F. Taylor and F. MacKenzie (eds.) *Development from Within: Survival in Rural Africa*, London: Routledge.

Collier, P. (1989) "Contractual Constraints on Labour Exchange in Rural Kenya," *International Labour Review* 128, 6: 745–68.

Cubbins, L. A. (1991) "Women, Men, and the Division of Power: A Study of Gender Stratification in Kenya," *Social Forces* 69, 4: 1063–83.

Davison, J. (ed.) (1988) *Agriculture, Women and Land: The African Experience*, Boulder, Colorado: Westview Press.

Harding, S. (1991) *Whose Science? Whose Knowledge? Thinking from Women's Lives*, Ithaca: Cornell University Press.

Hayes, J. J. (1986) "Not Enough Wood for the Women: How Modernization Limits Access to Resources in the Domestic Economy of Rural Kenya," Ph.D. Dissertation, Graduate School of Geography, Clark University, Worcester, Massachusetts.

Hoskins, M. (1983) *Rural Women, Forest Outputs and Forestry Products*, Rome: Food and Agriculture Organization.

Jiggins, J. (1988) "Problems of Understanding and Communication at the Interface of Knowledge Systems," in S. Poats, M. Schmink, and A. Spring (eds.) *Gender Issues in Farming Research and Extension*, Boulder, Colorado: Westview Press.

Land Adjudication Files, Government of Kenya District Offices, Embu (1987).

Lightfoot, C. and Ocado, F. (1988) "A Philippine Case on Participative Technology Development," *ILEIA* (Institute for Low External Input Agriculture) 4, 3: 18–19.

Munro, J. F. (1975) *Colonial Rule and the Kamba*, Oxford: Clarendon Press.

Munyao, M. (1987) "The Importance of Gathered Food and Medicinal Plant Species," in K. Wachira (ed.) *Women's Use of Off-farm and Boundary Lands: Agroforestry Potentials*, project report to the Ford Foundation, Nairobi: International Council for Research in Agroforestry.

Muthiani, J. (1973) *Akamba from Within: Egalitarianism in Social Relations*, Pompano Beach, Florida: Exposition Press.

Mwaniki, N. (1986) "Against Many Odds: The Dilemmas of Women's Self-Help Groups in Mbeere, Kenya," *Africa* 56, 2: 210–26.

Nairobi Standard (1989) May 11.

Ondiege, P. O. (1992) "Local Coping Strategies in Machakos District, Kenya," in D. R. F. Taylor and F. MacKenzie (eds.) *Development from Within: Survival in Rural Africa*, London: Routledge.

Ooki and Mbeo, M. (eds.) (1989) *Women and Law in Kenya: Perspectives and Emerging Issues*, Nairobi: The Public Law Institute.

Republic of Kenya (1977) *The Registered Land Act: Chapter 300*, Nairobi: Government Printers.

Rocheleau, D. (1984) "Criteria for Re-appraisal and Re-design: Intra-Household and Between-Household Aspects of FSRE in Three Kenyan Agroforestry Projects," Working paper, no. 37, Nairobi: International Center for Research on Agro-Forestry, November 1985.

—— (1989) "The Gender Division of Work, Resources and Rewards in Agroforestry Systems," in A. E. Kilewe, K. M. Kealey, and K. K. Kebaara (eds.) *Agroforestry Development in Kenya*, Nairobi: International Center for Research in Agro-Forestry.

—— (1990) "Gender, Conflict and Complementarity in Social Forestry Development: A Multiple User Approach," in *IUFRO: Congress Report, Vol. B*, Montreal: International Union of Forest Research Organizations, pp. 432–8.

—— (1991) "Gender, Ecology and the Science of Survival: Stories and Lessons from Kenya," *Agriculture and Human Values* 8, 1: 156–65.

—— (1992) "Gendered Knowledge and Resources in Kathama," in M. Jama, and L. Malaret (eds.) *Farmer/Researcher Collaborative Approach to Rural Development*, final project report submitted to the International Development Research Centre. Nairobi, Kenya.

Rocheleau, D. and Fortmann L. (1988) "Women's Spaces and the Role of Women in Food Production Systems," invited paper presented to the International Congress of Rural Sociologists, June, Bologna, Italy.

Rocheleau, D., Khasiala, P., Munyao, M., Mutiso, M., Opala, E., Wanjohi, B., and Wanjuagna, A. (1985) "Women's Use of Off-farm Lands: Implications for Agroforestry Research," project report to the Ford Foundation, Nairobi: International Council for Research in Agroforestry.

Rocheleau, D., Wachira, K., Malaret, L., and Wanjohi, B. (1989) "Local Knowledge for Agroforestry and Native Plants," in R. Chambers, A. Pacey, and L. Thrupp (eds.) *Farmer First: Farmer Innovation and Agricultural Research*, London: Intermediate Technology Publications.

Shiva, V. (1989) *Staying Alive: Women, Ecology and Development*, London: Zed Books.

Silberfein, M. (1984) "Differential Development in Machakos District, Kenya," in E. Scott (ed.) *Life Before the Drought*, Boston: Allen and Unwin.

Stichter, S. and Parpart, J. (eds.) (1988) *Patriarchy and Class: African Women in the Home and the Workforce*, Boulder, Colorado: Westview Press.

Swynnerton, R. J. M. (1954) *A Plan to Intensify the Development of Agriculture in Kenya*, Nairobi: Government Printers.

Thomas, B. (1988) "Household Strategies for Adaptation and Change: Participation in Kenyan Rural Women's Associations," *Africa* 58, 4: 401–22.

Thomas-Slayter, B. (1992) "Politics, Class, and Gender in African Resource Management: The Case of Rural Kenya," *Economic Development and Cultural Change* 40, 4: 809–28.

Thomas-Slayter, B. and Ford, R. (1989) "Water, Soils, Food, and Rural Development: Examining Institutional Frameworks in Katheka Sublocation," *Canadian Journal of African Studies* 23, 2: 250–71.

Thomas-Slayter, B. and Rocheleau, D. (1995a) "Research Frontiers at the Nexus of Gender, Environment, and Development: Linking Households, Community, and Ecosystem," in R. S. Gallin and A. Ferguson (eds.) *Michigan State University Women in Development Annual*, Boulder, Colorado: Westview Press.

—— (1995b) *Gender, Environment, and Development in Kenya: A Grassroots Perspective*, Boulder, Colorado: Lynne Rienner.

Tiffen, M., Mortimore, M., and Gichuki, F. (1994) *More People, Less Erosion: Environmental Recovery in Kenya*, Chichester, New York: Wiley.

Wachira, K. (ed.) (1987) *Women's Use of Off-farm and Boundary Lands: Agroforestry Potentials*, project report to the Ford Foundation, Nairobi: International Council for Research in Agroforestry.

Wangari, E. (1991) "Effects of Land Registration on Small-Scale Farming in Kenya: The Case of Mbeere in Embu District," Ph.D. Thesis, Department of Economics, New School for Social Research, New York.

Wanjohi, B. M. (1987) "Women's Groups' Gathered Plants and Their Potentials in the Kathama Area," in K. Wachira (ed.) *Women's Use of Off-farm and Boundary Lands: Agroforestry Potentials*, Nairobi: International Council for Research in Agroforestry.

7

DEVELOPING AND DISMANTLING SOCIAL CAPITAL

Gender and resource management in the Philippines

M. Dale Shields, Cornelia Butler Flora,
Barbara Thomas-Slayter, Gladys Buenavista

For island dwellers in the Central Visayas region of the Philippines, local social exchange networks are no longer as effective as they have been historically for gaining access to or conserving natural resources used for building livelihoods. Residents' use of these networks sometimes even contributes to environmental resource degradation. Women, in particular, have found it increasingly difficult to maintain their traditional networks, which allow them to earn income and sustainably manage the natural resource base. In this chapter, we seek an explanation for the current state of environmental deterioration in three communities in the Philippines.

Teams from the Visayas State College of Agriculture and Clark University conducted research in the communities of Napo and Tubod on Siquijor Island in 1991 and in the community of Agbanga on the island of Leyte in 1992.[1] The people in these communities are in the midst of a transition, complicated by population pressure and natural resource decline, from a predominantly subsistence economy to a market economy. Their livelihoods are increasingly tied to the market economies of neighboring towns and villages, the nation, and the world as a whole. The new economy is characterized by the need for cash and means of production, privatization, and competition for and degradation of the local natural resource base.

At the same time that people are involved in the market economy, they continue to participate in the subsistence economy based on gift giving and the exchange of resources between households. Through these exchange mechanisms, people build social capital between each other. Social capital is defined as reciprocity and mutual trust within a community (Flora *et al.* 1995; Putnam 1993). Many of the exchanges from which social capital develops involve products derived from the natural resource base or are built upon neighbors granting each other access to natural resources. Sharing and conserving natural resources and sustaining a material base for the community are thus an overall objective of social exchange mechanisms and the subsistence economy.

The new market economy[2] has brought about changes in the control and management of resources. Privatization, capitalization, and a new gender-based work specialization force men and women to reformulate how they work together. Villagers rely increasingly on the market for raw materials, to generate income and to ensure social security and they depend less and less on the local natural resource base and the social capital they establish at the community level. We compare how social capital has changed in these communities as they have undergone the transition to a market economy and how this has affected local ecosystems.

Analysis of the use of social exchange practices within the context of the new market economy, however, reveals a strong – yet at the same time ambiguous and contradictory – relationship between social capital and environmental sustainability. In some instances, there is a direct correlation between social capital and sustainable resource management. When social capital is developed, natural resources tend to be conserved; as social capital is dismantled, sustainable natural resource management practices tend to be abandoned.

There are, however, circumstances on both islands in which people's use of social exchange networks actually contributes to natural resource destruction. A person's desire not to risk social exchange linkages sometimes counteracts intentions to practice or move toward more sustainable resource management. The new marketplace has changed the context within which traditional resource exchange now takes place to such an extent that at times these practices, which once helped to conserve the local resource base, sometimes lead to degradation.

In this chapter, we compare how social capital is changing in these communities as they undergo the transition to a market economy. We explore the impact this transition has had on the environment and on women's position in the household and community. We hope this discussion will clarify the ways rural people and their organizations are " 'situated' in socioeconomic, political, and cultural structures that both enable and constrain as they construct their resource management strategies" (Bebbington 1993: 274). A viable plan for development must address the social relationships embedded within structural constraints.

COMMUNITY SUSTAINABILITY, RESILIENCE, AND SOCIAL CAPITAL

Community sustainability is based in part on the resilience of a community to respond to changes in conditions in the larger environment, and this in turn depends on the resources available to a community: financial and manufactured capital; human capital; natural resource capital; and social capital (Flora *et al.* 1995). Social capital refers to "features of social organization, such as networks, norms, and trust, that facilitate coordination and cooperation for mutual benefit" (Putnam 1993: 35–6).

Social capital has a variety of configurations, each of which has different implications for community sustainability. Social capital can be built horizontally or hierarchically. Horizontal social capital implies egalitarian forms of reciprocity between people from similar socioeconomic backgrounds. Hierarchical social capital, while it is also built on norms of reciprocity and mutual trust, involves vertical rather than horizontal networks. Individuals from different socioeconomic circumstances build, for instance, traditional patron–client relationships. Horizontal networks outside the sphere of the patron are discouraged, dependency is created, and mistrust of outsiders is generated. This type of social capital is prevalent in communities with persistent poverty (Duncan 1992).

Many communities suffer from the decline or absence of social capital. Such communities are characterized by extreme isolation, little trust, high population turnover, and high levels of conflict (Putnam 1993).

Relying on social capital is part of a survival strategy which is frequently, though not exclusively, gendered due to the different socially ascribed roles that women and men play in the private and public sphere. Whereas men in many regions tend to play a greater role in community politics and the cash economy, women are responsible for community management as a "natural extension of their domestic work" or reproductive labor. Community managing, according to Moser, consists of "work undertaken at the community level, around the allocation, provisioning, and managing of items of collective consumption" (Moser 1993: 34). Women can be found mobilizing the strength of kin and community networks around issues such as water, healthcare, education, garbage collection, community gardens, playground construction, Christmas bazaars, or altar guilds.

SOCIAL CAPITAL AND LIVELIHOODS IN THREE PHILIPPINE COMMUNITIES

Napo and Tubod are adjacent villages in the municipality of San Juan, one of six towns on the country's smallest island province of Siquijor. The island has a total area of 343.5 square kilometers (National Statistics Office, 1990) and is located some 600 kilometers south of Manila. A 1988 survey of Siquijor family mean annual income placed the province last among the nation's seventy-two provinces (National Statistics Office 1990). Far from the nation's central power and economic structures in Manila and with few natural or human resources of interest to the national government, most of the islanders have historically been marginalized in their access to politics and the national economy. The village of Napo is a hilly interior community of 313 hectares, 160 households, and over 600 people; Tubod is a slightly larger coastal village with similar demographic features.

The village of Agbanga is located on the southwestern coast of Leyte. The island is more than twenty-one times larger than Siquijor, with a land area of approximately 7,350 square kilometers (Spencer 1954). Although Leyte is approximately the same distance from Manila as Siquijor, its vast and rich land and coastal resources have historically positioned its people for much more participation in the national economy than has been possible for the people of Siquijor. Agbanga itself is a much smaller and more densely populated community than either Napo and Tubod. It has a population of 868 (National Statistics Office 1990) and 193 households living within only

250 hectares (Department of Agriculture of the Philippines 1987). It spreads across the same three agroecological zones found within Napo and Tubod: coastal, lowland, and upland.

As is true for much of the Philippines, overwhelming environmental degradation is evidenced in all three of these villages by the loss of most of the primary forest, declines in soil fertility, especially in the uplands, as well as reduced quality and quantity of fish (Buenavista and Flora 1993; DuBois 1990; Pomeroy 1989; Siquijor Provincial Development Staff 1983). The apparent decline is largely a consequence of two phenomena:

1 increasing numbers of people drawing from a limited, stagnant, or declining resource base; and
2 the integration of a historically subsistence-based economy into a rapidly expanding and competitive market economy.

Coping with an economy in transition

Capitalism has taken many forms as it has penetrated the subsistence economies in these villages. It has brought about major changes in the control and management of resources which have affected men's and women's ability to build systems of social exchange. Privatization of previously commonly used land has changed how people access and control natural resources. Agriculture and fisheries – previously subsistence-based livelihood activities – have been capitalized, increasing the scale of production and impact of these activities on the sustainability of the resource base. A gender-based work specialization has reshaped how households formulate their livelihood strategies. Competition for limited local resources and cash-earning opportunities has motivated large numbers of people to migrate.

Analysis of livelihoods in Napo, Tubod, and Agbanga reveals the different strategies that households and individuals employ in order to maintain or improve their economic and social situations in a new economy where rules, resources, and social relations are rapidly changing, forcing new circumstances upon them. Overall, livelihoods in these communities can be characterized by diversification. Household members – depending on their age, gender, and class status – utilize a variety of means and methods to mobilize resources for their families. Diversification is necessary due to several factors:

1 the seasonality of agriculture and fishing;
2 the risk of cyclical weather and pest-induced declines in productivity;
3 the high risk of failure in a volatile economy; and
4 the risk associated with a natural resource base in decline.

While traditional livelihood strategies were diversified because of the first two problems, the higher risk caused by the incorporation of the local economy into the national and international economies has made diversification an even more critical community-based risk reduction strategy.

People in Napo and Tubod have long been unable to support themselves solely on the fruits of their labor from the land and sea, yet they have few other economic alternatives. Local and regional entrepreneurs run a small number of industries, such as tourist facilities or cement-block making. Local and regional middlemen export livestock, coconut palm meat, and small quantities of products extracted from the remaining secondary forest and sea such as seashells, seaweed, and vines. Some people have participated in an export economy since the 1750s (DuBois 1990), but their activities have been limited by the small size of the island and the fact that the island's resource base does not offer anything unique to fill a special niche in regional markets. The economic problems of most islanders have also been exacerbated by issues of access and control. Local and regional entrepreneurs and middlemen have long monopolized the local economy. They hold tight control of both the supply and extraction of the island's export commodities, primarily by controlling access to transportation around and off the island.

Livelihoods in Napo and Tubod have subsequently built on a range of activities and resource bases such as agriculture, fishing, home-based income-generating activities, local off-farm employment, and out-migration. Each activity, however, is limited in its productive capacity by the communities' and the island's isolation from the national economy, the lack of natural resources of significant size from which to draw, and monopolization of the local economy by local and regional middlemen.

Households in Agbanga also juggle multiple livelihood activities day-to-day, combining agriculture, fisheries enterprises, public service, and other occupations within and outside the village. Their local economy has been more lively (compared to Siquijor's) due to the large size and rich nature of the land and coastal resource bases. Agbanga's historically large and productive lowland rice production and rich fisheries have long been the targets of government and private investments. Privatization has consolidated agricultural land and made it more efficient for modern agricultural technologies to be applied. Capital investments in agriculture include new breeds of livestock, new varieties of crops, and machinery; in fisheries they include motorized boats and new fishing methods, including dynamite fishing.

Resource exchange networks and social capital

Despite these pressures, livelihood strategies in these communities also remain embedded in the age-old practice of resource exchange. At the same time that families are involved in the market economy, they continue to participate in a subsistence economy based on gift giving and the exchange of resources between households. Gifts include raw materials such as corn, rice, vegetables, and fish – as well as items produced within the household such as cooked meals, processed food, and woven mats for sleeping. Resource exchange relationships involve the direct trading of these items between households. Neighbors share their harvest, catch, products, and even labor with one another. Both women and men share information about jobs in the local and regional centers to which their families migrate. Most exchanges are not written down, and strict accounts are not kept for future obligation.

In some cases, however, people carefully negotiate trades, such as exchanging seed varieties for planting materials and labor.

Through resource exchange, families not only diversify their sources of income and subsistence, but also build social capital. Social capital on the household level is the reciprocal trust members of a given household feel toward an individual, family, or another household. They know they can depend on each other and they also know who can depend on them in time of need. For the community as a whole, the benefits of social capital are not direct but additive and generally do not involve direct exchange. Resource exchange weaves an invisible web of relationships between individuals and families which binds a community together and contributes to its sustainability. One woman may give a sweet potato cutting to another, who gives fish to a third, who carries water for a fourth, who watches the child of the first woman.

Social capital also makes up the safety net for the village, and the resources that flow as it is built give each village member – particularly the most vulnerable – a sense of security and inclusion. It provides a mechanism for the most vulnerable to contribute resources to the community, as well as receive resources from it. In Napo and Tubod, the cultural value of social capital was expressed in a local phrase often used to insult a person who does not share – "*iya-iya, ako-ako*" or "what's yours is yours and what's mine is mine."

Gift giving and exchange relationships are built both within and beyond the boundaries of the community. They are established both horizontally between households of similar socioeconomic standing and hierarchically through patron–client relationships. Hierarchical relationships are found between landowner and tenant, livestock owner and sharebreeder, lender and borrower, financier and fisher, supplier of raw material and home-based manufacturer, employer and employee. The stuff of these exchanges most often involves locally derived renewable and stock resources as well as finished products.

Gender, class, and the changing nature of social capital

Although all members of these communities participate in exchange networks, gender and class emerge as important social variables that differentiate reliance on such networks. Within the family, all members have historically been responsible for networking. The gendered division of labor has structured to what extent men and women have dedicated themselves to building these networks and in this way affects the mix and scale of household livelihood strategies in general.

The gendered division of labor structures more time for adult men than women for building networks. Young adults, both male and female, must build networks which open up livelihood options for themselves and their siblings, both within and beyond the community. Women, with the help of young children, have had primary responsibility for reproductive tasks, but have also been expected to build networks through which they contribute to the household budget. They draw more heavily and directly than other household members upon the natural resource base for the raw materials

Plate 7.1 Women selling their husbands' fish at Sunday cockfight market

Source: Shields *et al.*

for gifts, exchange, and income generation. Social capital between neighbors is the means through which women gain access to the resources for carrying out their home-based enterprises. In this way, they have been able to make an important – and as we shall see below sometimes central – monetary contribution to the household's combined income.

The gendered division of labor has not historically dictated the kind of networks an individual could use or the type of resources to which an individual could seek access. Men and women develop the kind of networks which most efficiently provide them access to the resources they need. This has changed, however, with the market economy. Today, the gender-based work specialization of the marketplace forces households to deploy their members much more strictly along gendered lines. Male labor has been refocused toward the creation of vertical exchange and market relations. In the more stratified society, horizontal networks are neither as productive as previously nor as productive as vertical networks are today for gaining access to scarce resources.

The prioritization of men's social capital has limited women's access to productive resources. As we shall see, women have found it increasingly difficult to maintain the horizontal relationships that support home-based enterprises. As it turns out, many of these horizontal networks facilitated women's sustainable management of natural resources. When they have been abandoned, the natural resource base has suffered.

An individual's and household's reliance on social exchange networks is also shaped by the socioeconomic class in which the household is situated. Although people equate wealth and status with the capacity to access resources through social networks, our research revealed that there is no

Plate 7.2 Woman drying laundry, coconut meat, and starfish. She helped harvest the coconut and gathered the starfish herself and will sell them to local or regional middlemen for cash for the household She will control this cash

Source: Shields *et al.*

clearly defined positive relationship between status and participation in social exchange. In fact, middle-class families seemed to have more opportunity to create and maintain social exchange networks as a livelihood strategy. They were the most aggressive in using social exchange networks, and whatever other means possible, to gain access to resources and diversify their livelihood strategies. Upper-class households generally relied little on social exchange networks, possessing enough financial and manufactured capital so that they were more inclined to centralize their investments and labor. Lower-class households participated in exchange relations but at levels restricted by their lack of money, family labor, and sometimes health.

Overall, middle-class households were more dependent than other households upon social capital. Since they represent the biggest percentage of the populations in these villages, social capital takes on even greater significance in understanding the social and economic changes under way in these communities and the consequences for natural resource management.

WOMEN, SOCIAL CAPITAL, AND SUSTAINABLE NATURAL RESOURCE MANAGEMENT

The impact that the market economy has had on resource exchange networks and the connection between social capital and sustainable natural resource management can be seen particularly clearly in women's home-based enterprises. These are income-generating activities women perform within the sphere of the household, in conjunction with their reproductive work. They

allow women to perform tasks related to childcare and household maintenance and, at the same time, to generate new resources for the family.
Examples of such activities include share-rearing of livestock (*alima*), weaving
mats (*benig*), preparing and selling snacks and other foods, processing fish,
making roofing shingles from sago palm (*nipa*), operating general goods (*sari-
sari*) stores, and providing health services based on massage, prayer, and
herbal medicine. Local medicine is dominated by women in Agbanga, but
it is practiced by both women and men in Napo and Tubod.

Women have historically looked to the local natural resource base and
their neighbors for the raw materials and other resources required for many
of their home-based enterprises. They have relied upon their social capital
with their neighbors to gain access to these resources, and in turn have
relied upon their enterprises to build and maintain this social capital. In
some cases, the development of these exchange networks has given people
the opportunity to practice more sustainable natural resource management;
but in many other cases, women's intentions to more sustainably manage
natural resources have been undermined by forces in the market economy
and have jeopardized the natural resource base. Historical analysis of hog
raising and *nipa* roofing shingle production in our three communities illustrates both of these processes.

Hog raising: developing and dismantling women's social capital

Hog raising in Napo and Tubod provides an example of how women's social
capital can be developed in ways which are beneficial to the environment
and for women's livelihoods. In contrast, hog raising in Agbanga illustrates
how market forces have undermined women's social capital and livelihood
options in ways that have led to environmental degradation.

In Napo and Tubod, women in households without the capital to buy
their own livestock have historically utilized an age-old resource exchange
system called *alima* whereby they contract to raise and breed other families'
animals. The system depends on and builds social capital between individuals and households, both horizontally and hierarchically.

The household raising the animal is solely responsible for the expenses
of maintaining the health and well-being of the animal. Owners usually do
not interfere with their care and management. Exchanged animals are usually
female, which allows the families to divide equally the supplemental offspring
after the owner of the animal receives one offspring. In the case of draft
animals, the family raising the animal may – in lieu of receiving offspring
– negotiate for the right to use the animal for labor.

Women in Napo and Tubod have long depended on hog raising through
the *alima* system as a major home-based enterprise. Hogs are sold for cash
for household subsistence, to pay debts, or to finance other income-
generating activities; they may be slaughtered for annual celebrations; or
they may be bred. Some households even used *alima* to enlist other families to take care of their animals in times of stress, allowing them to avoid
selling their livestock.

Plate 7.3 Shaving the trunk of the banana tree, a local source of food for
indigenous hogs

Source: Shields *et al.*

In the not so distant past, women in Agbanga raised hogs as intensively
and in much the same manner as on Siquijor. Through *alima* or other means,
almost every household raised native hogs. The hogs were fed solely on a diet
of household and agricultural surpluses such as rice or corn bran, root crops,
thinly sliced banana trunk, and other locally available plants. Any sickness
among native hogs was treated with knowledge of traditional medicines developed from plants within the local ecosystem. Access to feed sources, indigenous medicinal plants, and knowledge about hog raising was readily available
through women's exchange relations with other households. Women's hog
production made a major contribution to the household economy.

Today, however, native hogs have almost completely disappeared from
the community and less than a third of all households are engaged in hog
raising. Most of the households raising hogs are from the upper class and
are raising the new "improved" hogs introduced by local agricultural agencies. Approximately 56 percent of the most wealthy households reported
raising pigs compared to 15 percent of the poorest households.

The new breed of hogs is bigger than the native hogs and therefore
promoted as "better" by local government agencies. They can produce more
meat and offspring per animal, increasing the country's overall agricultural
production. When the hogs became available in Agbanga through government programs, native hog-owning households were very quick to make the
switch. An easily accessible market for the pigs was already available in the
municipal center.

The unfortunate side effect of the promotion of these new hogs in
Agbanga, with the accompanying emphasis on production for the marketplace, is that it has undermined hog raising as a viable income-generating

option for women. Women engaged in hog raising through *alima* relationships with hog-owning households were unable to switch as easily to raising the finely bred hogs. Most Agbanga households cannot afford to raise them. Unlike the native hogs which had evolved to make the best of local feed sources and resist local diseases, the new hogs require expensive store-bought medicines and feed supplements in order to survive and grow.

Women in Agbanga now undertake hog raising primarily in times of stress, in a way which serves them more as a safety net than as a consistent source of income and social capital. This phenomenon was felt keenly in 1992, when women's hog raising increased dramatically as a means of livelihood after a typhoon destroyed the local fishing industry. As hog prices rose, women used any means possible, including *alima*, to gain access to hogs. But when fishing resumed in mid-1992, falling hog prices again squeezed most women out of production. Women found that "stockpiling" the hogs for emergency needs or to take advantage of higher prices was no longer cost-effective because of the high costs of maintaining the "modern" hogs.

Although the *alima* system was one of the ways women gained access to hogs during this crisis, the nature of the social exchange relationships involved in the contract had changed drastically. For the majority of women, hogs were only available through hierarchical relationships with upper-class households. In the past, when *alima* was used as part of a daily livelihood strategy, both horizontal and hierarchical sources were available to supply the physical capital, that is the hogs. Resources flowed both horizontally and hierarchically throughout the community, unlike today when resources tend more and more to be held at the top and flow down. Hog owners have become much more clearly differentiated, socioeconomically, from hog raisers, and potential hog raisers have fewer choices with whom to make their contracts.

The current priority of the government and the marketplace to maximize physical capital with respect to hog raising has all but eliminated opportunities for villagers to build horizontal social capital through hog raising. When not used, avenues for mobilizing resources horizontally grow vague and social capital is dismantled. Similarly, as native hogs disappear, avenues are also cut off for the continued transmission of locally based knowledge about native breeds. In time, the knowledge itself will disappear as well as the need to preserve the local plants which were used to treat the animals. Only hierarchical relationships will remain for generating livelihood options and for transmitting information.

Unlike Agbanga, the majority of households in Napo and Tubod still raise hogs. Hog raising has survived primarily because many people have persistently stuck to raising native hogs. Perhaps people recognized the futility of raising finely bred hogs given how quickly the island's markets are flooded and how little extra cash they had for purchasing anything, much less pig feed or medicines. Perhaps they saw they could not sell their hogs for a competitive price in provincial markets after paying for imported feeds and for transportation for the hogs to the markets. For a multitude of reasons, people have left the risk and expense of raising finely bred hogs to the wealthiest households and the government.

Plate 7.4 A man selling *kang-kong* (locally raised hog feed) to women in town center

Source: Shields *et al.*

One result of their determination has been the preservation of both the indigenous knowledge about how to treat hog illnesses with local medicinal plants and the conservation of these plants. Residents are still able to gather such plants from the local ecosystem.

The decision to raise native hogs has also had a ripple effect in terms of building women's social capital. With hog production so important to household economies, interest in and demand for locally based food sources for the animals has remained high. Women have recently recognized a market for the dark green leafy plant called *kang-kong*. Traditionally, it has been picked from riverbanks and rice paddies where it grows wild. It is used as a vegetable and as one of the many foods fed to the native pig.

Today, women in Napo and Tubod with access to rice paddies are cultivating small fields of *kang-kong* and selling it locally. When they are faced with a household labor shortage, women may plant the *kang-kong* in their rice paddies instead of the rice crop. *Kang-kong* saves them valuable time because it requires much less labor than rice to produce. They may plant it in the rice paddies during fallow periods. Women sell *kang-kong* to their neighbors or to households in the municipal center. It is also used to lower their own hog feed costs and to exchange with neighbors. And it gives them income which they can control.

Women's income-generating options have suffered in Agbanga from the capitalization of hog raising – one of their traditional home-based enterprises – by local government agencies. New breeds of hogs requiring new technologies have made hog raising so expensive that even through *alima*, which cuts the up-front costs of buying hogs, the enterprise is not viable for most women. Unable to engage in hog raising, women's opportunities to engage in social exchange and build social capital have been curtailed. Local knowledge about hog raising is rarely practiced now and the relevant medicinal plants are not longer being preserved.

Hog raising and *kang-kong* production in Napo and Tubod are examples of the way women can build their social capital in the community, earn income and preserve medicinal plants and indigenous knowledge in their communities. Women have utilized the marketplace in a way which helps them develop rather than dismantle their social capital. They have expanded their own local options for feeding their hogs, preserved local medicinal plants, and given themselves another source of income in the process. In addition, through this home-based enterprise, horizontal exchange relations are maintained, buffering the forces of stratification and sexism that come with the market economy; community sustainability is strengthened.

Nipa shingle production: Privatization, sexism, and degradation

Shingle production provides another example of how market forces are differentially affecting women's livelihoods, development of social capital, and sustainable management of the environment on Siquijor and Leyte.

In all three communities, roofing shingles are made from the long, slender frond of the native sago palm or *nipa*. *Nipa* is a 10–15 foot palm which grows in moist areas often adjacent to rice fields. Although many homes are now roofed with metal, *nipa* shingles are still widely used for houses, livestock shelters, and sheds. Many families make their own shingles, but those families who need to buy ready-made shingles place orders with the households that make them. Orders come from neighbors and surrounding villages.

In Napo and Tubod, women negotiate with the owners of the rice paddies for access to the fronds. Negotiations are often between neighbors who have longstanding relationships. Free shingles are often exchanged for the fronds. Secure access to the plants encourages their good care, and the supply of fronds is not a problem.

Women also gather the other raw materials needed for shingle production. Women must gather bamboo poles to use as the main shingle support and vines which they use both to sew the fronds together into a shingle and to lash the shingles to the house. Men harvest the *nipa* fronds from the plant and cut the thin bamboo rods to the desired length. Women then fold the fronds over the bamboo rods and secure them with the vines.

The shingle production process in Agbanga has been complicated as the market economy has expanded in the area. There, *nipa* now grows primarily in the coastal area on land which is primarily owned by the government. Little of this coastal land is owned by people in the area. *Nipa* used to be scattered throughout the village, particularly around the lowland rice paddies and along the riverbed, and was accessed as a common property resource. Most of these plants have been cut down in favor of lowland rice production or their use has been restricted by the privatization of the land. After the privatization of most of their supply of *nipa*, people quickly overused the few plants left accessible in common use areas. Today, shingle producers depend primarily on the access the government has granted them to the plants in the coastal area. At the time of this research, people were not engaged in replanting *nipa* anywhere in Agbanga.

The gender-based specialization of the market economy has brought about the prioritization of men's earning ability (and their social capital) in households in Agbanga. Women's home-based enterprises are not defined as an efficient use of male household labor. This has meant that male household labor is not always as available as in the past to support or backstop women's home-based enterprises. Women must now find the cash to hire men to harvest the *nipa* frond for shingle production.

The general degradation of the natural resource base throughout Agbanga has also meant that women must find the cash to purchase several of the other raw materials for shingle production from outside the village. Women must now collect the vine used to hold together the shingles from a nearby village or purchase a less sturdy substitute called *baliisan*. *Baliisan* can only be harvested in remote mountain areas and is generally purchased from an Agbanga resident.

The work specialization, natural resource degradation, and privatization experienced by households in Agbanga has put women's livelihoods and the local ecosystem at risk. Increasingly, the success of their shingle enterprises depends on suppliers of fronds and vines who come from outside the village. They also now need to have cash ready to pay the suppliers and the men who harvest the fronds. As women are forced to make increasing investments in their craft, their earnings are reduced.

As the women engaged in *nipa* production have become more focused on vertical or market relationships which connect them with external (i.e.. external to their local ecosystem) sources of these raw materials, their attention has necessarily moved away from their local environment. The local environment suffers when their attention is diverted from the local plants, their uses, and their protection. Plants such as *nipa* (or the medicinal plants used for treating hogs), once critical to their livelihoods, have already been or are in danger of being lost in the effort to survive in the new economy. Thus, the viability of women's home-based enterprises and the local environment in Agbanga have been threatened as the market economy has escalated pressure on the natural resource base and changed the nature of the community's social infrastructure in a way that no longer meets women's productive needs.

These changes represent a shift in the nature of social capital in Agbanga – from horizontal to hierarchical and market relations. This shift poses a serious threat to the perpetuation and stability of secondary incomes and women-centered work, with consequences for environmental sustainability and the continued transmission of locally based knowledge. When women lose opportunities for employment, they relinquish some control over the diversification of the livelihood of their households. In consequence, their economic autonomy and decision-making authority in the household are threatened. As women's productive work is historically essential to a household's livelihood strategy – both on a daily basis and in times of crisis – these losses present a serious threat not only to social capital but to the long-term prosperity and/or survival of the household and community in general.

On Siquijor, where women have been able to maintain the raw materials for their home-based enterprises as common access resources, they have been able to protect the natural resource base from overuse. Likewise, when they have been able to resist government capitalization programs and continue to raise native hogs, they have preserved horizontal social exchange, local knowledge about hogs, and local medicinal plant species in their community. The presence of social capital in Napo and Tubod – the horizontal social capital based on female exchange networks – facilitates resistance to the destruction of environmental resources and the community. Women's current distance from natural resources in Agbanga and the resulting decline in the social capital which protected them, has put those resources in jeopardy.

SOCIAL EXCHANGE NETWORKS FUEL NATURAL RESOURCE DECLINE

In all three communities there are, however, what appear to be exceptions to our observation that social capital enhances the preservation of environmental capital. There are circumstances at both sites in which residents' use of social exchange networks seems to have had the opposite and unintended effect of contributing to natural resource destruction. People in these villages on Siquijor do not invest in the long-term management of their land for they fear that if they restrict the way their land can be used, they may limit their own access to the resources of others. No one in these communities speaks out when livestock managed by men tramples the plants women use to weave mats whose production utilizes and builds social capital and whose sale brings a little more money into the household. In Agbanga, villagers accept free fish from dynamite fishers through the gift-giving practice of *mamarang*, despite the fact that this act of gift giving effectively garners support for the destructive fishing practice.

Land use practices and degradation in Napo and Tubod

Land use practices in Napo and Tubod provide an example of how farming families utilize social capital to gain short-term access to land resources, but in the process jeopardize its long-term sustainability. Almost all households in Napo and Tubod are involved in farming. Farms are highly fragmented, with an average of four lots per family, usually spread out on the landscape. The average plot size is an incredibly small 0.14 hectare (Shields and Thomas-Slayter 1993).

Both women and men build social exchange networks with the intention of locating plots of land which meet the family's needs in terms of agricultural production. The number of plots, their quality, and their location are balanced against the needs of the family and the labor pool available to work them. Both women and men may negotiate with a plot owner for access to the land or a resource on the land. Parcels are purchased, rented, borrowed, and mortgaged.

In tenancy arrangements, owners usually receive a 25 percent share of the harvest. Arrangements between owner and cultivator may involve specific cash agreements or they may be based on in-kind exchange agreements. Parcels are borrowed from neighbors or relatives on short- and long-term bases. Borrowing under these conditions does not usually require formal payment, but it is part of a continuing social exchange between families within the community. An extension of this approach to sharing resources is open access to fodder and fuelwood among neighbors. Rarely do neighbors require formal permission to gather these items, as long as the source is not completely depleted.

Households also organize access to land or other resources through *prenda*, a local variation on a "mortgage" or "pawning" negotiation. For example, the owner of a resource, such as a stand of coconut palms, may have a sickness in the family for which the household needs cash. Hospital bills drain the household budget and leave no funds for hiring labor to work the trees. Family labor is inadequate for the task. Instead of selling the land and the palms, the owner "mortgages" or pawns the stand of coconuts to a neighbor for a set price and, usually, for a set period of time.

Other studies of resource exchange practices in the Philippines have found arrangements similar to *prenda* to be exploitative (Brown 1990). Households which must resort to *prenda* are usually resource poor to begin with and, after "pawning" a productive resource, no longer have a way to generate enough cash ever to repay the debt.

In contrast, people in Napo and Tubod viewed *prenda* in a very positive light. As in the above case, the owner of the coconuts avoids having to sell critical resources, receives cash needed for the emergency, and maintains the land and coconut palms. The family gaining access to the land and coconuts through *prenda* temporarily expands the household's productive resource base. Both families benefit from strengthening the informal social ties between them. To reclaim the grove, the owner eventually repays the debt to cancel the mortgage. If a designated time period for buying back the grove lapses, and the owner does not yet have the cash, the neighbor's right to use the grove and the owner's right to buy it back continue indefinitely. There was no evidence that people were losing access to resources through *prenda*. In fact, as one woman commented: "When a family member is in trouble with debt we use our extra money to *prenda* their land rather than watching them sell it. Everyone wins when land is kept in the family."

Given the history of resource scarcity and competition in these communities, it may at first glance seem surprising, if not contradictory, that people in Napo and Tubod are seemingly so willing to provide their neighbors with access to the few resources they have. In fact, there is no contradiction. One of the most important ways people cultivate local exchange networks is by asking for and providing access to their physical resources such as fodder, livestock, and land. The owner of the resource is making a long-term investment in a relationship with the user.

The security provided by this type of exchange relationship is much more valuable for families in both the short- and long-term than the physical resource alone. The unfortunate side effect, however, is that the long-term viability of the physical resource is often sacrificed. In this context, households do not find it makes sense to make long-term investments in the physical

Plate 7.5 Drying strips of *romblon* which will be dyed and woven into *benig* or colorful sleeping mats

Source: Shields *et al.*

resources. Investing in the long-term productivity of land, for instance, could mean building fences or erosion control devices or planting trees. Actions such as these in effect restrict the way the land can be used by others. Severely restricting the use of one's personal resources cuts off one's access to the resources of others.

In the next section, we see how a social exchange arrangement may prevent women and men from speaking up to protect the *romblon* plant – the raw material for mat weaving – from destruction by livestock. In this case, men's social exchange practices allow them to pasture their livestock, but also lead to the depletion of a local plant. The destruction of the *romblon* plant leaves women without the raw materials for a home-based industry and effectively undermines their social capital.

Women and mat weaving

Weaving brightly colored sleeping mats is primarily a women's activity in Napo, Tubod, and Agbanga. Girls learn how to weave from the older women in their lives. Households weave mats for their own use, for gifts, and to sell or exchange. Wealthier households may buy the mats they need at the market or commission women outside their households to weave the mats. Mats are made out of the dried leaves of the *romblon* plant, which women have planted around their homes and on the boundaries of upland fields. Once abundant in all three communities and their leaves harvested freely, *romblon* plants are now in short supply. The supply is now so low in these communities that the continuity of this home-based enterprise is threatened.

The decline of the *romblon* plant can partly be blamed on the random staking and uncontrolled pasturing of water buffalo (in Agbanga) and water buffalo, cattle, and goats (in Napo and Tubod), primarily a male responsibility. *Romblon* is unpalatable to animals, so that livestock trample these and other types of plants as they graze. This problem was observed most often in cases where landowners allowed a community member to graze animals on their land through some sort of exchange mechanism. There was no perceptible challenge to this form of pasture management from the women or the men in the land- or animal-owning households.

It is not easy to understand why the members of resource-poor households do not take action to protect one of their scarce sources of livelihood. Part of the explanation lies in an understanding of the social dynamic between the men involved. The pasture owner – whether he is providing pasturing rights as a gift or through exchange, as a friend, a patron, or a client – does not want to complain about the trampling of the *romblon* because of the potential of jeopardizing his future social capital with the animal owner.

In addition, such exchange mechanisms – whether horizontal or hierarchical – embody the notion of "obligatory voluntarism" (Shigetomi 1992). Households feel an obligation to volunteer and give away resources to avoid the risk of damaging established community relationships. In this case, the landowning household is willing to overlook the destruction of its property, the *romblon*, in order to preserve social capital with the animal's owner. Providing assistance to prevent a neighbor's shortfall "minimizes costs in two ways: it reduces moral hazard, and avoids the waste of community resources" (Fafchamps 1992: 157). In hierarchical relationships, clientelism provides mutual insurance between patrons and clients. By providing clients access to productive resources, patrons become a source of insurance against collective risk.

Interviews with animal owners revealed that they do not fear that their destruction of the *romblon* will jeopardize their relationship with the landowner. Furthermore, although they are fully aware of the value of *romblon* to household economies, they do not take the time to protect it because they perceive they have more important and productive ways to use their time and energy. In an effort to gain access to resources in the competitive market economy, households have mobilized male labor in search of resources and wage-earning opportunities. There is no longer time for men or women to preserve the raw materials of women's mat weaving as the household struggles to shape a livelihood in the market economy.

For these same reasons, women in both land- and animal-owning households also do not speak up about the destruction of the *romblon*. If the *romblon* owner complains, the *romblon* damaging household could restrict access to resources which are more critical within the sphere of the household economy at the moment than mat weaving. By sacrificing their access to *romblon* (and their social capital), women support men's efforts to access other resources. As a result, women's social capital is dismantled and male social capital is enhanced.

The depletion of local supplies of *romblon* has forced many women to buy the leaves from sources within or outside the communities. This has required

Plate 7.6 Taking finished *benig* to town center to sell to vendors, friends, or door-to-door

Source: Shields *et al.*

women to develop more vertical rather than horizontal social relationships. Some weavers pay cash for the leaves but most enter into a labor sharing arrangement whereby *romblon* owners get a share of the product. This arrangement reduces the amount of income a woman is able to earn from her craft.

The degradation of *romblon* also represents a change in the nature of social capital and an overall loss of social capital, for women in particular and for the community in general. Women's social capital, generally built upon horizontal relationships, has been abandoned in favor of men's more vertical networks. In order to continue their craft, women have had to seek access to *romblon* through more vertical relations, especially through the marketplace.

Dynamite fishing and gift giving in Agbanga

People in Agbanga, through the age-old gift-giving practice of receiving fish from fishers, support the dynamite fishing industry in the area, and, in so doing, undermine the sustainable use of their coastal resources.

Fishing is a full-time activity and a primary source of income for many small-scale fishers in Agbanga. Approximately 30–35 percent of all but the top tier of the socioeconomic scale derive cash income from fishing. Fishing is also important as part of a diversification strategy for shareharvesters who are without subsistence resources between planting and harvest season.

While the fishery enterprise is heavily dominated by males, women do trade, process, and market fish. Women sort, process, and market fish and other seafood using informal networks that they have established. Women whose husbands are fishers or who work with other fishers have direct access to the catch. They bring raw fish to the local markets to sell. Other small-scale traders sell fish to women processors by going door to door. The women then dry or salt the fish. Processed fish is a source of food for the family; it can be stored for a time or given away to neighbors in exchange for other foods or raw materials. Women also sell processed fish, providing the household with additional income, especially during seasons when the fish catch is low or when sales from fresh fish are less.

Once rich and productive, Agbanga's fishing grounds are increasingly characterized by rival consumption, uncontrolled access, and declining fish yields. A growing number of men and women have come to depend on fishing for primary or secondary employment as paid job opportunities wane in agriculture and population pressures on natural resources escalate. The absence of collective authority in Agbanga leaves indiscriminate exploitation of resources unchecked. Guidelines for fishing practices and licensing of fishers are not enforced. Local and national ordinances have been enacted to control how the resource is exploited, yet enforcement vacillates.

One consequence is that the use of explosives as a quick and successful (though destructive) method has grown quite popular. Some 23 percent of the households interviewed reported engaging in dynamite fishing, probably a conservative figure since others may have felt hesitant to report such illegal activity. From March until May, when anchovy are abundant, dynamite fishing is at its peak. Explosions can be heard and felt physically through the ground as they rock the earth and ocean throughout the day and night. Fishers sink homemade explosives in glass bottles under the surface of the water and blast fish to the surface where they are scooped up into baskets.

Dynamiting has had both environmental and health consequences for the people of Agbanga. The process has led to the overfishing of the most sought-after fish species in the area and a general depletion of fish stocks, and it has indiscriminately destroyed other parts of the ecosystem such as coral and other fish. Pomeroy (1989) reports that dynamite fishing has destroyed the original coral reef at Matalom, another fishing village on Leyte.

Dynamiting also jeopardizes the health of fishers and their families. Dynamiting has been responsible for multiple deaths and maimings of fishers in Agbanga. Fishers and their families routinely expose themselves to the dangers of working with explosives as they mix the fertilizer and gasoline in empty Coke bottles and then store the "dynamite" in their homes until needed. Although fishing is an important source of social capital for women and cash income for the household, many women with husbands involved in dynamite fishing expressed serious concern for their husband's and their family's safety. Some were instrumental in getting their husbands to stop the activity and were willing to risk losing a certain amount of social capital in order to cease this dangerous practice, but only a few.

Dynamite fishing is a capital-intensive fishing technique. Practiced in the area since the 1930s, it usually requires the financing of a patron. Commonly, the patron or financier owns both the fishing gear and the craft. At the end of a period set by the financier and the crew members, the financier pays the crew members – deducting expenses for materials such as explosives, gasoline, oil, kerosene, and, in some cases, food. Dynamite fishing symbolizes the tendency of fishers and their financiers toward maximization of benefits using short-term strategies at the expense of resource sustainability, long-term societal good, equitable access, and fishers' safety (Pomeroy 1989).

Dynamite fishing and Agbanga's fisheries in general, however, like its agricultural system, are not solely driven by economics. Despite capitalization by financiers and the intensification of the market economy, many male fishers still use social capital to set up exchange relationships which gain them access to the labor, fishing craft, and gear they need to catch fish. Fishers still work cooperatively to produce a means of subsistence and economic gain from a shared resource.

In addition, despite the increasing need to commodify fresh fish in order to generate cash and participate in the market economy, fishers still practice *mamarang* – the act of giving away portions of their catch. Villagers from all social classes gather on the beach at the end of the day to meet the fishers when they land. More than 70 percent of the fishers give away a portion of their catch. Before docking their boats, fishers separate their fish into piles – fish for the market, fish for the men who help carry their boat to shore, fish for home consumption, and fish for those waiting on the beach.

Although *mamarang* is still practiced as an important social support mechanism in Agbanga, fishers note that it is more of a strain to maintain the practice today. Declines in the quality and quantity of fish and an increase in the number of people waiting at the beach no longer allow them to give away fish at the level they did in the past. Now, even children are sent to receive a portion, whereas once only adults went.

The only time there is enough fish for everyone in the *mamarang* exchange is when the fishers return from dynamite fishing. Villagers receive a handful or more of the dynamited fish. By accepting a gift of fish from fishers engaged in dynamite fishing, however, people effectively support a fishing practice that is destroying the ocean ecosystem, their community's environmental capital.

Villagers are well aware of the long-term consequences of accepting these fish through *mamarang*, yet they see no other way of gaining access to the food they need as they struggle to make ends meet in the new economy. Rules, resources, and social relations (roles) are rapidly changing in the transition to the market economy, forcing new circumstances upon them. Many depend on *mamarang* as a safety net as population pressure, natural resource degradation, and capitalism make other livelihood options less viable. Yet old methods of gaining access to resources are not what they once were. Traditional social exchange mechanisms sometimes have new, unintended negative consequences for neighbors or the environment. Sometimes it is difficult for people to deal with these new consequences of their actions. People feel a significant sense of despair as to how to keep up with the market economy, much less how to stop dynamite fishing. This despair in some cases leads to inaction.

Dynamite fishing was controlled during the Martial Law period because of the imposition of a "shoot to kill" ordinance, which legitimized the Navy shooting on sight anyone deemed to be engaged in illegal behavior such as dynamite fishing. As the use of explosives has once again intensified, many people see the reimposition of such drastic measures as the only way to control this destructive and dangerous technology – a typical response to a decline in social capital, according to Putnam (1993).

Efforts by local villagers to control the practice have been squashed by the intimidating methods of goons employed by the financiers of dynamite fishing. Local officials who have tried to stop it have had their lives threatened or have been successfully bribed by financiers to stay out of the issue.

Despite the intimidation, fear, and danger associated with the destructive technology, fishers continue to engage in dynamite fishing. The technique provides for the fishers' overwhelming need for cash and a means of production. More than 50 percent of respondents in each class cited "easy and fast cash" from dynamite fishing as the primary motivation for the persistence of this destructive practice. Many of the lower- and middle-income households engaged in dynamite fishing said that they had no other equivalent means of livelihood. A few individuals who were interviewed had tried dynamite fishing and then had given it up "because of the danger to their lives or because of the environmental impact," but they were all from wealthy families and had many other livelihood options.

The infusion of financial capital and new technologies into the fisheries has changed the nature of social capital in Agbanga in ways that have had a negative impact on both social and environmental capital. There is an overall loss of social capital as fishing becomes less about working cooperatively as fishers (through horizontal resource exchange) and as community members (through *mamarang*), and more about maintaining good hierarchical and market relationships with powerful capitalist financiers. As horizontal relations surrounding the fisheries deteriorate, coastal resources have no protection.

CONCLUSION

Capitalism has brought about significant changes in the nature of social capital in these three communities. The marketplace has prioritized physical, environmental, and human capital over social capital. Privatization, degradation, capitalization, and the gender-based work specialization which characterize the market economy, have created an atmosphere of competition among people for scarce productive resources. Over time, these same forces have also concentrated resources in the hands of the wealthy (Agbanga) and/or in the hands of the government (Siquijor).

In the new marketplace, households are abandoning many horizontal social exchange arrangements in favor of hierarchical networks and market relations. In the more stratified society, the latter are much more productive for gaining access to and control of scarce resources. As seen above, as households redeploy their labor along these lines, women have found it increasingly difficult to maintain the horizontal relations which allow them to earn income. Many of the horizontal relationships which have been abandoned once facilitated women's conservation of the natural resource base.

The market economy has changed the context in which social exchange is now practiced to such an extent that, at times, people's use of social capital directly or indirectly leads to natural resource degradation. When this happens, people are fully aware of the ramifications of their actions. Most often, however, they are not willing to invest their social capital in order to preserve environmental capital. In this way, people's desire not to risk social exchange linkages sometimes counteracts intentions to practice or to move toward more sustainable resource management.

People are loath to give up their social exchange networks, because, in order to operate as efficiently as possible as a household in the market economy, they may have already abandoned many of the relationships that formed their safety net. People are now unwilling and/or unable in some cases to give up the last few linkages they may have left through social exchange (primarily gift giving and men's hierarchical networks) in order to sustainably manage natural resources. Gift giving provides families with a safety net and men's hierarchical networks are critical for linking poorer households with the resources of more wealthy households. Villagers in large numbers perceive the loss of social capital as an equal or greater threat to household security than the loss of environmental capital.

In making these decisions, it is true that people are not using environmental preservation as a gauge in evaluating whether they should or should not abandon the networks. They are putting their own short-term security above the long-term management of environmental resources and their long-term security. However, they perceive that they have very little choice but to compete as well as they can for the natural and other resources which population, natural resource degradation, and the capitalist economy have made so difficult to obtain and yet are so critical for survival. In the past, the social system may have made it possible for them to combine their concern for the environment with their concern for social exchange, but this is not so easily accomplished now that competition, not cooperation, is the mode of economic operation. Capitalist forces have increased stratification in these communities and decreased community social capital as a potential tool for sustainable natural resource management.

That there are circumstances in which people's use of social exchange leads to natural resource destruction should not be used as grounds for labelling rural people as "destroyers of the environment." We have seen numerous examples of how social capital has the ability to make other forms of capital, including environmental assets, more efficient. We have also shown how in the process of building social capital people have become empowered to preserve the natural resource base. The potential for sustainable natural resource management on Siquijor is greater than on Leyte, where the deterioration of, and changes in, social capital (from gift and horizontal exchanges to vertical exchanges) have decreased community-based options for long-term economic and natural resource security.

Likewise, we have observed how various modes of capitalist penetration have destroyed social capital by infusing communities with financial and physical capital and extracting environmental, human, and local physical capital, particularly in Leyte. These forces have undermined people's efforts to maintain social exchange networks and their intentions to move toward more sustainable resource management.

Rural people's social exchange practices have the potential for both sustainable environmental outcomes and negative or harmful environmental impacts. What is clear, however, is that women's social capital as it has been used to sustain gender-based livelihoods, has suffered and declined in the face of a growing commercialization. This commercialization has strengthened opportunities for men in the broader market economy. In this process, both community and environmental sustainability are diminished.

NOTES

1 See the individual case studies for field methodologies used, or refer to Thomas-Slayter *et al.* (1993).

2 In the first part of this chapter, we use the term "market economy" to refer collectively to the wealthy class of patrons, local financiers, and middlemen who have – in cooperation with local, regional, and national government representatives – promoted and benefited greatly from the capitalist economy in these communities. In the second part of the chapter, however, we give this "market economy" faces and refer to specific actors in our examples.

REFERENCES

Bebbington, A. (1993) "Modernization from Below: An Alternative Indigenous Development," *Economic Geography* 69, 3: 274–92.

Brown, E. (1990) "Tribal Peoples and Land Settlement: The Effects of Philippine Capitalist Development on the Palawan," unpublished Ph.D. Thesis, State University of New York at Binghamton.

Buenavista, G. and Flora, C. B. (1993) *Surviving Natural Resource Decline: Exploring Gender, Class and Social Capital in Agbanga, Philippines*, An ECOGEN Case Study, Blacksburg, Virginia: Virginia Polytechnic Institute and State University.

Department of Agriculture of the Philippines (1987) *Socioeconomic Survey of the Municipality of San Juan, Province of Siquijor*, Province of Siquijor, Republic of the Philippines.

DuBois, R. (1990) *Soil Erosion in a Coastal River Basin: A Case Study from the Philippines*, University of Chicago Geography Research Paper No. 232, Chicago: Committee on Geographical Studies, University of Chicago.

Duncan, C. M. (1992) "Persistent Poverty in Appalachia: Scarce Work and Rigid Stratification," in C. M. Duncan (ed.) *Rural Poverty in America*, New York: Auburn House.

Fafchamps, M. (1992) "Solidarity Networks in Preindustrial Societies: Rational Peasants with a Moral Economy," *Economic Development and Cultural Change* 41: 147–74.

Flora, C. B., Kroma, M. and Meares, A. (1995) "Indicators of Sustainability: Community and Gender," Proceedings from Indicators of Sustainability Symposium, Athens, Georgia: Sustainable Agriculture and Natural Resource Management–Collaborative Research Support Program (SANREM-CRSP).

Moser, C. (1993) *Gender Planning and Development*, London and New York: Routledge.

National Statistics Office (1990) *Siquijor Provincial Profile*, Manila: National Statistics Office, Republic of the Philippines.

Pomeroy, R. (1989) "The Economics of Production and Marketing in a Small-scale Fishery: Matalom, Leyte, Philippines," unpublished Ph.D. Thesis, Cornell University.

Putnam, R. D. (1993) "The Prosperous Community: Social Capital and Public Life," *The American Prospect* 13: 35–42.

Shields, M. D. and Thomas-Slayter, B. P. (1993) *Gender, Class, Ecological Decline, and Livelihood Strategies: A Case Study of Siquijor Island, the Philippines*, an ECOGEN Case Study, Worcester, Massachusetts: Clark University.

Shigetomi, S. (1992) "From 'Loosely' to 'Tightly' Structured Social Organization: The Changing Aspects of Cooperation and Village Community in Rural Thailand," *Developing Economies* 30, 2: 154–78.

Siquijor Provincial Development Staff (1983) *Siquijor Province 1983: Socio-economic Profile*, Province of Siquijor, Republic of the Philippines.

Spencer, J. E. (1954) *Land and People in the Philippines: Geographic Problems in Rural Economy*, Berkeley and Los Angeles: University of California Press.

Thomas-Slayter, B. P., Esser, A. L. and Shields, M. D. (1993) *Tools of Gender Analysis: A Guide to Field Methods for Bringing Gender into Sustainable Resource Management*, Worcester, Massachusetts: Clark University.

8

"OUR LIVES ARE NO DIFFERENT FROM THAT OF OUR BUFFALOES"

Agricultural change and gendered spaces in a central Himalayan valley

Manjari Mehta

INTRODUCTION

Over the past few decades, the relationship between women and the ecological systems on which their subsistence activities depend has undergone a dramatic change in many parts of the Global South. This is most keenly manifested in rural women's diminishing control over and access to cultivable pasture and forest lands, and increasingly arduous conditions under which they collect fuelwood, fodder, and water.

The complex linkages between economic impoverishment, diminishing resources, and environmental degradation, on the one hand, and gender inequities, on the other, are particularly pronounced in the mountains where subsistence activities are tightly interwoven with the renewability of surrounding natural assets. However, this close and symbiotic relationship, developed over many generations, is under assault on many fronts. Deeply embedded in the economies of the industrialized lowlands, mountain communities have long been net exporters of natural resources and labor to the plains, as well as bound to development policies that offer them little say in decision-making about the control and management of their local natural resources. Today, integration into the global economy is occurring at ever faster rates as roads and mass markets transform the small-scale farming and bartering systems of once-remote mountain areas.

These collective assaults on the environment and the attendant processes of economic impoverishment and social marginalization have had a big impact on mountain communities as a whole. Their consequences have, however, been particularly severe for women whose roles as producers and household providers are now increasingly carried out in physical and social landscapes altered by privatization, diminishing access to agricultural, pasture lands, forests, and water resources, and the emergence of the institutions and personnel of the market economy.

As this case study from the Indian Himalaya illustrates, women are now being confronted with new challenges arising from the adoption of commercially oriented agricultural practices developed in the plains. Relying heavily on production inputs, credit, and other institutions of the market economy, these practices are intensifying ecological vulnerability with the erosion of traditional genetic diversity and the gradual loss and devaluation of a wide repertoire of indigenous knowledge on which the sustainable use of resources has traditionally drawn. Women in this part of the Himalaya have long had to carry the burden of family survival; however, these new agricultural practices are creating yet another frontier of gender conflict by overemphasizing men's participation in the cash-based economy and rendering invisible the full extent of women's critical roles.

TRANSFORMATIONS IN MOUNTAIN FARMING

During the 1980s villagers in certain agro-ecologically rich regions of the Indian Central Himalaya began to diversify their agricultural regimes, combining vegetable cash crops cultivated for market sale with the traditional subsistence grains. This trend, along with increased levels of male migration into urban labor markets, has affected women's work and women's lives in a number of ways. Drawing on field work in a small, thirty-one household Brahmin village in a valley in the north Indian state of Uttar Pradesh,[1] this chapter focuses on:

1 changes in the type, amount and value of women's labor;
2 women's ability to exercise some measure of control in various aspects of agricultural decision-making and resource management;
3 the erosion and devaluation of knowledge systems that women have drawn on to conduct their work;
4 women's dependence on men to perform specific gendered tasks; and
5 how these factors, in combination with increasingly commercialized relations and values, are reinforcing gender asymmetries and changing rural landscapes.

The region's participation in the market economy – both due to men's involvement in off-farm employment and the more recent introduction of cash cropping into the once subsistence agrarian regime – have affected women's agricultural roles and their ability to manage the land effectively. This discussion revolves around two broad time periods.[2] The first includes the years through the late 1960s: the local agrarian economy was still largely subsistence-oriented, agricultural production was dependent on the use of locally generated inputs, and the concept of surplus was defined in terms of production over and beyond domestic consumption requirements. As recently as thirty years ago, male migration into the plains was more the exception than the rule. Although some men sought employment outside the area (typically in the army or in domestic service), most remained in the village, going to the plains only intermittently. Finally, the extent and intensity of subsistence activities demanded considerable flexibility and coordination among all household members, young and old, female and male. While certain activities in the domestic and agricultural realms were designated as strictly "male" or "female," in the days when there were few local or nonlocal employment opportunities available, men assisted their womenfolk in a variety of tasks.

The second period begins in the early 1970s, with the construction of local and regional roads, and comes to a head in the early 1980s when a majority of the households in the area were cultivating cash crops (potatoes, peas, and beans) for sale in urban markets. This initiates a "transitional" period in the agrarian economy which is now characterized by a dual agricultural regime based on both subsistence and market-oriented cash crops. By this time agricultural inputs were more dependent on the market, while the notion of "surplus" was increasingly being defined by market demands. Male migration was on its way to becoming a key economic strategy to bridge the growing gap between household subsistence production and consumption needs, while a few who remained in the village found local employment in public service enterprises such as the Forest Department and the Public Works Department. The absence of men for long stretches of time, coupled with increasing emphasis on sending boys (as future income earners) to school, has eroded the complementarity of male and female domestic and agricultural work roles and responsibilities and has created a pronounced gendered dichotomy between the farm and off-farm domains. The one constant between these two periods, however, is the implicit ideology that women's labor is elastic and can be infinitely expanded to meet household labor demands.

By the late 1980s, village households were closely integrated into and dependent on local and external markets, with increasing access to disposable cash incomes. Not only was money critical to daily production and consumption strategies; it had become an important measure of individual worth and standing in the community.

These factors have contributed to the widening chasm between women's and men's lives. Money has always been considered "men's responsibility." Now, however, men have far greater opportunities to earn cash incomes while women continue to be inhibited by cultural and structural prohibitions from gaining access to new opportunities generated by the burgeoning cash economy. In a sense, men's highly valued participation in the off-farm, income-generating domain has come at the cost of the growing invisibility of women's contributions in various domains of the agrarian economy and limited participation in decision-making. These increasing spatial and gender divisions are located in a set of interlocking structural and ideological "impediments" which both literally and figuratively serve to distance women's work from the market/cash domain.[3]

POLICY IMPLICATIONS

This locale-specific discussion is grounded in a well-established feminist critique of the gender-differentiated effects of growth-based development policies and environmental degradation throughout the Global South. Examples from South Asia, Africa, and elsewhere demonstrate how the commercialization of agriculture strains the ability of many rural households to meet subsistence needs, and often negatively affects women's roles and expertise in food production, processing, and the renewability of biomass systems. Studies have also described how uneven access to and control over resources and increasing segmentation of rural labor markets reinforce gender-differentiated access to productive resources, markets, and money. In many instances, these factors underscore the structurally dissimilar positions occupied by women and men within rural households, while skewed access to institutions and resources – both within local and external economies – often serve to replicate gender asymmetries.[4]

The need to apply this level of gendered analysis to the study of mountain farming systems has yet to be fully acknowledged. In part this is due to the marginal position mountain farming concerns and needs occupy among national agricultural policy and research priorities which are more influenced by lowland agro-ecological systems.[5] Neglect of highland agricultural issues has been further exacerbated by the fact that farming continues to be largely conducted on the periphery of the wider cash economy, only minimally dependent on market inputs and cash exchanges. In addition, compared to rural communities in the plains, gender relations and divisions of labor in Himalayan communities are relatively egalitarian. Consequently, there has been a tendency to minimize the myriad ways in which contemporary development interventions are affecting the lives of mountain women and men in ways reminiscent of patterns established over longer periods of time in the plains.

Against the backdrop of the economic crisis of the 1980s and the growing connections among poverty, the feminization of agriculture, diminishing levels of food security, and environmental degradation in the 1990s, these oversights are troubling. Economic need, coupled with educational opportunities and improved conditions for occupational mobility, are siphoning males away from farming activities at increasing rates. In this context, women (and their daughters) remain closely tied to the maintenance of farms and households, while at the same time lacking access to the resources, information, and services on which agricultural production and household provisioning depend. In short, the intensification of women's traditionally high profile in agriculture is not emerging as a positive development. Rather, women's increasing marginalization from agricultural decision-making, lack of voice in shaping work agendas, dependence on men for cash and access to the market, and the erosion and devaluation of knowledge systems upon which subsistence practices depend, contribute to a growing dissonance between women's roles as agriculturalists and the social recognition accorded to them. This situation has particularly troubling implications for household food security, since the main responsibility for this lies in women's hands.

THE SETTING

The village is located in Saklana, a south-facing valley in the outer hills of the sub-Himalayan range in Tehri Garhwal District, one of the eight mountain districts in the north Indian state of Uttar Pradesh. It rises from an elevation of approximately 350 m (1,150 feet) in the south to 2,450 m (8,000 feet) along its northern boundary, covering a total of 1,357 hectares. There are over forty villages scattered along the notch of the valley ranging in size from hamlets of fifteen households to communities of over fifty households. Some of these are single-caste villages, while the majority have multi-caste populations. The valley is endowed with a varied natural resource base that ranges from rich oak and pine forests to sparsely covered slopes. It is located in the outer hills, exposed to the full force of heavy seasonal monsoon rains and in close proximity to the catchment of the Song River. This provides the valley with access to plentiful water, distinguishing it from the drier agro-ecological zones which lie only a few miles away on the other side of the ridge. This easy access to water has encouraged the adoption of cash crops and the emergence of a relatively dynamic agricultural base in recent years.

THE SUBSISTENCE AGRICULTURAL REGIME

As recently as two decades ago, a system of mixed farming was practiced in the valley, based on the cultivation of coarse-grained millets, wheat, irrigated and rain-fed paddy, a variety of legumes, and potatoes.[6] Forests supplemented agricultural activities and animal husbandry, providing compost for the fields and fodder grasses and leaves for the maintenance of livestock, as well as providing fuelwood for domestic, social, and ritual purposes, timber for construction, food, and other minor forest products.

Animal husbandry has also been an intrinsic part of the mountain agricultural regime, providing traction, farmyard manure (the sole source of fertilizer), and milk. In earlier decades the size of animal holdings also served as an important indicator of household wealth and status. The seasonal transhumance of animal herds was an important part of local livelihoods: performed exclusively by men, this entailed moving cattle and goat herds to the upland homesteads (*chaans*) or farther south into the flatlands to take advantage of milder climates and good pasture lands.

Over the past few decades cows (once the dominant animal stock) have been replaced by buffaloes. According to villagers, this shift is due to declining forest and field fodder resources, as well as to changes in the composition of household demography and availability of labor. The transition from cows to buffaloes has had a major impact on gendered divisions of labor. Cows used to be grazed by men and boys. Buffaloes, on the other hand, are stall-fed (since they are unable to negotiate steep mountain paths) on forest leaves and agricultural residues collected by women and girls. While maintaining buffaloes close to homesteads has facilitated collection of manure, the overall effect of this shift in animal stock has made women's workdays more arduous.

As in many highland areas, subsistence farming has been a highly labor-intensive and low-productivity enterprise. The successive subdivision of land through inheritance has produced small and fragmented landholdings, most of which are less than a hectare (2.4 acres) and typically scattered at various elevations. Three types of land were cultivated, their quality identified in terms of access to irrigation, proximity to forests, types of soil and availability of farmyard manure: irrigated fields (*kyare*), typically located in the valley flatlands; terraced rain-fed fields (*ukhar*) near villages in the middle-level lands; and *ukhar* upland fields located near *chaans* (the temporary dwellings located on the margins of forests which used to be part of the pastoral cycle). Few

Plate 8.1 Selecting wheat seeds in preparation for the forthcoming planting season

Source: Manjari Mehta

households were able to produce a surplus because of the poor quality of the soil and their heavy dependence on rain-fed cultivation.

Traditionally, the spring festival of *Besant Panchami*, celebrated between the end of February and early March in accordance with the lunar calendar, ushered in the start of a new agricultural season. During the cold winter months agricultural work came to a standstill: animal husbandry was the main activity, and portions of households with large holdings would migrate to warmer lowland areas to pasture their herds, returning to their villages in the following spring.

The extent and intensity of the myriad subsistence activities that households were engaged in demanded a certain flexibility and coordination of all household members, irrespective of age or gender. While plowing was strictly a male activity and women had primary responsibility for the innumerable domestic tasks, there was considerable overlap in the activities women and men performed. Older members of the community say that a relatively egalitarian gendered division of labor was commonplace up until a generation ago when there were few off-farm employment opportunities available; before men started migrating in large numbers into the plains, they had little alternative but to assist women in various subsistence tasks.

The primary unit of production was the extended family unit, while the larger kin group and fellow villagers were also periodically mobilized to provide assistance for the more labor-intensive activities. One common form of labor mobilization was *padiyaal*, a system of reciprocal interhousehold labor exchange in which women participated for weeding millet seedlings, transplanting paddy, and planting and harvesting potatoes. Other interhousehold forms of assistance were observed during weddings, funerals, and other ritual occasions when both women and men helped to cut and collect fuelwood, and prepared and cooked food. Intervillage service exchanges between high and low castes (who live in an adjacent village) also provided an important source of labor assistance.[7]

As recently as the late 1950s, markets and money played a minor role in the lives of villagers. This was particularly the case for agricultural strategies: seeds were selected and stored from one harvest to the next, farmyard manure was stored until required for application on fields, and shortfalls were met by borrowing or trading from fellow villagers. While migrants inevitably served as a channel through which new types of seeds and techniques of production entered the community, these did little to alter the face of farming activities.

Similarly, markets played a minimal role in households' consumption strategies. The local market consisted of a single shop which doubled as a post office and carried basic items of consumption such as wheat, rice, sugar, tea, and cooking oil. It used to be a matter of pride for households to subsist on whatever their land produced, and it was not uncommon to be able to produce up to ten months' worth of grain requirements. Most of the agricultural produce was absorbed for immediate household consumption, while a small amount of grain was set aside for social and ritual exchange within the village community. Women, who controlled the consumption and distribution of food within their households, also played an important part in the distribution of grains for social purposes. The small agricultural surpluses were also used as barter within the community and, occasionally, in intervillage exchanges. The absence of roads prevented most households from transporting their produce (mainly potatoes) to marketing centers elsewhere in the hills or farther beyond in the plains. Even today, it is common to hear elderly men recount stories about the two-day pony trek over narrow and poorly maintained bridle paths, a trip which can now be made in a matter of hours by bus.

THE "NEW" AGRICULTURE

By the end of the 1970s old patterns of cropping and divisions of labor were undergoing considerable transformation. The main impetus for these changes was the construction of the link road connecting Tehri, the administrative center of the district, and the important market center of Mussoorie (about 40 kilometers/25 miles west of the study area) along the northern ridge during the 1960s. In the mid-1970s, this was followed by another road which cut through the valley along its north–south axis. Thus, for the first time villagers had relatively easy access to nearby urban centers in the hills and the towns and cities in the plains. By 1988 the northern ridge road was a major thoroughfare into the interior hill areas, well-trafficked by commercial and private vehicles. Government-sponsored tourism, facilitated by the recent construction of a five-star hotel along the ridge road, was also attracting large numbers of tourists from the plains.

Road construction and the expansion of transportation facilities played a key role in the diversification of the valley's agricultural regime. By the late

1970s government and private agricultural extension agencies were playing a pronounced role in the area, introducing new vegetable crops such as peas and beans to communities along the ridge road. This was part of a wider hill policy initiative designed to help regenerate local village economies by providing local income-earning opportunities and help stem the flow of migration. Although villages in the valley lacked direct access to extension support services, in a matter of a few years through an indirect process of diffusion virtually all the valley villages were also cultivating these potentially lucrative crops (and potatoes) for wholesale markets in the plains.

At first glance, the present agricultural regime – combining both the old subsistence grains with the new vegetable crops – appears to be a success. Villagers, both women and men, of high and low castes, tend to agree that the sale of cash crops provides their households with a much needed source of income. Households now have the means of earning the money required to bridge the growing gap between declining subsistence production and a wide array of monetized needs (e.g. clothing, medicines, education, foodstuffs). Purchasing power has increased and standards of living have improved in terms of the variety of foods people can now afford, the clothes they wear, and their greater mobility.

A casual visitor to the area would be struck by other apparent signs of prosperity. Through much of the year the fields are lush. During the peak agricultural season from June until September, each day as many as seven trucks carry potatoes from the local commercial center (a twenty-minute walk downhill from the village) to wholesale markets in Dehra Dun, some 20 miles south of Mussoorie. There are also visual signs of conspicuous consumption: a common sight in certain parts of the valley is cement houses complete with the flat roofs typical of the plains (which are highly inappropriate for this high-rainfall area) and occasionally even sporting television antennae.

There are also less visible expressions of this newfound prosperity. Some households have invested surplus cash in trade, and a number of local traders have become seed merchants and/or marketing middlemen; a few have even managed to expand their roles by investing in trucks to transport goods and produce. Most commonly, households invest resources (money, kinship, and other contacts) to establish their men in urban-based jobs and to provide their sons with access to schools (occasionally even extending to college educations in the plains).

With closer scrutiny, however, the complexities of this market-propelled agricultural change become more apparent. Compared to agricultural changes characteristic of farming systems in the plains, technological transformations in the mountains are minor. A newcomer to the area expecting to find new systems of agricultural husbandry would be disappointed. There are no tractors plying consolidated tracts of land, no apparent evidence of canal-based irrigation or contract systems of labor deployment – all images associated with agricultural diversification based on green revolution technologies in the plains. On the contrary, in many respects the local agricultural repertoire, activities performed and by whom, continues to exhibit a remarkable similarity to patterns that prevailed twenty, forty, even a hundred years ago. Nevertheless, significant changes are occurring, even though their full implications will become apparent only in the longer term.

1 The widespread adoption of cash cropping and rising land values have led to both intensive and extensive land utilization patterns in all villages. As a consequence, traditional fallow cycles are declining and being replaced by a continuous agricultural cycle. Thus, instead of the earlier harvesting of two crops over a one-and-a-half year period, now three and four crops are taken from the land, which depletes soil fertility and ultimately reduces yields. Although a few households still observe *Besant Panchami* in the early spring, the festival has lost all but its ritual significance in the minds of villagers.

2 An important characteristic of the new agriculture is the virtual absence of modern cropping technologies, with the exception of the use of some new seeds and sporadic use of chemical fertilizers. As a result, increased productivity is dependent on intensive uses of land and labor (rather than a function of improved production inputs), thereby increasing overall agricultural workloads. Most of this work burden is concentrated in women's and girls' hands. A large portion of male household labor has been siphoned away into education and off-farm employment, and there is limited use of hired labor.

3 In most parts of the valley subsistence crops are being displaced as households devote better quality land to cash crops. This has affected availability of agricultural residues which are an important source of animal fodder. This compels women to rely more on the use of forest resources. Shortfalls in foodgrains (millets and traditional varieties of wheat) are also forcing households to purchase a greater proportion of consumption items, once available from the land, in local markets. Market dependencies are also being tightened as a result of changing food preferences.

4 The new crops are locking households into new market dependencies as producers: nowadays a growing proportion of their agricultural inputs (seeds and, on occasion, chemical fertilizers) must be purchased from seed merchants and the local agricultural extension agent. Because money is rarely available at critical stages of the agricultural cycle, and because of the absence of a regularized credit system, virtually all households are tied into relations of dependency to money lenders (some of whom also double as the local seed merchants) and are forced to sell their produce at disadvantageous prices.

5 Given the agro-ecological and infrastructural constraints of farming in this highland environment, there are few incentives for villagers to invest in improved farm technologies and inputs. Apart from purchasing better quality seeds, chemical fertilizers and, occasionally, livestock, the general absence of avenues for productive investment has encouraged a high level of conspicuous consumption. One expression of this is the observation of dowry (replacing the traditional custom of bride-price) patterned after the increasingly commoditized practices of the plains. This has resulted in rising levels of indebtedness: with overall expenses often as high as 50,000 rupees – at the time of the study about US$2,500 – girls are now bemoaned as drains on their households' resources and only secondarily recognized as valuable workers.

6 This era of commercial opportunities has coincided with a trend toward dispersed settlements. A growing number of households have now taken up semi- and even permanent residence in their upland *chaans* or closer to the road because they are easier to maintain, are of better quality than village lands or, because their location closer to the road offers easier access to transportation and marketing networks. Spatial fragmentation of village communities has particular significance for women who, in their day-to-day activities, can no longer rely on certain labor networks.

Villagers are well aware that dependence on externally acquired production inputs and fluctuations in market prices, information, transportation facilities and access to markets has opened up new insecurities for them. There is widespread cynicism that while money is easier to acquire now, the continuous credit and debt cycles in which households are embedded are increasing the pressures of daily life. It is common to hear villagers say that "As expenses increase so do our worries" and "However hard we try it still isn't enough because all the money goes into our stomachs." This highlights the way people view their increasing vulnerability to the larger economy.

Villagers also feel that the central and regional governments, as well as private research and extension institutes in the area, are not committed to making the new cash crop regime sustainable over the long term. Many say, "Our lives are being consumed by peas and potatoes," and "These cash crops have betrayed us." The elderly often compare the present with times in the past when the ingredients for successful agriculture included access to quality land, an adequate labor force, and a large measure of luck thrown in: "*Vikas* (development) has brought us access to more money but agriculture remains dependent on the skies."

Older members of the village are particularly critical of the new emphasis on money and how it has affected the moral fibre of the community and relations between people. A concern commonly voiced amongst the elderly is that whereas "in the past there was only one *hookah* (waterpipe) and people (men) would congregate around it, now everybody has money to buy their own. In the old days, people had some thought for the community but nowadays people are only concerned about finding employment for themselves and the well-being of their immediate families." One man in his late seventies who had spent many years employed in domestic service in the plains said: "In the past we had less: people only wore *khadi* (homespun cotton) and had one extra set of clothes, but even so everything was all right. People say they have more money now, but the truth is there is never enough because our needs keep increasing and life is deteriorating. I ask you, how is it possible to save anything when it all goes into the stomach and to pay off loans?" Yet another expressed his opinion poignantly: "Oh, why ask what it was like in the past; that was another world."

STRUCTURAL AND CULTURAL CONSTRAINTS ON WOMEN'S WORK

Long-established gender divisions of labor are reinforced by women's increased work burdens due to cash cropping and men's withdrawal from farming. Contemporary patterns of socioeconomic change, however, render women far more vulnerable than men. Like their men's lives, mountain women's lives have also been affected by the policy- and market-driven processes which have drawn their communities into wider state and market economies. For the younger generation of women this has meant certain benefits: some small measure of schooling, access to medical centers to facilitate difficult pregnancies, the prospect of purchasing foods and other goods in local markets, occasional travel into towns, and glimpses of other lifestyles disseminated through the media.

Despite these changes, women's work, access to resources, life options, and constraints remain firmly embedded in a complex ideological and structural matrix which is shaped by the patriarchal extended family household, a system of patrilineal descent and inheritance, and patrilocal residence and kinship relationships. As in the past, men continue to hold ultimate authority over the material resources of the household, the land, and the labor of junior members and women. Women are thus excluded from exercising direct control over land and other productive resources.[8]

Contemporary patterns of engagement with the market economy have further accentuated women's contingent relations to property. This has occurred through processes which deny women access to occupational mobility or opportunities to earn incomes in the local economy, and which limit their control over money and decision-making. As the following discussion suggests, these contingent relations have been further exacerbated by households' dependence on multiple sources of income which encourage the retention of the extended family system.[9]

SECLUSION IDEOLOGIES

Women's materially dependent status is ideologically reinforced through socially legitimized forms of submission to men. Expressed through seclusion ideologies (*purdah*) and operationalized by what Sharma (1980: 198) has called an "etiquette of public invisibility," these shape the content and value of women's work and interactions with different members of the village community at different stages of their life cycles. Seclusion ideologies also define the unique relationship women have to the geographical spaces in which their lives and work are embedded.[10]

Unlike rural women in the plains, women of the Indian and Nepal Himalaya are not confined by the formal observations of *purdah*.[11] The ecological exigencies of mountainous terrain compel women to be extremely mobile and visible in public areas: most of their daily work is performed outdoors, often in forests and fields located at considerable distances from their homesteads. Seclusion ideologies are also attenuated by the unique form that caste relationships assume in the mountains, whereby the ritual distinctions between high and low castes play a relatively minor role in differentiating between the types of work different categories of women do.

One way that high-caste households resolve the contradiction between status ideals (e.g. that women should not work) and material realities that demand that women play a prominent role in the agrarian economy is by extending the notion of the domestic realm to include family fields. In this way households can retain the belief that they observe a modified form of seclusion by limiting women's labor to domains which do not involve cash transactions.[12] Conceptualizing fields as external extensions of the domestic sphere in effect ensures that the agricultural work women do brings them neither credit nor prestige. As Sharma notes (1980: 123–4): "Not being paid, the agricultural work of the family laborer is no more 'proper' work than washing the dishes. ... So if there are social or moral satisfactions to be got from working one's own land, these are much diminished for women."

The association of women's work with the subsistence and non-monetized domain, men's de jure ownership of resources, and the constraints women experience in getting access to inputs and services in the market economy feeds into a widely held view that women are not farmers and managers of the land but merely workers. The "infinitely expandable" and service nature of women's roles in the agrarian economy is captured in the remark that "women are our arms and legs."

SPACE AND POWER

Women now find themselves in an untenable position. Increasing monetization of the local economy is reinforcing long-held cultural expectations regarding the availability of women's labor to meet the needs of their natal and affinal households. Women and their daughters are expected to absorb a greater share of domestic and family farm labor demands in order to free up male labor for the off-farm sector, and are tied to the land through lack of alternatives. At the same time, however, women have limited access to the highly valued monetized domain. This prevents them from broadening their roles in agricultural management by denying them access to new systems of knowledge and spheres of exchange.

Village women's access to and exclusion from certain physical spaces is shaped both by kinship and family structures and processes through which the village system (and individual households) have been incorporated into larger institutional structures. If "[c]ommand over space is a fundamental source of social power" (Enslin 1990: 6) then it follows that lack of or limited access to certain spaces can play an important part in disempowering certain individuals or groups of people relative to others. This is true throughout rural India, where space as a geographical reality has different meanings for women and men as a result of their differential access to and uses of it. In this sense, women's access to and exclusion from certain spaces also make an important statement about the differential exercise of social power between the sexes.

The village is embedded within a complex of "natural" and built spaces. The former include the village and homesteads, agricultural fields, *chaans* (upland grazing/agricultural lands), common pasture lands, irrigation channels, and forests, all of which to varying degrees are crucial to the village's agrarian economy. In addition, a number of more recent spaces have been created as a result of developmental interventions over the past three decades, including roads, markets, schools and various governmental institutions (a bank, post office, and the agricultural extension office).

Like other aspects of social life, the meanings imbued in these spaces have been changing in response to the wider community of villages' inter- actions with the wider political economy over the past few decades. Previously, the distinctions between women's and men's access to and uses of these spaces were relatively minor. While marketing and exchange were the respon- sibility of men, and men were permitted a greater freedom of movement and leisure than women, the commercial domain did not occupy a promi- nent part in the life of villagers.

The flexibility of gender divisions of labor ensured that both women and men participated more or less equally in the various subsistence domains, be they within the household, village, fields, common pasture lands, or forests. Since women were primarily responsible for gathering activities, common grazing and forest lands were regarded primarily as the domain of women. This in turn contributed to women's intimate knowledge of the environ- ment, much of which continues to be expressed in folklore and songs.

Forests and other common lands have also served an important nonutil- itarian function in a social context where interactions between young and old, women and men are closely structured around gender and generational hierarchies. The commons offered a milieu in which women of all ages could seek a measure of privacy, daughters-in-law finding a place to socialize while collecting fodder and wood away from their mothers-in-law's watchful eyes and sharp tongues, older women to talk more freely away from men.

RECONFIGURATION OF VILLAGE SPACES

In the past few decades, various factors have contributed to the expansion or shrinking of the natural and built spaces in which the village economy is embedded. These changes have affected women and men in different ways, reflecting their different positions within the domestic economy, their work obligations and responsibilities, and their access to newly emerging spaces within the market economy.

In 1953, in an effort to improve management of forest resources and check soil erosion, the central government passed legislation nationalizing all forest lands including all uncultivated lands surrounding villages. Although this legislation did not prohibit use of forest resources, it established tighter restrictions by redefining the terms and content of usufructuary rights to forests and their produce and prohibiting extension of landholdings.

Other external interventions also exacerbated the growing trend toward privatization. Since the 1960s, the development of markets and the construc- tion of roads have altered the layout of large areas of agricultural fields and forests, and rerouted access between certain villages and forest lands. The largest amounts of forest cover were lost in the process of constructing roads during the late 1970s when the valley road was built and cost large tracts of forest, pasture, and even agricultural lands. The local commercial area underwent considerable expansion: once a forested bowl in the valley, it consisted of little more than a single all-purpose retail shop until the early 1960s; by the mid-1980s it was a flourishing commercial area. The most recent transformation of space has occurred with the introduction of cash crops which has resulted in a gradual displacement of subsistence foodgrains.

Fragmentation of landholdings has also encouraged a process of land use conversion from grazing to cropland. Much of this particular form of privatization occurred at the expense of the common pasture lands (*gochar*) and waterways (*guhls* and *nallas*) in the lower-lying part of the village. More recently a few individual households have illegally appropriated certain sections of forest land by constructing walls beyond the limits of their property. More commonly, the promise of high returns from cash crops has encouraged households to establish new agricultural fields illegally in areas beyond their main holdings.

GENDER IMPLICATIONS OF SPATIAL TRANSFORMATIONS

Men have either been relatively unaffected by many of these changes or, as in the case of development interventions which have expanded markets and occupational mobility, overtly benefited from them.

The "gendering" of geographical spaces has had a much more tangible, and negative, impact on women's lives. In many instances the creation of forest enclosures has narrowed women's access to fodder leaves, grasses, and fuelwood, compelling some to walk considerably longer distances in order to meet their households' daily needs. The shrinking of common lands and processes of privatization have also eroded the small measure of control and freedom of movement that women once exercised in defining their work agendas independent of men. The growing importance of both local and nonlocal markets in household consumption and production strategies, and the diminishing ability for subsistence agriculture to meet household requirements, has contributed to a decline in the *perceived* importance of the spaces – fields and forests – in which much of women's work is conducted. Today, one of the biggest challenges village women face is having to operate in a social environment in which they have unequal access to the very geographical spaces that are vital to the accomplishment of many of their daily tasks. In short, while men's spaces are expanding (if not necessarily literally, then in terms of the importance associated with them), women's spaces are shrinking without enabling them access to the new arenas of prestige.

The cyclical pattern of migration practiced here and the tendency toward extended multigenerational households, also reinforce the devaluation of women's spaces and the heightened value of men's spaces. Both factors provide a partial explanation for why the "feminization of agriculture" has neither proved to be a positive development for women nor contributed to the emergence of what Epstein (1987) has called "matri-focused" households.

Compared to more interior distant regions of the Uttar Pradesh Himalaya, the villages of the upper Saklana Valley have relatively easy access to the towns and cities of the adjacent plains where most men work. As a result, men characteristically return to the village at least once during the year. At the time of the study, twenty-two households of a total of thirty-one in the village had at least one male migrant and over half of these individuals were able to return to the village regularly throughout the year. There were also two households in which the brothers – all employed in the same cities – took turns coming back to the village every alternate month.

Inevitably, some men are unable to return to the village on a regular basis due to job-related demands, distance, or lack of funds. In such cases, the multigenerational extended household typically enables other male relatives (elderly fathers, retired or locally employed brothers, or even younger brothers still at school) to attend to specifically "male" designated tasks: plowing, purchasing production inputs and household provisions, paying off credit, negotiating bank and private loans (with seed merchants and fellow villagers), and attending to the marketing and exchange of the agricultural produce.

Women who are the de facto heads of their nuclear households (there were five at the time of the study, one of whom was a middle-aged widow who had no sons) experience the temporary absence of men from their lives – along with the effects of gender-based differential access to resources – in particularly harsh ways. This is rooted in the structures of ownership in which men are the de jure heads of households and owners of all alienable property. In practical terms, this means women are unable to get access to the collateral necessary to raise cash in financial emergencies. Because bank loans are only made to formally designated male heads of households, women cannot get formal credit administered through the local banking system nor negotiate bank or even personal loans from within the village community. At the time of the study, only one woman in the village attempted to address the problem of lack of money by raising goats. While she was aware that "without money we are nothing," she was also concerned that social norms would prevent, or at the very least make it difficult for her to initiate distress sales of these animals on her own. Although she had never actually had to consider the prospect of selling a goat, she hoped she would be able to appeal to one of her husband's kin to assist her should such a situation ever arise.

MALES SPACES, FEMALE BEHAVIOR

The commercial area, with its numerous shops and flow of vehicular as well as pedestrian traffic throughout the day, has in a relatively short period of time developed into the local center of trade and socializing. It is also very much a man's world. At tea stalls and merchants' shops, men have an opportunity to exchange local news and gossip. Radios, newspapers, travelers recently returned from the plains, and truck and bus drivers coming in from other parts of the hills and plains, also serve as conduits for local, regional, and national news. Much like their *pahari* (mountain) counterparts in other areas, men from the villages of upper Saklana are extraordinarily well versed in a wide range of issues. At virtually any time of day, it is common to see groups of men, sitting over cups of sweet, milky tea and the occasional hard-boiled egg, hotly debating the state of the national economy, talking about price hikes, production gluts, and strikes. They draw connections between national-level political corruption and the difficulties local men face in finding jobs without having to pay bribes, and they argue the pros and cons of various types of hill seed potatoes versus the plains varieties. Thus, within the relative confines imposed by its mountain environment, the local commercial center offers men the opportunity to participate in social networks that connect them to communities well beyond their membership in particular villages.

Even though the area is not formally closed off to them, women experience the local market in a completely different way. The intrinsic "maleness" of this area is palpable. Even women whose days are spent traversing long distances between forests, fields, and homesteads are affected by this. Women cannot use public spaces of the market in the same casual manner as men; rather, their interactions are restricted to practical activities and shaped by a series of unspoken "rules" which dictate how they should behave. These constraints have far-reaching implications both for their actual and perceived economic and political roles within the local community.

It is, for instance, not appropriate for women to sit in tea shops: not only do women admit to feeling uncomfortable in such a milieu, but they also consider the idea of paying for something which can be prepared at home an "unnecessary" and "pointless" expense (it is worth noting that men do not think of refreshments as an extravagant luxury). Similarly, it would be highly improper for women, after making their purchases, to linger to gossip with traders; the only women from whom such behavior is tolerated are those whose status as low castes or widows enables them to transgress certain mores without inviting the sharp criticism that would inevitably be heaped on younger, higher-caste, and married women.

These seclusion norms have practical consequences for all women. At a superficial level, they deny the vast majority a voice in exercising choice over matters such as selecting saris, bangles, and other articles for their personal use, items which are typically purchased by husbands and fathers-in-law. On occasion women's inability to interact freely with "public" institutions has more serious consequences – as when the absence of male kin prevents women from taking advantage of medical services provided by men or even taking the initiative to see the midwife or take sick family members to seek medical treatment in the plains.

In terms of their roles as agriculturalists, women's circumscribed access to the central artery of the village economy is particularly significant. Limited mobility affects women's ability to attend to domestic provisioning and agricultural needs. Women's limited ability to initiate commercial interactions and negotiations, along with their limited control over money, affects their ability to participate actively in various aspects of agricultural decision-making, devalues the knowledge systems upon which subsistence works (for which women continue to be responsible), and reinforces their dependence on men to perform key tasks related to women's own work.

Illiteracy imposes an additional constraint on women's ability to participate effectively in the marketplace. As one woman admitted: "Without a man life is very difficult. It is not possible for a woman to open a bank account on her own ... of course I see traders write in their books how much I have purchased, but I do not know how to read so how do I know how much they are charging?"

Plate 8.2 Women farmers taking a break during the potato planting
(the main sowing season is February to March)

Source: Manjari Mehta

EXCLUSIONS IN THE GENDER DIVISION OF LABOR

In many respects, village women's status is more closely defined in terms of
the tasks from which they are excluded rather than those they perform.
Three of the main ones – observation of the plowing taboo, limited access
to the market, and lack of control over money expenditures – create a twofold
dependence on men for cash and labor that all women experience. Women
cannot initiate the agricultural cycle on their own, cannot acquire and handle
cash independently of men (and thus are unable to hire labor to plow the
fields), do not market their own produce, and have minimal voice in dictating
how money will be spent.

Women's exclusion from handling the plow is also reinforced by gender
ideologies regarding the types of tools women and men have access to. For
instance, while men both *own* and *control* tools based on animal energy,
women only have *access* to these tools whilst largely relying on the physical
energy derived from their own bodies. These gendered uses of energy also
serve to legitimize the belief that what women do is somehow less productive than men's work. Although detrimental to all women, this situation can
be particularly harsh for women who cannot rely on their husbands' or
other male kin's assistance in preparing the fields.

Women's dependence on men to initiate the agricultural season is mirrored
at the other end of the agricultural cycle by constraints which affect their ability to market the produce independently. Although there are no formalized
barriers to women's participation in the local market, gender ideologies and
expectations affect the way in which women relate to it. Men disparagingly

talk about women's inability to negotiate loans, and their general illiteracy and "irresponsibility" over money matters. Both women and men agree that it is inappropriate for women to deal with nonlocal men. Women have also internalized these gender ideologies, agreeing that "women know nothing about money matters" and that "accounting is men's work."

Under these circumstances, it is not surprising that women do not regard going into the commercial area as a measure of autonomy. Echoing a common sentiment, one middle-aged woman remarked: "We have so much work to do already – why is there any need to go if there are others to go for us? The *bazaar* (market) is a place for the men: don't women have enough to do without also having to take responsibility for that work?" Another woman said: "Money is a matter for men; what do women know about it?" One old woman expressed amusement when asked why women never carried the harvested produce to the market: "*Boris* (gunnysacks) of potatoes and peas easily weigh between forty and fifty kilos; why would we want to do such work?" When reminded that women routinely carry equally heavy headloads of fodder, fuelwood, and densely packed agricultural straws and harvested foodgrains from forests and fields back to their homesteads in the course of their daily work, she shrugged saying: "What we do is women's work, but working in the market and keeping *hisaab* (accounts) is the responsibility of men."

Most women in the village live in households with older children (boys and girls) and/or have a man available for attending to market-related tasks. Those who cannot rely on male affines or youngsters, however, have a hard time negotiating on their own. All women, regardless of age, have reservations about going into the market and dealing with men outside their immediate kin group and village. For some the problem is compounded by their youth since young women carry particularly heavy labor obligations, yet must conform to norms of deference in dealing with actual and fictive kin within the village, non-kin and strangers.

Many young women who are de facto heads of households – and whose children are too young to provide much help – have little choice but to go into the *bazaar* frequently to take grain to the electric mill, buy provisions, and, on occasion, to purchase agricultural inputs. They regard these periodic trips as unwanted additional pressures to their already tight schedules. Despite their regular commercial interactions they do not feel they are getting anything positive from participating in what they regard as "men's work" in "men's spaces."

The presence of nonlocal and transient bus and truck drivers also presents concerns to women. One woman explained why she was uncomfortable about having to deal directly with traders over matters above and beyond straightforward transactions: "Earlier there was more respect for women; today men are *batemeezi* (rascals), and I have seen in the plains (*desh*) how easy it is for women to lose their *izzat* (honor)." Another remarked: "So far our men are all right, although I have heard that in other places they drink a lot and then beat their women. Even here, I can tell you that this sort of thing is beginning to happen as the younger generation of men are picking up bad habits from the city."

WOMEN'S RELATIONSHIP TO MONEY AND DECISION-MAKING

While women's reluctance to take on additional responsibilities is under-standable, their exclusion from marketing and exchange activities takes on particular significance from an economic perspective. Women's current limited access to money underscores the general lack of control they exer-cise over the product of their labor as well as the erosion of their roles as active participants in agricultural and resource management decision-making. So deeply internalized is this gender ideology that even women who have no choice but to engage routinely in commercial transactions and bargaining are adamant that they know little about the handling of money. It is all too common for women to deny any knowledge of how much things cost, what the proceeds of the day's harvest are, and to claim ignorance of how much (or rather how little) money there might be in the house.

Typically older women talk about the gender division of labor as repre-senting the economic interdependence that exists between members of the household and specifically between husbands and wives. Many argue that it matters little who actually handles money or activates commercial nego-tiations since women, as members of households, ultimately benefit from the goods bought by their men.

Increasingly, however, women are beginning to recognize the link between their limited freedom of movement and their ability to control money in any meaningful way. As one older woman put it: "Yes, it is true that I keep the money, but what is the point if I or my *bahu* (daughter-in-law) cannot go into the *bazaar* (market) to spend it? We (women) look after the money when it is in our homes but it is the men who will be able to use it."

The increased emphasis on money to meet domestic consumption needs and men's roles in acquiring it has also contributed to an erosion of the social power older women once had within the domestic domain. Many older women refuse to acknowledge that their work is even connected to the cash nexus and even imply that this is the reason why their participation in vari-ous types of decision-making has eroded. A comment by one older woman that "those who earn the money should decide how it is to be spent," poignantly acknowledges the harsh reality that men, as the owners of pro-ductive resources and as earners of money, are imbued with a source of power to which even senior women cannot hope to aspire. The devaluation of women's contributions was starkly expressed by women who asked: "What good is our work without the money with which to purchase oil and salt?"

EROSION OF LOCAL KNOWLEDGE SYSTEMS

Growing emphasis on cash and market-based transactions is also contributing to the gradual erosion of local knowledge systems, a development which also has gendered consequences. A common theme noted in much of the liter-ature is that "[w]omen's specialization as the managers of subsistence production does not counterbalance male specialization or translate into a broader parity with men in family or commercial affairs" (Bourque and

Warren 1981: 124). Traditional systems of knowledge have been undergoing transformation over the past three decades largely due to the introduction of hybrid varieties of seeds and the loss of local genetic variation. In recent years the new cash crops coupled with the growing value of exogenous scientific knowledge systems has further eroded the importance of local knowledge in many people's minds. This tension between *old* and *new* knowledge systems has particularly negative consequences for women who draw on these practices in their daily work and who, in any case, lack access to new information and products.

Most subsistence tasks continue to rely on technical expertise (in the selection and storage of seeds, systems of composting farmyard manure, and various activities related to animal husbandry) of which women are the main repositories. In this sense, locally based knowledge is viewed as "female," even though some older men in the village are knowledgeable about the tremendous local ecological diversity. Botanical information and resource management skills are passed down from one generation of women to the next, with young girls learning by observing and helping their mothers and grandmothers. Older women in the village have a remarkable facility for identifying different types of grasses, trees, shrubs, and plants or naming their unique qualities, uses and relative nutrition. With a casual touch they can tell whether farmyard manure has matured sufficiently for application on the fields. A few women still collect herbs believed to have curative properties which are used by local *dais* (midwives) for women's ailments. Older women talk with pride about how it was their expertise in differentiating the various qualities of seeds, of knowing which ones should be saved for the next season's sowing and which used for consumption, that often helped their families through difficult times. Their knowledge constituted a major contribution to production and was once acknowledged as such.

Not surprisingly, the younger generation of men who seek to emulate the ways and knowledge of the outside world tend to disparage what they perceive as signs of a "backward" way of life. Whereas many of these men are best situated – because of education and/or well-placed jobs in the plains – to serve as disseminators of new information and production techniques, few see any value in pursuing agricultural livelihoods and most consider subsistence technical expertise to be inferior, and to contribute to lower levels of productivity. During the time of the study, there were only two households in which young men were actively involved in agricultural activities.

The situation is somewhat different among older men. Many have, since their retirement from wage labor, begun to play a more direct role in their households' agricultural affairs (even though this is often restricted to managerial rather than labor activities). Nevertheless, even these men often have little detailed knowledge about the new inputs (e.g. optimum levels of chemical fertilizer applications, improved composting techniques, correct spacing for seed potatoes, depth of trenches, etc.). Many men freely admit that they have little real knowledge about more effective techniques of commercial agricultural production, that they have learned new practices through costly trial-and-error processes, and that learning is simply a matter of *andaaz* (approximation). Most older men and women consider chemical fertilizer

Plate 8.3 Selecting seeds for storage after harvest

Source: Manjari Mehta

(*khad*) to be a mixed blessing, believing that its promise of short-term gains must be weighed against the longer-term impact on soil depletion. This belief is also gaining prominence among some agricultural research scientists and extension workers. Yet traditional composting methods based on farmyard manure (*gobar*) are now viewed less favorably.

Various factors contribute to women's diminishing participation in certain aspects of agricultural management. The marginalization of the traditional crops now compels households to purchase a large portion of seed stock (both vegetables and foodgrains like wheat) in the local market, an activity generally performed by men. In addition, with the growing commercial value of potatoes, men are now playing a greater part in the selection and preparation of seed potatoes. As a result, women's role in the selection and storage of seeds has become more or less restricted to the less valued millets.

The growing use of chemical fertilizers (*khad*) has also eroded certain aspects of women's control of soil management. The purchase and application of *khad* is considered the exclusive responsibility of men. Women's inability to participate in these activities reinforces the view that men's roles are indispensable to good farm management and high agricultural productivity, while women's work is "old fashioned."

Older women are aware that their roles are no longer as respected as they once were. One elderly woman remarked as she pored over a heap of many-colored millet grains: "We are brought up to believe that women do not have the same abilities as men, and that we cannot work as hard as they do. You will even hear it said by some women that what we do is not as difficult as what the men do. But I cannot believe this: it takes a sharp eye, a sensitive hand and a lot of patience to tell the difference between these seeds. But these

Plate 8.4 Pounding millet (*jhangora*) after harvest

Source: Manjari Mehta

are not the things that are honored any more." Nowadays, few of the younger women in their twenties and thirties, much less young girls, have the "feel" and breadth of knowledge exhibited by their mothers-in-law, mothers, and grandmothers.

CONSEQUENCES OF GENDERED ACCESS TO INFORMATION

The emerging gender monopoly over information confirms an unspoken local belief that those who work the land are not necessarily considered "farmers" – highlighting the chasm between women's laboring roles and men's managerial roles. It also serves to reinforce women's dependence on men to act as mediators to critical resources, services, the banking system, and to various personnel (e.g. the agricultural extension agent) of the market economy. These levels of dependency not only have material consequences, but also legitimize women's ideological inferiority to men by reinforcing the cultural norm that "money is a matter for men" and that "women do not know anything about it." On a practical level, this situation also undermines women's ability to serve as "substitute" decision-makers for migrant men. Women's inability to acquire necessary information also affects household agricultural strategies and threatens to lower agricultural productivity over the longer term.

Gendered access to the market and information affects all women, particularly those living in nuclear households. While men's participation in social networks in the local market and urban migration make them privy to sources of information vital to efficient farming practices, this is not necessarily shared with women. Few women in the village are able to use relevant information in ways to balance their workloads or to ensure higher agricultural returns independent of men. This limited access to market-derived information is particularly striking in the case of women who, as de facto agricultural managers, must make important day-to-day decisions on which crops to plant, in which and how many fields, and what types of labor assistance to draw on.

The case presented below illustrates some of the predicaments arising from this lack of communication. Meera, a woman who at the time was in her late forties, was her household's main agricultural worker. Her husband paid little attention to the land, while all the children attended local schools. She had never been to the market and was completely dependent on her husband and her sons for whatever outside financial and technical information they deemed significant to share with her. Meera was very open about how little she knew about market matters, laughingly saying that the various production inputs, qualities of different seed varieties, new cropping technologies and pricing differentials were issues she did not understand, that "these are the *zimedaari* (responsibility) of men."

The contradiction of being the household's main worker, who nevertheless lacked access to relevant information, was sharply illuminated during 1988–9. Heavy monsoon rains in the summer of 1988 rotted the potato crop at the same time as a countrywide glut of the crop led to a precipitous drop in the wholesale price of potatoes. At this time, a common topic of conversation among men in the commercial area was how to strategize to minimize losses. A few households in the village decided to cut their losses with the rotting potato crop and began preparing for an early sowing of the much more lucrative pea crop. Meera, however, knew nothing about these developments; although she agreed that peas brought in better money and were a much less time-consuming crop than potatoes, she did not feel entitled to take the initiative to alter the household's cropping regime on her own.

Meera's situation is revealing on a number of levels. It illustrates that carrying the workload does not entitle women to consider themselves as full agricultural partners of their men. It also reveals the costs to women of being unable to move, even in groups, outside the periphery of the wider community of villages unless accompanied by men. It is a revealing indicator of these village women's isolation that, given the restrictions on women's access to market services, there were no informal women's networks through which relevant information was disseminated. When asked if women could benefit from such organizations, Meera shrugged her shoulders and replied: "Women do not go to community meetings . . . we do not have the time because there are so many things to do." Then, almost apologetically, she said: "Women here are very timid: we do so much work and yet we would never travel or live alone the way you do. It is quite funny really."

Intrahousehold gender hierarchies also function to undermine women's authority by pitting their needs against men's interests. This is illustrated by the situation of Kusum, who was the household's de facto head. Her husband spent the greater part of the year out of the village, both her in-laws were elderly and frail (an acknowledged informal transfer of authority from the older to the younger woman had already taken place), and she had four young daughters below the age of twelve. As a result, Kusum was responsible for hiring the scheduled caste labor to prepare the fields for each new season; arranging for the harvested produce to be taken to the market; taking most of the decisions regarding purchases for household consumption; and, because a new family house was under construction, spending long hours cooking not only for her family but the laborers as well.

However, even though her father-in-law was no longer able to participate actively in market-related work, much less in field work, his word carried greater weight than hers in all agricultural matters. At the time, because all of her efforts were concentrated on attending to basic subsistence tasks and preparing meals, very little agricultural work was being done. Given the pressure on her time, Kusum's preference was to devote as much land as possible to peas since they are an easy crop to manage, require minimal labor and, despite marketing constraints and price fluctuations, provide the highest returns of all the cash crops. Her father-in-law, however, was opposed to this, insisting instead that she plant more millets; according to him, it was unnecessary to plant peas since his son had a high-paying job and cash was not a driving concern for the family.

Kusum offered a logical defence of her position: "My *sasur's* (father-in-law's) decision would be good if we needed more *jhangora* and *mandua* (millets) – they are good for our animals, they fill our stomachs, and can be stored for a long, long time. But what need is there just now with only two buffaloes? [The bullocks had been sold to raise money to help defray the costs of construction.] Even on my own I can easily collect grasses from the (fallow) *chan* lands. You have helped me do the *gudai* (weeding) so you can see what heavy work it is. How can I do this on my own, when my children are still so young? Peas are so easy to grow and they bring in money, but he [father-in-law] will not listen to me." She was not able to persuade him to support her point of view, and the old man's word prevailed over his daughter-in-law's needs.

WOMEN'S WORK, WOMEN'S WORTH

It would be inaccurate to suggest that only women have been affected by the increasing presence of cash and new values in the village economy. Older men, especially those with little or no experience of urban living and off-farm employment, similarly complain that younger men's access to education, jobs, and money challenge their status which was once unquestioningly conferred by seniority hierarchies. What is significant in the case of women, however, is that access to money is structurally limited through a combination of gender hierarchies, the gender division of labor, and differential access to crucial resources. A major impediment to village women's ability to accumulate cash is the absence of local opportunities to earn money. This gender-based disadvantage extends even to the scheduled caste women from other villages who work on higher-caste villagers' lands. They typically receive in-kind payments such as grain, vegetables, and rights from high-caste women to cut grasses and collect wood from private woodlots. However, even when paid in cash, these women earn less than men.[13]

Men's control over money concretizes gender ideologies which claim that "men are gods," "they are superior to us whatever they are like," and "nothing can be done here without men." Gender-differentiated control over money diminishes women and reinforces the view that men are the natural heads of households and, hence, the *real* managers of the land and all matters pertaining to family sustenance. This perception, in combination with men's considerably easier mobility and access to information networks, lends increasing support to the widespread belief that women's primary role is to provide labor while men's roles are mainly managerial.

Today, the patriarchal ideology of male supremacy within the household is intensifying. Boys, not girls, are socialized from a young age to develop the requisite skills that enable them, in the absence of their fathers and older brothers, to participate in the marketplace, bartering, purchasing, selling, and negotiating. However imperfect their familiarity with markets may be, they grow up to be adults who consequently have very tangible powers to exercise over women. This raises the less visible but critical issue of gender equity and distributional justice at the intrahousehold level: at a time when households have greater access to cash than ever before, gender entitlements to resources are becoming more asymmetrical.

CONCLUDING REMARKS

It is still too early to visualize the full ecological and socioeconomic impact of cash cropping in this particular mountain valley. Nevertheless, this snap-shot of a community undergoing rapid transition from a subsistence-based to a market-oriented economy offers some insights into the dilemmas facing mountain regions throughout the world. Mountain communities remain ill prepared to respond to changes induced by often poorly conceived development initiatives, extractive industries, and the social upheavals induced by migration. It should come as no surprise that mountains are home to disproportionately large numbers of the very poorest people who find themselves, increasingly, economically and politically marginalized.

It is likely that economic impoverishment coupled with rising expectations created by the mass media and market interactions will continue to compel large numbers of men to seek employment opportunities far from their villages. This situation in which women have to shoulder the responsibility for subsistence provisioning and farming activities has the potential for enabling them to expand their participation in the agrarian economy. However, given their lack of independent access to productive (and, by extension, status-enhancing) resources such as land, jobs and money, this is an unlikely outcome. To the contrary, such a scenario may actually bolster traditional sexual asymmetries by reinforcing women's dependence on men for access to institutions and personnel of the market economy.

The fragmentation of village communities, on the one hand, and women's "primary but subordinated" status, on the other, have disturbing consequences for the well-being of the local ecology and household food security. This is particularly problematic given the absence of infrastructural and institutional development in most mountainous areas in the Indian Himalaya that prevents even relatively accessible mountain communities from receiving adequate access to production inputs and information. In such contexts the loss of viable aspects of indigenous knowledge, techniques, and ways of knowing derived from a broad base of knowledge passed down from one generation of women to the next paint a dim picture for sustainable agricultural production.

The time has now come to move actively beyond the "gender blindness" that has typified too much of the discourse on mountain development and to make connections between macro development processes, the transformation of subsistence and food production systems and gender relations. It is, however, inadequate simply to recognize that gender-balanced approaches have to undergird development initiatives and to seek to improve women's access to information, resources, and education. Instead, there has to be a deeper understanding of how intrahousehold dynamics shaped by gender and generational hierarchies, on the one hand, intersect with cultural systems of entitlement and institutional constraints, on the other, to perpetuate women's limited access to institutions, personnel, and services of the wider political economy.

The Mountain Agenda, the blueprint for action which was included in Agenda 21 at the Earth Summit in 1992, offers an important new approach to address some of these challenges. Its recommendations are wide ranging, urging reformation of national policies to represent better the interests of mountain communities, decentralizing systems of management and decision-making, recognizing local land and resource rights, and protecting the indigenous knowledge systems and the resource management systems which they have helped sustain. Equally critically, it seeks mechanisms for incorporating the perspectives of the least powerful and most vulnerable social groups, typically women, the elderly, and those belonging to the lowest classes and castes. It remains to be seen how effective this agenda for action can be in a context where highland–lowland interactions are now increasingly being shaped by the dictates of the global economy and structural adjustment policies. Nevertheless, it offers an important opportunity to begin to work innovatively to create equitable and authentically sustainable forms of development which address local priorities in their myriad forms.

NOTES

1 The study was part of doctoral research which I conducted in the village from mid-1988 until early 1990.

2 These time phases are to some degree arbitrary for the groundwork, since present-day hills–plains interactions extend back to the nineteenth century with the advent of British colonial rule in the mountains, as discussed by Guha (1989) and Saklani (1987) among others. Nevertheless, I have selected them because they are significant contemporary watersheds when a coalescence of policy interventions, market forces, and a variety of pressures generated from within village communities –

including migration, land fragmentation, declining subsistence returns and higher educational levels – have helped to forge new patterns of interaction with the external market economy.

3 This notion of cultural, economic and political "impediments" comes from Bourque and Warren's (1981) comparative study of rural women in the Peruvian Andes.

4 See, for instance, Agarwal (1991); Sen and Grown (1987); Rodda (1993); Sontheimer (1991); Dankelman and Davidson (1988).

5 This invisibility of mountain issues is slowly changing as the development and environment communities become more aware of the special needs and constraints of highland areas. Since 1990 this concern has been formalized with the establishment of the International Union for the Conservancy of Nature (IUCN) Commission on Mountain Protected Areas and, in 1992, the inclusion of Article 13 ("Managing Fragile Ecosystems: Sustainable Mountain Development") into Agenda 21 at the Earth Summit.

6 Rhoades (1990) discusses how the potato was introduced into the middle Himalaya in the second quarter of the nineteenth century by European explorers. Saklani (1987: 5) provides evidence that potatoes were being cultivated on lands surrounding the newly created hill town of Mussoorie by the last years of the nineteenth century.

7 In comparison to agrarian relations in rural communities in the plains, however, hired low-caste labor has historically played a minor role in mountain agrarian economies. For further detail see Guha (1989); Rawat (1989); Saklani (1987); Sanwal (1976).

8 As Joshi (1929) and Berreman (1963) illustrate, the rights of the affinal household to women's reproductive and productive labor were so total that communities observed the practice of levirate (whereby a widow was married to her husband's brother) to keep the woman within the patrilineage.

9 As a result, members of a single extended family household often have multiple class identities: white-collar workers, those engaged in manual off-farm labor, and unpaid, household agriculturalists.

10 In South Asian rural communities, in addition to defining the nature of women's work, seclusion norms are also closely associated with overall household status. See Horowitz and Kishwar (1984); Jacobson (1974); Papanek (1982).

11 At the same time, studies indicate how cultural norms limit mountain women's movement and access to resources (Acharya and Bennett 1981). There are, for instance, communities where – even though it is common for women to participate in weekly markets – the extent of their mobility is either shaped by life-cycle issues or dictated by their caste status. Typically, however, lower-caste, older married, divorced, or widowed women are less confined by *purdah* norms than young and married women.

12 The mechanisms of this process in Nepal are elaborated in greater detail in Acharya and Bennett (1981), while Kemp (1987) offers an illustration of similar issues operating in the rural plains.

13 The situation here contrasts with many plains areas where economic imperatives compel low-caste and even impoverished high-caste women either to hire their labor out within the village or to migrate to urban areas in search of employment. As Acharya and Bennett (1981) note, a similar situation prevails in many Tibeto-Burman communities in Nepal. See also Mies (1986); Sharma (1989); and Jeffery *et al.* (1989) for conditions in the rural Indian plains.

REFERENCES

Acharya, M. and Bennett, L. (eds.) (1981) *The Rural Women of Nepal: An Aggregate Analysis and Summary of 8 Village Studies. The Status of Women in Nepal*, Vol. II, Part 9, Kathmandu: Tribhuvan University, Center for Economic Development and Administration.

Agarwal, B. (1988) "Neither Sustenance nor Sustainability: Agricultural Strategies, Ecological Degradation and Indian Women in Poverty," in B. Agarwal (ed.) *Structures of Patriarchy: State, Community and Household in Modernizing Asia*, New Delhi: Kali for Women.

—— (1991) "Engendering the Environment Debate: Lessons from the Sub-continent," *Center for Advanced Study of International Development, Distinguished Speaker Series*, No. 8, East Lansing: Michigan State University.

Berreman, G. D. (1963) *Hindus of the Himalayas*, Berkeley: University of California Press.

Bourque, S. and Warren, K. B. (1981) *Women of the Andes: Patriarchy and Social Change in Two Peruvian Towns*, Ann Arbor: University of Michigan Press.

Dankelman, I. and Davidson, J. (1988) *Women and Environment in the Third World: Alliance for the Future*, London: Earthscan.

Enslin, E. (1990) "Recovering the Commons: Gender, Space and Power in Chitwan, Nepal," Paper presented at the American Anthropological Association Meetings, New Orleans.

—— (1991) "Remapping Feminist Theories: Reflections on a Woman's Campaign for Community Space in Nepal," based on presentations at the University of Iowa and Cornell University, April.

Epstein, T. S. (1987) "Cracks in the Wall: Changing Gender Roles in Rural South Asia" in J. W. Bjorkman (ed.) *The Changing Division of Labor in South Asia*, New Delhi: Manohar Publications.

Guha, R. (1989) *The Unquiet Woods: Ecological Change and Peasant Resistance in the Himalaya*, New Delhi: Oxford University Press.

Harcourt, W. (ed.) (1994) *Feminist Perspectives on Sustainable Development*, London: Zed Books.

Horowitz, B. and Kishwar, M. (1984) "Family Life – The Unequal Deal: Women's Condition and Family Life Among Agricultural Laborers and Small Farmers in a Punjab Village," in M. Kishwar and R. Vanita (eds.) *In Search of Answers: Indian Women's Voices from Manushi*, London: Zed Books.

Jacobson, D. (1974) "The Women of North and Central India: Goddesses and Wives," in C. Mathiasson (ed.) *Many Sisters: Women in Cross-Cultural Perspectives*, New York: The Free Press.

Jeffery, P., Jeffery, R. and Lyons, A. (1989) *Labor Pains, Labor Power: Women and Childbearing in India*, New Delhi: Manohar.

Jiggins, J. (1994) *Changing the Boundaries: Women-Centered Perspectives on Population and the Environment*, Washington, D.C.: Island Press.

Joshi, L. D. (1984) *Tribal People of the Himalayas* (first published 1929), New Delhi: Mittal Publications.

Kemp, S. (1987) "How Women's Work is Perceived: Hunger or Humiliation," in J. W. Bjorkman (ed.) *The Changing Division of Labor in South Asia*, New Delhi: Manohar Publications.

Mies, M. (1982) "Purdah: Separate Worlds and Symbolic Shelter," in H. Papanek and G. Minault (eds.) *Separate Worlds: Studies of Purdah in South Asia*, New Delhi: Chanakya Publishers.

—— (1986) *Indian Women in Subsistence and Agricultural Labor*, New Delhi: Vistaar Publications.

Rawat, A. S. (1989) *History of Garhwal, 1358–1947*, New Delhi: Indus Publishing.

Rhoades, R. E. (1990) "Potatoes – Genetic Resources and Farmer Strategies: Comparison of the Peruvian Andes and Nepali Himalayas," in K. N. Riley *et al.* (eds.) *Mountain Agriculture and Crop Genetic Resources*, New Delhi: Oxford University Press and IBH Publishing.

Rodda, A. (1993) *Women and the Environment*, Women and World Development Series, London: Zed Books.

Saklani, A. (1987) *The History of a Himalayan Princely State: Change, Conflicts and Awakening*, New Delhi: Durga Publications.

Sanwal, R. D. (1976) *Social Stratification in Rural Kumaon*, New Delhi: Oxford University Press.

Sen, G. and Grown, C. (1987) *Development, Crises and Alternative Visions: Third World Women's Perspectives*, New York: Monthly Review Press.

Sharma, M. (1980) *Women, Work and Property in North-West India*, London: Tavistock.

——(1989) "Rural Women in Rajasthan," *Economic and Political Weekly*, Review of Women's Studies, 24, 17: 38–44.

Sontheimer, S. (ed.) (1991) *Women and the Environment: A Reader*, New York: Monthly Review Press.

Part IV

GENDERED KNOWLEDGE

9

GENDERED KNOWLEDGE: RIGHTS AND SPACE IN TWO ZIMBABWE VILLAGES

Reflections on methods and findings

Louise Fortmann

Political ecology has brought home the importance of understanding local resource users and their decision-making environment in the analysis of natural resource use and management (Blaikie 1985; Blaikie and Brookfield 1987; Sheridan 1988). Feminist political ecologists have emphasized the need to understand both gender differentiation of natural resource use and management, and how broader social relations affect women's use of the environment as compared to men's (Walker 1995; Carney and Watts 1990; Jackson 1993; Rocheleau 1995; Schroeder 1994). Researchers have explored these questions at a variety of scales from the village to the city.

This chapter explores methods of obtaining data to address the queries of feminist political ecologists as well as recent questions being raised among feminist scholars on methods.[1] On a more fundamental level, however, it is about generating and sharing knowledge and expertise. It presents research findings and describes how the knowledge generated and revealed through a research project came to be "owned" by the study village through the use of participatory methods, and the deliberate and systematic empowerment of both the village research team and other villagers. Because the story concerns a personal intellectual journey as well as more traditional schol-arly matters, I tell it in my own voice and include my own reflections on the process.

The concerns which motivated my choice of methods and the writing of this chapter are encapsulated in an episode from my 1994 graduate seminar which was studying Elinor Ostrom's (1990) analysis of the management of a southern California water basin. It was clear from the text that one of the communities was singular in an unidentified way. I asked a student who had lived in the area what she knew about the region in general and that town in particular. "Oh," she said, "I only lived there a few years, so I really don't know very much about it." "In international development," I quipped, "a few years would make you an 'expert'."

Expertise, in short, is problematic. In development circles – including the arena of women and environment – who is an "expert" and whose exper-tise "counts" is often shaped by the unsavory forces of elitism, racism, and neocolonialism. For example, women from urban-industrial places may find themselves sitting in judgment on the relative sustainability of a resource management system developed through long local experience. Or they may be asked to introduce "environmentally sound" agricultural practices to seasoned farmers without any basis on which to compare customary practices to new techniques. Similarly, the "professional" criteria of researchers may fail to capture subtle but crucial distinctions used by local people. Or scientists may document, present, and use the empirical and analytical insights of rural people with at most a cursory mention buried in the acknowledgements section. The development literature and the oral histories of field scientists and planners are replete with these kinds of "expert" behavior by women and men.[2]

Needless to say, when in 1991 I undertook field research, I hoped to avoid all of these sins. The research itself explored the intersection of tree tenure, gender, and tree planting and use in two villages[3] located in two different ecological zones within 100 kilometers of Harare, the capital of Zimbabwe. Residents of both sites depend on rainfed agriculture and jobs in town for their livelihood. Trees play a crucial role in the livelihood strategies of the two study sites. They provide domestic and commercial sources of food, medicine, browse, poles, fuel, mulch, and wood for carving as well as serving ecological and religious functions. The research took three complementary forms – a standard random sample survey, participant observation, and a series of participatory methods.

Having been trained as a rural sociologist, one of the ways I approached this question was through the use of a survey of a stratified random sample of households. Since my Shona language skills never progressed beyond the "me-go-store-now" variety, I hired a team of seven villagers to help with the administration of the survey.[4] There were four women in this research team, three of them between the ages of 36 and 52, middle-aged women like me. They provided an entry to the situated knowledge of middle-aged and elderly women – their own knowledge and that of their friends.[5]

I have always been adamant about returning data to the village, both making sure the village gets copies of any written reports and always going back to discuss my findings with villagers. That is what I originally intended to do in this case. But perhaps because this was the first time in many years that I was actually living in my research site,[6] or due to the obvious excellence of the team and the equally obvious need for the villagers to be able to do their own research, I realized that I had to give them my own skills.

How do we leave our skills behind? Obviously formal survey methods which require paper, duplication facilities, and statistical analysis are of little use to the average village. In contrast, participatory methods not only got me useful data, but they were skills and techniques that villagers could learn and replicate whenever they needed similar sorts of information.

To develop a participatory research process, I returned to the lessons I had learned at Cornell University in the 1970s from Ivette Puerta, a Puerto Rican doctoral student working with Latinas. Wanting to find a way of building community through the research process, Ivette Puerta developed the strategy of using community members as a research team (Puerta and

Plate 9.1 Patricia Matsuimbo instructs the author in how to use bark as rope to tie up a bundle of firewood

Source: Louise Fortmann

Bruce 1972). The idea was that as community members gathered information through a survey, three things would happen. First, people would develop a consciousness about their problems as they talked about them in the interview. Second, the research team would become experts on community problems and could become community spokespeople. Third, they would through their interviews develop a network which could be mobilized later on.

I had used Ivette Puerta's methodology in my dissertation. At the end of the survey, the five women (all welfare mothers, none with a high school diploma) who worked with me took the Dean of the College of Agriculture on a rural poverty tour of the county with articulate and pointed commentary throughout. He was in shock for quite some time.

So in Zimbabwe, I returned to my methodological roots after two decades. I decided to adapt my methods so that the villagers would really "own" this research. And I wanted to empower the women at the same time I was learning from and with them. This took five basic forms: Foxfire books, resource mapping, the questionnaire survey, wealth rankings, and public presentation of the research by the research team.

FOXFIRE BOOKS

As I began rethinking my methods, I ran into Dianne Rocheleau, the feminist geographer, who reminded me about the Foxfire books. So I explained to the headmaster of the village primary school about these books which were written by rural children and teenagers in the Appalachian Mountains in the United States about their own culture and environment. The young authors of the original Foxfire books set the precedent: the children of the village could interview their own parents or grandparents or describe something they knew themselves. But since I wanted this to be their book, not my book, I stayed out of the process.

The first set of essays I got were a terrible shock. Most of them had been copied (two sets were *identical*) or paraphrased from a book on trees. Each said at the bottom: "Warning: District Council Prosecutes Illegal Tree Cutters." This is the persisting legacy of British colonialism in which education is based on memorization and regurgitation. The notion that people's own knowledge in their children's own words could have any value is quite inconceivable in that system. I went to the headmaster and said, "There are a few problems with these essays." And he, fortunately, said, "Yes, I thought you would say that." So I wrote some titles: How My Grandmother Uses Trees, My Favorite Tree, and so on. And we started again.

The final book consisted of all of the second set of essays plus salvaged ones from the first round, if the author had not written in the second round. They were typed and bound, and every child who wrote, plus several of the village dignitaries, got a copy of the book at the farewell ceremony described below. The headmaster of the secondary school where some of the children were now enrolled came to pick up their copies. A few children read their essays aloud. Everyone was incredibly proud! They had written their own book which they could use in school. They could read it to themselves and to others. It captured their own knowledge and information. The headmaster thought that it could perhaps be published and distributed all over Zimbabwe.

I did not use the Foxfire books as a data source on gendered knowledge and tree management, although with more careful planning I might have. But that makes them no less valuable. It is important to remember that not everything in our research needs to be for us.

RESOURCE MAPPING

Mapping is just what it sounds like. You ask people to draw a map. You can have them map anything you need to know about – the village, the wealthy, water, markets, and so on. I asked people to map where they got tree products. For this you need a long stick and a lot of branches. The stick is for drawing the map on the ground. The little twigs represent trees. Rocks and other props are also useful. For example, one group of men filled a cup with water for a dam. A group of women fashioned a wonderful wind-mill out of cornstalks. In another group, we were sitting under a mango tree which kept bombarding us with hard green mangoes. Finally one woman put a twig in the map and announced it was the tree which was attacking us. Drawing on the ground with a stick (or many sticks as people get into it) avoids all the school and learning connotations that pen, pencil, and paper carry. But there is a certain vulnerability involved. For example, the agricultural extension agent drove his motorbike right through the village research team's practice map.

Plate 9.2 Hilda Chitsa adds "trees" to a village forest resource map

Source: Louise Fortmann

After people had done their maps, I asked them what it looked like in 1970, in 1980, and what they would like it to look like. Not surprisingly, everyone said there used to be more trees and they wished there were more once again.

It is very important to do this sort of thing separately for women and men. For starters, women and men map things differently, putting different elements in their maps in different order, with different degrees of detail. Both began their maps with the two rivers which bound the village. Women then drew a very detailed map of "domestic" social space, household by household, sometimes with details of particular houses – extra windows, tin roofs, and so on. Men, in contrast, concentrated on public and production spaces – the roads, the grazing areas, the shops.

But in order to tease out these different spatial perceptions, you must allow women their own arena for mapping. On two occasions women were included in the same group with men. Perhaps the most striking example of gender hierarchy occurred in the village research team. They had insisted on doing a "proper" map with pencils instead of the "childish" map in the dirt. So I bought a large piece of poster board and they did a second map. Each researcher had an individual pencil and eraser so everyone could draw. In the group that day were the three middle-aged women, a 20-year-old woman, and a 19-year-old man. Who drew the map? The young man, of course! In this case, the older women were far above the young man in the age hierarchy and were confident and assertive, at least in the research arena. So they were quite vocal about what should go on the map and made the young man erase things and redraw them to their specifications.

Other women were less able to make their voices heard in a mixed group. Women who participated in the grazing committee mapping exercise

Table 9.1 Spaces used for selling by women and men respondents and women and men fruit sellers[a] (1990–1)

Sales Spaces	Men (N = 48)	Women (N = 105)
Inside community	17 [62]	19 [57]
Nearby areas	2 [8]	7 [20]
To outside buyers	10 [38]	3 [20]
Urban market	13 [46]	2 [6]

Source: Fortmann and Nabane (1992a)

Notes: $x^2 = 10.89$, 3 df, significant at .01 level
    a: Figures in brackets are for fruit sellers only

primarily muttered only an acquiescent "*ndidzodzo*" (it's OK) as the men drew the map. In response to a direct question, one woman did indicate specific termite mounds that were good places for collecting fuel wood. But on the whole, the shy silence of these women was a striking contrast to the assertive laughing women of the single-gender groups.

The maps provided a useful visual indication of resource clusters. They also were the most accurate check on elite myths, in particular the myth of the community woodlot. The chairman of the Grazing Scheme proudly took all visitors around the village woodlots, one area of substantial eucalyptus and one of regenerating indigenous woodland. This, the story went, was where villagers came for poles – everyone used it. Indeed, the village had won several prizes for their wonderful woodlots. But (with the exception of the elite grazing committee) when villagers drew their maps of where they got tree products, the village woodlots were conspicuous by their absence. Survey data confirmed that only the rich used these woodlots. But their absence from the maps tells an even more powerful story – most villagers do not even think of them as an available resource. No survey can tell you that quite so clearly.

SURVEY FINDINGS ON GENDERED KNOWLEDGE AND SPACE

The survey of gendered space complemented and confirmed the resource maps with a more quantitative "map" in numerical form. The research showed that knowledge about trees is highly gendered. As revealed in key informant interviews, women not only knew many more uses of specific trees, but they were especially knowledgeable about medicinal uses of trees (see Chidari *et al.* 1992 for details).

As revealed in the participatory mapping exercises described above, the construction and use of space were also highly gendered. Not only did women and men use the same space differently, they used different spaces.

Table 9.2 Gendered tenure in tree source areas (1990–1) – percent of tree species locations[7]

Reported uses of trees	Women	Men
Fruit to eat $x^2 = 23.54^c$ 1 df	N = 1321 53% Individual 47% Communal	N = 649 41% Individual 59% Communal
Firewood $x^2 = 6.43^b$ 1 df	N = 971 20% Individual 80% Communal	N = 357 14% Individual 86% Communal
Medicine $x^2 = 21.18^c$ 1 df	N = 563 59% Individual 41% Communal	N = 359 43% Individual 57% Communal
Browse $x^2 = 17.33^c$ 1 df	N = 399 32% Individual 68% Communal	N = 309 18% Individual 82% Communal
Poles $x^2 = 28.43^c$ 1 df	N = 375 38% Individual 62% Communal	N = 231 10% Individual 90% Communal
Agricultural implements $x^2 = 2.23$ ns 1 df	N = 78 19% Individual 81% Communal	N = 121 12% Individual 88% Communal
Household utensils $x^2 = 4.7^a$ 1 df	N = 94 24% Individual 76% Communal	N = 129 13% Individual 87% Communal
Edible insects $x^2 = 15.30^c$ 1 df	N = 153 29% Individual 71% Communal	N = 126 10% Individual 90% Communal
Fertilizer $x^2 = 0.05$ ns 1 df	N = 172 34% Individual 66% Communal	N = 37 32% Individual 68% Communal

Note: ns not statistically significant
a: significant at .05
b: significant at .01
c: significant at .001

Source: Fortmann and Nabane (1992a)

This was particularly striking in spaces in which women and men sold tree products. Women were most likely to sell in the village and nearby areas. Sales to outside buyers and in the urban market were primarily made by men. This is demonstrated for fruit selling in Table 9.1. In this table, percentages are based on all respondents. Figures in brackets are percentage of fruit sellers only. In 1990–1 there were thirteen men and thirty-five women who sold fruit.

As shown in Table 9.2, survey findings ran counter to the received wisdom on women's use of space, showing that women were more likely than men to name privately controlled spaces as the location of the tree species they used. A moment's reflection on the spatial distribution of a woman's workday in the family fields and doing domestic work and the expeditionary nature of much of men's resource collection will indicate why this would be so.

This is particularly interesting in light of the detailed mapping of "domestic" space done by women.

WEALTH RANKING

Participatory methods for wealth ranking involving card sorting are pretty standard. I sat down with my research team and said, "Tell me what rich people have, tell me what poor people have." I wanted a five-point scale. They pushed me to six. Their scale included the usual cast of variables – cattle, house type, employment. But also included among their variables was one I had never thought of – secondary education: where children went to school (in the village or in town) and how continuously, if at all. They laid out categories of people who were dependent on other people for livelihoods. There was a huge argument over the importance of owning the means of production (plows, cattle, fields) versus owning consumer durables (fancy house, radio).

Then I had each village researcher rank all our respondents leaving blank anyone they could not or did not want to rank. What I found particularly interesting in these rankings was that they ranked a number of widows much lower than I would have ranked them. Why? They ranked widows on what they themselves personally controlled/owned as opposed to what their children were able to give them. Since children might withdraw their favors or be run over by a bus, their largesse did not count in this exercise. There was a very strong sense of vulnerability in these criteria for wealth and well-being.

The results of the ranking exercise were eventually used in statistical analysis. They are clearly related to the traditional measures of wealth (see Table 9.3). But they reflected the nuances of local reality far better. And the research team got practice for future use in thinking through categories of people and how they were differentially affected by various things happening in the village.

REPORTS BY THE VILLAGE RESEARCH TEAM

If the villagers are to own research, they ought to be able to use it right from the start. So I asked the team if they wanted to present the results to the village. They were enthusiastic. Each person picked a topic. I tabulated and printed the data for them. Each person wrote a speech in English. I checked it. They translated it into Shona. My colleague, Nontokozo Nabane, checked it. I gave them tips on public speaking. And then every Tuesday at 10 a.m. for ten weeks we practiced.

At the beginning I was a little worried. These were youths and middle-aged women – people whose role in meetings is generally to listen respectfully to the older men. We had gigglers. We had clutchers. Who knew what would happen on the day!

The day came. The meeting began with a very long and eloquent prayer about trees. The school choir sang about trees. And then the research team gave their speeches. The gigglers and the clutchers had turned into seven

Table 9.3 Traditional wealth indicators (%) and the wealth ranking scale

Wealth levels	0 $N = 9$	1 $N = 28$	2 $N = 23$	3 $N = 38$	4 $N = 7$	x^2 4 df
Indicators						
Own cattle	22	46	83	82	100	25.34[a]
Own Scotch cart[8]	0	11	35	61	71	26.17[a]
Have tin roof	33	21	35	66	100	22.69[a]

Source: Fortmann and Nabane (1992b)

Notes: 0 = Poorest, 4 = Least poor
a: significant at .001 level

confident, polished, authoritative speakers. They were dynamite!! They were proud!! And everyone listened! "We never thought," said the chairman of the Grazing Scheme in his concluding speech, "that we would learn something from a woman, but we have."[9]

This sort of thing is the best way that I know of that will enable villagers to "own" the research.[10] Any time someone wants to know something, they can just pop round and ask their neighbor. The community can replay that meeting or parts of it any time they want to. And there are now women who can speak for themselves, for women, for their village. And, of course, they could do some research for themselves.

PAYING IN OUR OWN CURRENCY

Our own currency has nothing to do with foreign exchange. It has to do with credit for ideas and knowledge. A classic example can be found in another twenty-year detour to an acknowledgment I read in the mid-1970s in a development book. It went something like this: I would like to thank my wife who accompanied me in the field, assisted in the interviews, typed my field notes every day, coded the questionnaires, helped with data analysis, read and commented on all my drafts, and typed and proofread the final manuscript. "And why is she not the co-author?" I muttered for the next week (and indeed for the next 20 years). I have long forgotten what the book said and even who wrote it. I have never forgotten that acknowledgment.

In my survey, I asked what trees people used and for what. We ended up with a list of 122 indigenous trees. There were also a large number of exotics. The next task was to get physical specimens since the same tree has different names and different trees have the same name in Shona. And some people simply made names up.[11] For example, around the school grounds grew clumps of Banket Hedge, named for the town of Banket where the headmaster got a cutting and brought it back to the village.

So one very hot day, the research team and I sat under the shade of a mango tree and went through the list. They quickly grouped the trees by habitat: trees that grow in rocky places, trees that grow in fields, trees that

Plate 9.3 Patricia Matsuimbo interviews a farmer during a break from ground nut planting

Source: Louise Fortmann

grow by the river. One "tree" that grew *in* the river turned out to be a water lily – another reason for doing this exercise. Then we set out and in four hours we had collected specimens of ninety-five different species of indigenous trees. Not only did my team know the habitat for each species, they knew the location of individual trees. It was a staggering exhibition of botanical expertise.

There were, of course, duplicates. Some, like the water lily, were not trees at all. And some trees turned out to be located in distant places where people had worked so we could not collect them. In the end we had a list of 114 different tree species, 99 identified in the Herbarium by botanist Robert Drummond.[12]

Of course, all this brought back that acknowledgment that has been irritating me for so long. That acknowledgment reminded me of the acknowledgments that I see in so many works utilizing local knowledge. And it came to me, that if local knowledge really *is* important and not just something we pay lip service to, then we should pay for it in our own currency – not with an offhand acknowledgment but with full-blown academic credit.[13] If we have relied on their knowledge, then they should be co-authors. And so Chidari *et al.* (1992) "The Use of Indigenous Trees in Mhondoro District" was born. The villagers, as they should be, are the senior authors.

This booklet was also presented at the farewell ceremony. Every member of the research team, every headman, the headmaster, the chairman of the Grazing Scheme, the Forestry Commission representatives, the District Officials all got individual copies. Everyone was proud. This was their knowledge. And it had the potential to be used in future development efforts.

Subsequent thought and conversations and correspondence with people from all over the globe have convinced me that paying in our own currency *must* be a model for the way we publish our research whenever that research depends on local people's knowledge. We would not use our colleagues' ideas without acknowledgment. The same principle applies here. To do anything other than paying with academic credit is probably unethical and certainly colonial.

CONCLUDING THOUGHTS

The ability to articulate needs for and rights to natural resources to government agencies and nongovernmental organizations (NGOs) is an increasingly important skill for villagers as pressure on these resources rises and government and national and international NGO penetration of rural areas increases. In this context the need to use participatory research methods is obvious. They are empowering if they are done right. They provide a forum in which people learn and share knowledge together. This in turn provides a common basis for certain kinds of planning and decision-making. In particular, it means that right from the start researchers are plunged into people's own categories and language, and this is essential for getting accurate maps and inventories of resources, rights, and practices. In short, participatory methods can serve both the researcher and the researched.

The take-home message from this chapter is that researchers have ethical responsibilities to ensure that not just they but also villagers benefit from and own any village-based research. The academy must never be stingy with its ability to do research. Rather it is incumbent upon us to nurture in others the skills to do research, for they may have to depend upon themselves. We must draw the circle of expertise larger – truly honoring their expertise and enabling them to add our expertise to their repertoire.

ACKNOWLEDGMENTS

I gratefully acknowledge the field assistance of Nontokozo Nabane, and the comments and editorial hand of Dianne Rocheleau.

NOTES

1 For additional sources on feminist methods, see among others: Nielsen (1990), Reinharz (1992), Roberts (1981), Staeheli and Lawson (1994), Stanley (1990), and Wolf (1993).

2 I must tell this story as an illustration of the persistence of these pernicious practices. In February 1995, I witnessed a woman explaining a very complex agricultural livelihood system to a young economist who was considered a development expert because he had once spent roughly a month in the country he studied. After she had explained the cultural, social, political, and labor considerations that shaped the system, he said (and I am *not* making this up), "Well, all you have to do is get the prices right." We have a very long way to go!!

3 For purposes of anonymity, these two villages are labelled on any maps within this volume with the pseudonyms Chamitimirefu and Mombe.

4 I interviewed about ten women from the community. They did practice interviews in Shona and conversed with me in English and were chosen for their interviewing skills.

5 The research team also had three men on it. The women interviewed female respondents and the men interviewed the male respondents.

6 Obviously the sine qua non of participatory methods is that you have got to be there. For a number of years I had been asking questions that were best answered through archival data or could be answered (whether best answered is another issue) through mail surveys or through short-term field stays to do key informant interviewing.

7 It is essential to understand the unit of analysis, tree species locations, precisely. Respondents were asked what species of trees they used, if any, for fifteen different purposes. They were then asked where they got each species listed ten years before and in 1990–1. This generated a list of tree species locations with some species having more than one location. Thus, a respondent might have reported using *muhacha* located in her homestead and in the grazing area. This would constitute two tree species locations, in this case, one under individual tenure and one under communal tenure. It is essential to bear in mind that this table does *not* tell what percentage of the products came from which location. Nor does it tell us what percentage of individual trees were located where. But it does indicate the location of species and some suggestion of where people are getting certain products.

8 Scotch cart is a local name for an animal-drawn cart, usually with two wheels.

9 This statement was made in the context of a farewell speech to me and hence might refer to me. But its application to the women on the research team is equally obvious.

10 The Centre for Applied Social Sciences at the University of Zimbabwe has replicated this method and found it effective in communicating research findings.

11 I am sympathetic to this since I grew up eating Newcastle Sandwiches. These were ordinary grilled cheese sandwiches named by our family for the town where my father's amateur softball team played in the state championships and where we kids first ate them.

12 His extraordinary patience with my inexpertly preserved specimens is herewith gratefully acknowledged.

13 This is not to suggest that village researchers should not also be paid living wages. Obviously they should.

REFERENCES

Blaikie, P. M. (1985) *The Political Economy of Soil Erosion in Developing Countries*, New York: Wiley.

Blaikie, P. and Brookfield, H. (eds.) (1987) *Land Degradation and Society*, London: Methuen.

Carney, J. and Watts, M. (1990) "Manufacturing Dissent: Work, Gender and the Politics of Meaning in a Peasant Society," *Africa* 60, 2: 207–41.

Chidari, G., Chirambaguwa, F., Matsvimbo, P., Mhiripiri, A., Chanakira, H., Chanakira, J., Mutsvangzwa, X., Mvumbe, A., Fortmann, L., Drummond, R., and Nabane, N. (1992) "The Use of Indigenous Trees in Mhondoro District," Centre for Applied Social Sciences, University of Zimbabwe, Natural Resource Management, Occasional Paper 5.

Fortmann, L. and Nabane, N. (1992a) "Fruits of Their Labours: Gender, Property and Trees in Mhondoro District," Centre for Applied Social Sciences, University of Zimbabwe, Natural Resource Management, Occasional Paper 7.

—— (1992b) "Poverty and Tree Resources in Mhondoro District: A Research Note," Centre for Applied Social Sciences, University of Zimbabwe, Natural Resource Management, Occasional Paper 8.

Jackson, C. (1993) "Environmentalisms and Gender Interests in the Third World," *Development and Change* 24: 649–77.

Nielsen, J. M. (1990) *Feminist Research Methods: Exemplary Readings in the Social Sciences*, Boulder, Colorado: Westview Press.

Ostrom, E. (1990) *Governing the Commons: The Evolution of Institutions for Collective Action*, Cambridge: Cambridge University Press.

Puerta, I. and Bruce, R. L. (1972) "Data Collection with Low-Income Respondents," Paper presented at the Adult Education Research Conference, Chicago, April.

Reinharz, S. (1992) *Feminist Methods in Social Sciences*, Oxford: Oxford University Press.

Roberts, H. (ed.) (1981) *Doing Feminist Research*, London: Routledge and Kegan Paul.

Rocheleau, D. (1995) "Gender and Biodiversity – A Feminist Political Ecology Perspective," *Institute for Development Studies Bulletin* 26, 1: 9–16.

Schroeder, R. A. (1994) "Shady Practice: Gender and the Political Ecology of Resource Stabilization in Gambian Garden/Orchards," *Economic Geography* 69, 4: 349–65.

Sheridan, T. E. (1988) *Where the Dove Calls: The Political Ecology of a Peasant Corporate Community in Northwestern Mexico*, Tucson: University of Arizona Press.

Staeheli, L. A. and Lawson, V. A. (1994) "Women in the Field – The Politics of Feminist Fieldwork – Discussion," *Professional Geographer* 46, 1: 96–102.

Stanley, L. (1990) *Feminist Praxis: Research, Theory and Epistemology in Feminist Sociology*, London: Routledge.

Walker, P. (1995) "Synopsis and Annotated Bibliography of Political Ecology," unpublished paper.

Wolf, D. L. (1993) "Feminist Dilemmas in Fieldwork: Introduction," *Frontiers: Journal of Women Studies* 13, 3: 1–8.

10

FROM FOREST GARDENS TO TREE FARMS

Women, men, and timber in Zambrana-Chacuey, Dominican Republic

Dianne Rocheleau, Laurie Ross, and Julio Morrobel
(with Ricardo Hernandez, Cristobalina Amparo, Cirilo Brito, Daniel Zevallos, the staff of ENDA-Caribe and the members of the Rural Federation of Zambrana-Chacuey)

PRACTICING FEMINIST POLITICAL ECOLOGY

While political ecology and feminist scholarship have been associated primarily with critique, there is a real need to move beyond critique and to transform practice in the context of land use change and resource management. In addition to the issues of gendered resource access and organization, this chapter addresses the questions of how we acquire knowledge and how we use it. We explore the relations of power embedded in a rural reforestation project in the Dominican Republic, moving from a broad regional and national context to a case study analysis based on field work. A short discussion of research methods summarizes our own tools for learning. As a direct outcome of a feminist political ecology analysis we then offer a series of practical recommendations for a particular project in one region. We table specific suggestions for technology and tree species choices, tenure innovations and organizational changes. By summarizing both the process and practical applications of feminist political ecology in one place we hope to illustrate the possibilities for critical engagement in the everyday work of resource management.

GENDER, FARMS, AND FORESTS IN REGIONAL CONTEXT

The prevailing national and international images of rural women in Latin America and the Caribbean as housewives and "not farmers" has masked their extensive involvement and their very diverse economic, cultural, and environmental interests in agriculture and forestry (Townsend *et al.* 1993; Arizpe *et al.* 1992; Ronderos 1992; Urban *et al.* 1994; Flora 1986; Katz 1993; Momsen 1993; Guzman *et al.* 1991; Silva 1991).[1] The extent and nature of women's actual participation in forestry and farming ranges from provision of family labor for men's enterprises, to processing, marketing, and management of the products at household level, to full identification of women as farmers by themselves and others. Women farmers and farm managers may take direct responsibility for household food and/or cash crop production, including commercial tree products from farm and forest. Women often manage and collect fuelwood and water supplies and in many places they gather medicinal herbs and wild foods from the forest. They may also use the forest or its products for religious observances, particularly where the practices of peoples of African descent and those of indigenous people persist on their own or within syncretic religious practices in the Catholic Church (Menchu 1989; Rocheleau *et al.* 1995b). However, women's knowledge and practice of agriculture and forestry have remained largely invisible to national and international environmental and development agencies.

The discrepancy between outsider gender stereotypes and daily life is especially relevant to the experience of the women and men of the Zambrana-Chacuey region in the Dominican Republic. The region provides examples of both deforestation and intensive tree crop horticulture, and of edible forests[2] as well as clear-cut hillslopes and cash crop farms. The region has been the target of national military anti-deforestation campaigns – complete with helicopters – as well as the privileged site of a forestry enterprise pilot project. Zambrana-Chacuey epitomizes the simultaneous pursuit of deforestation and reforestation by a highly differentiated set of actors, both women and men. They use the trees as both tools and sites of struggle from the household to the national level.

In this chapter we focus on the gendered interests at stake in changing livelihood systems and landscape patterns. We do this through the prism of a successful timber cash cropping program jointly sponsored by the Rural Federation of Zambrana-Chacuey and ENDA-Caribe, a regional branch of an international environment and development organization. The case study explores the problems of deforestation and reforestation in national, regional, and local contexts. We examine gendered knowledge, interests, rights, responsibilities, and institutions in the process and outcomes of reforestation.

The divergent experience of people within specific smallholder households and within the Federation demonstrates the diversity of actors and circumstances and the gendered nature of landscapes, livelihood strategies, and institutions within the region. Their stories also make visible the gendered dimensions of reforestation not yet represented in the summary numbers and the district maps of forestry-as-usual.[3]

NATIONAL AND REGIONAL CONTEXT FOR THE CASE STUDY

Forestry and rural people

The Dominican Republic has a long history of direct state intervention in forest management and regulation, from the colonial period to the present (Betances 1994). In 1967, the government passed the Forestry Act (Law 206), which effectively placed not only forests, but all "trees" under the protection and regulation of the state (Veras 1984). It prohibited the felling of any tree without the express permission of the newly formed, militarized forestry service (Direccion General Forestal - DGF). This law extended the authority of the state into the lands of every resident and property holder, and into the daily lives of every tree user in the nation. In practice the forestry service (DGF) selectively enforces tree-cutting bans against poor smallholders, wood workers, and charcoal producers and awards permits primarily to commercial producers and local elites.

A recent turn toward "Green" discourse on forests has captured the public imagination and the attention of Dominican political and technical professionals (Lynch 1994). Forest management policy and enforcement of forest laws have generated dramatic and politically charged environmental conflicts. Forest department campaigns to stop deforestation have employed helicopter surveillance and periodic raids in rural communities, and many rural people have reported abuses by forest guards and accompanying police or soldiers (Lynch 1994; Rocheleau and Ross 1995a).

At the same time that the DGF has campaigned to protect forests, other state agencies have promoted the expansion of "nontraditional" agricultural exports, including pineapple, citrus, and root crops, at the expense of coffee and cocoa stands, pasture, and forest lands (Lynch 1994; Raynolds 1994). This is part of a regional trend throughout Latin America and the Caribbean, promoted by international banking and development institutions, to increase and diversify agricultural exports.[4] Given the increased competition for land between and among largeholder and smallholder farmers and international agribusiness corporations, these export crops have often displaced forests and diverse multicropping systems. This trend in agricultural development constitutes an important part of the context for deforestation and reforestation in the region.

The institutional actors in these contradictory trends in land use change include national and regional environmental NGOs, national and international corporations, government agencies and popular organizations, each representing different interests on any given issue. Between the conservation concerns of a privileged few and the monolothic control of forest resources by the state, lies the complex terrain of economic development and forest use and protection in the public interest. This is all the more complicated by the existence of many publics, with distinct interests and aspirations. Likewise, the promotion of nontraditional cash crops finds most people in the middle of a battle between powerful competing interests. Both women and men, across classes, in rural and urban contexts, have a stake in national debates over forestry and agricultural policy and are making everyday decisions on land use, environmental management and access to land, trees, and other forest resources.

Gender dimensions of forestry in the Dominican Republic

Throughout the Dominican Republic both men and women are gardeners, farmers, livestock keepers and breeders, forest managers, forest gatherers, drawers of water, food processors, market vendors, and keepers of the "natural" and built environment. Yet, in every region of the country women and men differ with respect to specific divisions of labor, responsibilities, interests, and control in agricultural and forestry production and resource management. Their knowledge, experience, constraints, and opportunities are in many ways distinct, not by biological necessity, but by custom and current practice.

Ethnicity, race, class, and locality all shape constructs of gender and frame the terms of women's everyday participation in farming and forestry. For example, in the Sierra, on the northern slope of the Central Mountains (Cordillera Central), the gender division of labor reflects a pervasive Spanish influence, with women identified more as housewives or as farmworkers (such as coffee harvesters) and not as farmers. While men may be the exclusive tillers of the land in much of this area, women do participate in both subsistence and commercial production as food processors, wood and water collectors, and as keepers of small livestock (primarily hogs and chickens for both sale and home use). They often also act as farm managers and supervise sons or hired men in male-identified tasks such as land preparation and cultivation. Most Serranas also engage in agriculture within patio gardens and some have cultivated in women's group gardens (Flora and Santos 1986).

In contrast, in the hills of Zambrana-Chacuey near Cotui, gender identity is strongly shaped by African as well as Spanish cultural influences. Men and women share farm tasks under a more flexible ethos which leaves very broad scope for choice by individual women and their households, with many women identified proudly as farmers and others as housewives (Field interviews 1992–3). Yet women's and men's knowledge, access to resources, and organizational affiliation are clearly gendered, reflecting flexible complementarity of labor and authority under uneven relations of power (Rocheleau and Edmunds 1995). The current distribution of power clearly favors men with respect to land ownership, control over crops, trees, water, and livestock, and institutional linkages to technical support from national and international agencies.

Throughout the country the work, knowledge, interests, and ideas of both women and men matter for resource management, and, conversely, natural resource policy and technologies matter to women's and men's daily lives and their possible futures, both separately and together. The story of the people of Zambrana-Chacuey and their encounter with "sustainable development" within the Forestry Enterprise Project illustrates several ways in which gender affects resource management and conversely, how resource management technologies and policies affect men's and women's lives differently.

The region

Zambrana-Chacuey is a hilly farming and forest region at the margins of the fertile Cibao Valley. It encompasses an area of 250 square kilometers, with elevations ranging from 100 to 600 meters above sea level. The zone was once covered almost entirely in humid and very humid subtropical lowland forests. Currently the region exhibits both the scars of deforestation and the colorful brushstrokes of cocoa forests topped by the bright blossoms of the Amapola tree (*Erythrina poepoegiana*). Throughout the area edible forests flourish alongside pastures, croplands, and riverine forests. Current smallholder crops include coffee, cocoa, citrus, tobacco, cassava, yams, sweet potato, taro, pigeon pea, beans, maize, squash, and a host of other fruits and vegetables.

There are well over 10,000 inhabitants living in Zambrana-Chacuey and population density averages seventy inhabitants per square kilometer.[5] The majority of the people are smallholder farmers engaged in a tenuous mix of subsistence and commercial agricultural and forest production with off-farm labor. Large cattle haciendas held primarily by absentee owners in New York or Santo Domingo, and expanding agribusinesses – from Dole pineapple to citrus companies – compete with smallholders for land. Many smallholder plots are "microfundias" (as opposed to "minifundias") and are no longer divisible into smaller units for the next generation.

The region once constituted a frontier for smallholder farmers in the 1960s, and now, due to land scarcity, it is a "sending area" for migrants to the capital, other cities, and to the new agricultural frontiers. To keep their land and roots in Zambrana people engage in a combination of subsistence and commercial strategies. Many farmers rent land for food or cash crops while others engage in wage labor – on farms, in factories or in service industries. Home-based enterprises (food processing, crafts, and furniture) provide another option. Trading of agricultural and other goods is also gaining in importance. Finally, the Rural Federation, ENDA-Caribe and their joint initiatives are integral components of many people's livelihood systems.

THE FORESTRY ENTERPRISE PROJECT: THE ACTORS, THE STORY, THE STUDY

The Rural Federation of Zambrana-Chacuey

The Federation is an enduring grassroots organization created by small farmers in the region to advocate for their social, political, and economic rights. The current members (approximately 800 people in 500 households) belong to fifty-nine local associations of farmers, women, youth, and, most recently, wood producers. When one considers informal connections (family ties and other social networks) at least 4,000 people benefit directly from the Federation and many more benefit indirectly (Ross 1995; Rocheleau and Ross 1995). The Federation owns land on which it has constructed a meeting place with cooking facilities, a primary school, and a cement block construction workshop. It operates a rural medical clinic staffed by a doctor and several nurses paid by the state, as well as Federation volunteers. In a separate locale it has constructed an agricultural implements workshop, a small furniture factory, and a sawmill operated by ENDA and the Wood Producers' Association (WPA) (Ross 1995).

The regional organization is affiliated with the Confederation Mama Tingó (consisting of seventeen other rural federations). It has thirty years of organizational experience and a highly successful track record of securing land for landless farmers in the region. While land rights continue to be a theme, conflicts between the state and the rural poor in Zambrana-Chacuey increasingly revolve around resource rights.

The foundations of the Federation can be traced back to the mass evictions of peasant farmers from the fertile valleys of the Dominican Republic by commercial agriculture, industry, and the state in the 1940s and 1950s (Rocheleau and Ross 1995). Toward the end of the Trujillo era, and after the dictatorial president's assassination, peasants in Zambrana-Chacuey organized to reclaim their territory.[6] A variety of community organizations, from women's clubs to marketing cooperatives emerged and confronted the state over the right to organize, to make demands on national agencies, and to table grievances against local elites. The first decades of the movement were spent largely underground or in direct conflict with the military and law enforcement officials. The Federation challenged local elites over land claims, terms of employment, and issues of local governance. In 1974 several community-based organizations converged to form the Rural Federation of Zambrana-Chacuey in order to protest the expansion of the Rosario Dominicana gold mine in their region (field work 1992; Lernoux 1982).

The Federation was founded on and still represents a coalition of three distinct currents within the broader rural movement. Liberation theologians focus on human rights, social justice, and class struggles and see the Federation as a catalyst and agent of social change. The cooperative enterprises sector promotes the association of producers and consumers into groups to protect and promote their interests as small commercial farmers and traders in the local and national market place. The traditional church-based membership views the Federation as representing local interests in securing infrastructure, services, and other "basic needs" including land, from the state as well as acting as a buffer against outside encroachment. All three groups agree in principle on the use of civil disobedience and nonviolent protest although they may disagree on the most appropriate strategies and tactics in a given circumstance.

The leadership of the Federation has encompassed a broad spectrum of people with respect to class, gender, ideologies, occupation, and locality. The commonly acknowledged founders and the current leadership include both women and men from all three "wings" of the Federation. Among the early advocates of local human rights one Catholic sister explicitly linked women's rights to human rights and encouraged women to form their own organizations and to table women's specific concerns within the larger movement. Women have played a strong and visible role in the governing council and have also led vigorous popular struggles to secure underutilized state lands for smallholder farmers, to seek compensation for communities displaced by large dams, and to protest contamination of air and water by hazardous wastes from the Rosario Gold Mine.

Most recently women have played a key role in environmental, development, and health projects, often gender-divided in their structure and content. They have worked to integrate these apparently separate sectors of development activity into their communities in a way that is consonant with their own gendered experience of environmental politics and ecological science in everyday life. Women herbalists and midwives thus serve as council members in community governance, as promoters of forestry, agricultural, and agrarian reform and health innovations, and as key figures in community

Plate 10.1 Zoila and Alfonso Brito with family and friends in the *ramada* (porch) built with *Acacia Mangium* timber from their farm, under the new accords between ENDA and the DGF

Source: Dianne Rocheleau

religious life. Women members of housewives' associations and farmers' groups, as well as women married to farmers' association members have played a major role in shaping the landscape as farmers, as gardeners, as nursery managers, as livestock managers and as gatherers and managers of fuelwood, medicinal herbs, and water supplies.

The farmers' associations, predominantly men's organizations, most often focus on marketing and technical assistance for cash crops. Both farmers' and women's associations have established tree nurseries for fruit trees, coffee, and cocoa and some have organized cooperatives to buy and sell agricultural inputs and to maintain shared wells. Many groups secure cropland plots for joint cultivation of food crops for home use, sale and/or group fund-raising. The "housewives" associations also often function as formally constituted, broad-based, mutual support groups between neighbors, providing everything from informal savings-and-loan opportunities and income-generation activities to assistance with child care and health problems.

ENDA-Caribe

ENDA (Environment and Development Alternatives) is an international NGO headquartered in Senegal. It engages in development programs throughout the world, with a branch in the Caribbean (ENDA-Caribe) since 1980 and in the Dominican Republic since 1982. ENDA works with local communities in both rural and urban areas, with an emphasis on indigenous plants, as well as local knowledge, practices, and participation in agriculture, forestry, and health programs.[7]

In Zambrana-Chacuey, ENDA-Caribe's medicinal plant research/action project brought the Federation and ENDA together. They focused primarily on the work and knowledge of a handful of women herbalists connected to the local clinic. This early effort led to a series of Federation–ENDA collaborations over the next decade under the umbrella of a regionally integrated rural development project.

The Forest Enterprise Project

In 1984, ENDA and the Federation initiated a forestry and agriculture project. They combined timber production, agroforestry, agriculture, livestock production, soil conservation, and gardening activities. Each production technology was treated as a somewhat separate enterprise with a distinct "target" group (timber for men, gardens and small livestock for women).

ENDA and Federation researchers established experimental nurseries of timber, fruit, and soil improvement trees, primarily exotic species and tested more than sixty tree species. Research focused increasingly on *Acacia mangium*, the project's Australian "miracle tree" which produced timber for milling within six to eight years. Excitement about the trees, particularly the *Acacia*, spread rapidly.

By 1993, a growing number of Federation households were planting rows and blocks of *Acacia mangium* and other timber and fruit trees in connection with the Forestry Enterprise Project, a timber cash cropping initiative that evolved from the original ENDA–Federation collaboration. There were eighty-seven community nurseries and more than 300 household nurseries for timber and fruit trees. Overall, the project had planted 800,000 timber trees, 40,000 fruit trees and had 250,000 seedlings in nurseries (Valerio 1992). The project drew its strength from the broad base of Federation participants and the role that ENDA played as a "techno-political" broker with powerful national institutions, particularly with the DGF (Rocheleau and Ross 1995). Special accords between ENDA and the DGF allowed the farmers to harvest the trees that they planted and permitted the farmers to transport, process, and market timber derived from *Acacia*. As a result of these accords, many Federation members were able legally to sell logs, small stakes, and poles at the farm gate, within a context where tree harvesting, processing, and sale were otherwise prohibited.

Tied to the widespread adoption of the *Acacia* as a cash crop, ENDA encouraged and assisted members of the Federation to form a Wood Producers' Association (WPA). The organization was constituted as a semi-autonomous affiliate of the Federation. In 1993 ENDA and the Wood Producers' Association constructed a community-based sawmill, the second such institution in the country granted cutting and processing rights by the DGF. Their goal was to make the sawmill a self-sufficient, community-managed enterprise within two years.

The ECOGEN Research Project

When we began our study the Forest Enterprise Project was at a crucial stage. Tens of thousands of *Acacia mangium* trees on farm were ready for

Plate 10.2 Residents of Tres Bocas beneath the 4–6-year-old project trees planted by a
member of the Wood Producers' Association

Source: Dianne Rocheleau

harvest and milling at the new ENDA-WPA sawmill. The same practices,
species and policies were about to be replicated at four other sites in the
country. Furthermore, ENDA planned to turn over the sawmill and other
project activities to the Wood Producers' Association by 1996. There were
many outstanding questions about the appropriate division of labor and
power between the WPA and the Federation.

These questions warranted research on their own merit, in addition to
the policy significance of the project as a major pilot effort at national and
international level. The ENDA project director asked us to document the
extent and nature of the project's effects and the social differentiation of
responses to timber cash cropping. We were uncertain of the gender issues
which might surface, but noted that women were participating in the project
nursery work as well as in agroforestry practices.

Our research team[8] used a multi-method approach to understanding the
region, project, local organizations, people, and issues that mattered most
to the distinct land user groups involved in or affected by agroforestry practices and related project activities. We combined several data collection
activities: attendance at formal meetings; group interviews; focus groups;
household histories, labor calendars and mapping exercises; key informant
interviews; personal life histories; and a formal survey (including sketchmaps
and plant inventories) of a random sample drawn from the adult members
of the Federation.[9] Our research methodology mixed recent innovations in
feminist ethnography (Katz 1993; Katz and Monk 1993; Behar 1993; Moore
1988) and participatory sketchmapping with the strength of quantitative and
formal surveys (Rocheleau *et al.* 1995; Rocheleau 1995).[10]

Most of our interviews combined an introduction of ourselves and our own research objectives with open-ended questions to individuals, families, and groups about their overall history, land use change, their previous and current practices of agroforestry, as well as their participation in development projects. We included questions about the gender division of labor, knowledge, organization, and decision-making in different domains of land use. We explored the effects of family composition, land tenure, age, and source of income on agroforestry practices and project participation. We also sought opinions about species, practices, project function and structure, and the role of ENDA, the Federation, and the wood producers, as well as suggestions for the future.

We visited thirty-one of the fifty-nine organized groups within the Federation, and traveled to sixteen of the thirty-one communities in the initial reconnaissance, including some chosen explicitly for their remote location, marginal status or negative experience with the project. We specifically sought out and arranged later interviews with women heads of household and/or women members of the farmers' associations and their families. ENDA social promoters and Federation leaders identified several people with unique knowledge of Federation or project history, or people representing distinct circumstances. We also continued to identify people in group meetings who seemed to represent groups so far under-represented in our interviews and discussions (e.g. younger families, near-landless people dependent on off-farm work.)

The random sample for the survey (forty-five people, over 6 percent of the adult Federation membership) at the last stage of field work made us keenly aware of the number of younger families living on small residential plots, dependent on off-farm work, trading of local crops in the capital city, or on access to family (parents'), rented, or sharecropped land to produce crops. We also encountered a number of households connected to the Federation through only one person. All of these latter groups had some stake in the tree planting project, but were generally invisible both to ENDA staff, to Federation leaders and, initially, to us.[11]

GENDER, CLASS, AND THE FORESTRY ENTERPRISE PROJECT

Differences influencing receptivity to the *Acacia*

While a history of cooperation and struggle runs deep through the Federation, its membership is neither socially nor economically homogeneous. The same highly skewed patterns of social and ecological stratification that occur in the region and the nation (Vargas-Lundius 1991; Sharpe 1977) are repeated within the Federation, with class and gender pre-eminent among them. What is distinct is the Federation members' ability to recognize and build upon affinities within the groups and to maintain solidarity while engaged in struggles with the state, commercial interests, and local elites. The Forest Enterprise Project presented a challenge to the membership to maintain that solidarity in the midst of an apparently successful project which affected people differently based on class and gender.

While Federation members readily acknowledged the class differences among them, they often articulated differences between men and women in terms of complementarity – of activities, responsibilities and domains of authority (field interviews 1992). While difference does not necessarily imply dominance, the uneven relations of power between men and women have shaped the terms of the gender division of work, resources, responsibilities, and rewards in households, communities, and regional institutions. Many people in the region did not wish to identify these relations as conflictual, and empha-sized the cooperation of women and men within the household and in the Federation. However, many others recounted the ongoing struggles of women in the region to change overall conditions as well as to protect and advance their own interests within their homes, communities, and local institutions.

A combination of gender and class differences accounted for members' *differential receptivity* to the *Acacia* as a timber cash crop, as well as for the *distinct outcomes* of commercial forestry at individual and household level. Three categories of gender and class differences most clearly influenced the adoption and effects of timber cash cropping: land tenure, labor and liveli-hoods, and terms of affiliation with the Federation.

Land tenure: differences between households

Land is not equally distributed throughout the Federation's membership (Figure 10.1). For example, within the Federation, 82 percent of the members live on land holdings of less than 5 hectares. Rather than *minifundias* (small-holdings), their 32 percent of the land area is mostly divided into *microfundias*, plots consisting of less than one hectare. Within the Federation membership there are no *latifundias* (very large holdings), yet 18 percent of its population owns 62 percent of the land. Thus, while the extreme differences in land hold-ing size that exist on the national level are not completely mirrored within the Federation, there are significant differences among the membership.[12]

On average, farmers own 1–1.5 hectares of land which is usually divided among two or more plots. Forty-two percent of Federation members rent or borrow additional land for cash and subsistence cropping from neighboring largeholders who generally are not in the Federation. A sizable percentage of farmers in Zambrana-Chacuey are on state owned or non-titled land; thus, they own not the land but the value of the *mejora*, or "improvements" such as cleared croplands as well as coffee, cocoa, citrus, pastures, fencing, buildings, and more recently, "legal" timber such as *Acacia*.

Prior to the ENDA–DGF accords farmers often invoked a local saying: "To put trees on the land is to put the land in chains." The legal permission to harvest, process, and market the *Acacia* has reversed the role of trees, from tenure liabilities – signs of abandonment – to tenure assets – as indicators of investment. This has encouraged many Federation members, especially large-holders, as well as nonmembers, to establish plantations of *Acacia* and other recognized commercial timber trees[13] on their state and non-titled land. The trees strengthen their land claims and tenure security as well as their income.

The 87 percent rate of timber planting among Federation members sug-gests that landholding size had little effect on the decision to plant *Acacia* and

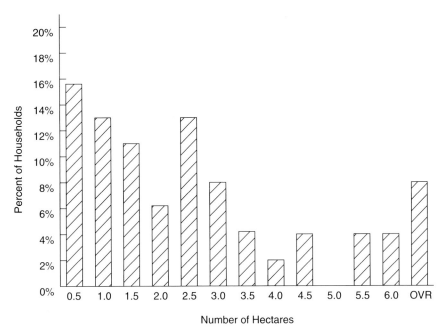

Figure 10.1 Distribution of landholdings

Source: Rocheleau and Ross (1995)

other timber trees. Yet the project was designed for people who could incorporate blocks of 50–1,000 monocropped *Acacia* trees on their own farms. Landholding size limited many smallholder farmers to far fewer trees (2–20). If they had been involved more fully in project planning many near-landless and smallholder farmers said they would have insisted on tree species more compatible with existing land use on very small plots – timber trees that could be intercropped, smaller trees for stickwood and poles, and more fruit trees.

As a result of their exclusion from the mainstream of the project, these households are very vulnerable to loss of plant diversity, especially tree species. The near-landless are faced with trade-offs between their most diverse plots – patio, canada (riverine forest), cropland, and tree crop stands – against timber blocks. In contrast, largeholders can trade off monocropped tobacco or cassava for timber and still retain their forests, gardens, and multiple food crops.

Gender, trees, and tenure: differences within households

Like near-landless families, women face a land tenure barrier when attempting to control the nature and benefits of forestry production. Women voiced an interest in timber trees amenable to intercropping as well as more fruit trees, and trees that produce smaller, more portable, and readily harvested products to sell as needed for quick cash. Yet, comparable to the case of people with very small farms, women have largely been left out of species and technology choices.

Many widows and some divorced or separated women own and manage their own farms, while others with absentee or wage laborer husbands are the

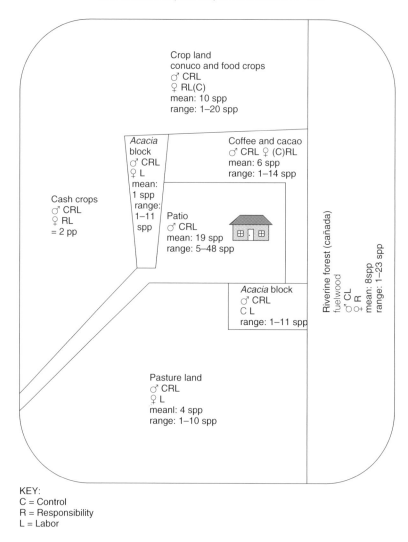

Crop land
conuco and food crops
♂ CRL
♀ RL(C)
mean: 10 spp
range: 1–20 spp

Acacia
block
♂ CRL
♀ L
mean:
1 spp
range:
1–11
spp

Coffee and cacao
♂ CRL ♀ (C)RL
mean: 6 spp
range: 1–14 spp

Cash crops
♂ CRL
♀ RL
= 2 pp

Patio
♂ CRL
mean: 19 spp
range: 5–48 spp

Acacia block
♂ CRL
C L
range: 1–11 spp

Riverine forest (cañada)
fuelwood
♂ CL
♀ R
mean: 8spp
range: 1–23 spp

Pasture land
♂ CRL
♀ L
meanl: 4 spp
range: 1–10 spp

KEY:
C = Control
R = Responsibility
L = Labor

Figure 10.2 Sketchmap of crop and tree biodiversity in gendered domains on the farm
Source: Rocheleau *et al.* (1995) and Ross (1995)

de facto farm managers (a total of 20 percent according to a Federation survey in 1991). Yet most of the women in the region (approximately 95 percent according to several surveys and 100 percent in our own sample) live in households legally headed by men and on land owned by husbands or male relatives.[14] The situation of women with resident husbands was strongly influenced by intrahousehold distribution of control over land and resources.

Many women noted that they could not plant *Acacia* and other timber trees on household lands for lack of clear tenure rights. Some of them had already tried and encountered direct opposition from their husbands. One young woman who was a member of the Women's Association in her community planted several Acacias on the patio, near the house. Her husband, who was not a Federation member, cut them down with his machete. She

Plate 10.3 Leonida's family in their patio – a forest garden with project trees interspersed

Source: Dianne Rocheleau

expressed a clear sense of injustice over her inability to control even the patio, which was commonly accepted as the woman's domain on the farm. While women control this space, they do so without legal rights.

In other cases the women had successfully established up to twenty trees on the patio or perhaps had even negotiated permission to plant part of the plot boundaries to timber. However, many women whose husbands were not Federation members or not actively involved in the tree project voiced an interest in planting blocks of timber, and were frustrated at not being able to control more than the patio and perhaps a piece of the property line.

Conversely, there were women whose husbands *have* planted *Acacia*, perhaps without consulting them, at the expense of women's land or plants. In some cases women's vegetable gardens had been replaced by blocks of *Acacia*. In other cases the multistory, diverse stands of fruit, timber and cash crops on the patio or *conucos* (diverse food crop plots) were being overtaken by the very aggressive *Acacia*. These women expressed skepticism about the timber production enterprise: "we have seen this before . . . peanuts, tobacco, now this *Acacia*, they all take over our croplands and reduce the food which we can grow for ourselves." Several women stated that they would far prefer and might not resist a tree that was more amenable to intercropping (Rocheleau and Ross 1995).

Overall, women's inability to control the *Acacia* (whether they wish to cultivate it or to exclude it) has rendered their authority over all farm land

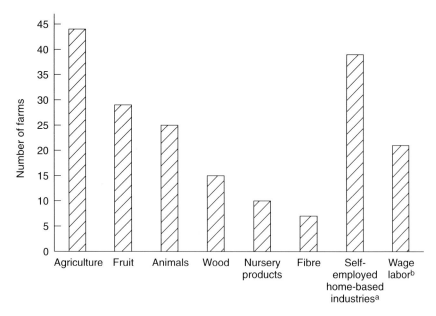

Figure 10.3 Household income sources

Source: Rocheleau and Ross (1995)

Notes: a Examples from this category are local business, sales and trading, carpentry, artisans, food processing, and other services

b Examples from this category are work on citrus plantations, caretakers for large landholders, and casual labor on other farms

– including the patio – increasingly vulnerable. Their inability to influence decisions about land use in regard to the *Acacia* has implications for species diversity on the farm as well. The patio is the area on the farm with the highest rates of species richness (Figure 10.2). This is highly significant for women's plants, for their participation in timber cash cropping and for the future of forest and garden ecosystems in the region.

Livelihood strategies: household differences

In response to unreliable markets, declining yields of cash and staple crops, and new employment opportunities, people in Zambrana-Chacuey have developed diverse livelihood strategies to produce subsistence goods and earn cash. While households vary substantially in the exact combination of income sources, production activities, and occupational specialization, agriculture plays a key role in almost all of the households, as does wage labor and/or trade (Figure 10.3). The nature of these strategies affects the interest and the capacity of households to engage in timber cash cropping or alternative forms of forest and agricultural production and influences the choice of species, planting arrangements and land use practices.

The major cash crops in the zone are coffee, cocoa, tobacco, cassava, citrus fruits, and, increasingly, timber. The majority of families relied in part on the sale of cash crops for regular income as well as on small livestock

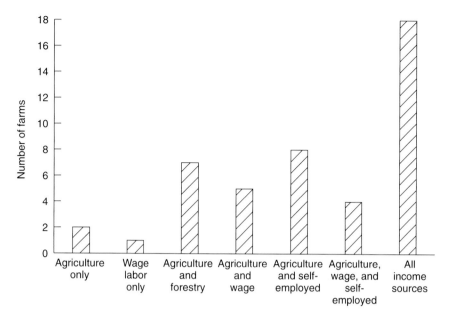

Figure 10.4 Combinations of household income sources

Source: Rocheleau *et al.* (1995) and Ross (1995)

sales (chickens and pigs) in times of cash shortages or emergencies. Yet only 5 percent of surveyed households relied solely on agriculture for their income and another 5 percent did not receive any income from agricultural products. The overwhelming majority relied on a combination of regular income from wage labor, home-based production enterprises, and on the produce and income from their crops and livestock.

People earned cash through nonagricultural means by producing goods for sale in home industries, including food and crafts. Even prior to the forestry project, many households engaged in tree product enterprises such as charcoal, woodworking, and tree nurseries. One third of the households in the Federation had at least one adult involved in trade (*compra-venta*). In most households some adults also engaged in wage labor – both on and off farm. The majority of households exhibited a high degree of economic diversification (Figure 10.4), yet there is a widespread sense of economic marginalization among farmers and rural residents.

Household livelihood strategies, especially the extent of off-farm employment, trading, and forest products industries, have important and convergent implications for the expansion of smallholder timber cash cropping in the region. Timber constitutes a very promising "nontraditional cash crop" for smallholders in this region for two reasons. Timber is a more lucrative cash crop per unit of land than any other currently available to smallholders and it is low in both labor and capital input requirements. In addition, unlike charcoal production, it is legal and has DGF support.

This new commercial option also offers an alternative to land sales and migration for smallholder households otherwise unable to maintain their holdings. As noted by several farmers, timber cash cropping would allow many farm households to move some members into towns or cities to obtain access to factory employment, secondary education, and health services as desired or required. This would permit the continued involvement of thousands of smallholders in crafting the patchwork of plant communities in the region, as opposed to concentrating land and land use decisions in the hands of largeholders and agribusiness interests (Rocheleau *et al.* 1995). Yet the degree of project specialization in *Acacia* blocks may leave some households locked out of this economic activity, and could result in the sale of some smallholdings to local largeholders – keeping land in local hands but increasing local land concentration.

Gendered division of labor and management within households

The cultural construction of gendered labor among Federation members is flexible and does not preclude the active participation of women in agriculture and forestry and their self-identification as "farmers" (Rocheleau and Ross 1995). What differs between households is the way that tasks are divided, and the degree of women's versus men's authority and control in various activities. Survey results indicated that married women were more likely to be "in charge" of activities in home and patio spaces and to "help out" in other spaces (Rocheleau *et al.* 1995; Ross 1995).

In general, women are in charge of the activities that take place in the patio – including household activities such as cooking and cleaning, all or part of the processing of cash crops such as coffee, cocoa, and tobacco, and the management of small livestock (particularly goats, hogs, and chickens). Women usually supervise the collection of fuelwood and water. Often they plant crops, such as pigeon peas, or have gardens in what are considered "men's" *conucos*. They "help" men prepare crop land for cultivation, take hot food to the fields, and often stay to "help out" with tasks in progress. Most women and girls also harvest agricultural products and tree crops, and some women market cash crops (coffee, cocoa, fruit). Some married women own or manage their own coffee and cocoa stands inherited from their parents. Women heads of household, those who take a special interest in farming or those whose husbands work off-farm often take charge of all agricultural activities and identify themselves as "farmers."

Both women and men reported that women work with all types of trees – from fruit, coffee, and cocoa to timber trees – and participate in all stages of production, including plant propagation nurseries, planting, maintenance, harvesting, processing, and marketing (Rocheleau *et al.* 1995; Ross 1995). *Acacia* (as a monocropped timber) stands out from these other tree crops in that women's labor is largely restricted to nursery and planting tasks, and they are virtually excluded from processing and marketing.

The project's ideology of men's and women's labor in agriculture and forestry differentiated *Acacia* from other trees and discouraged women's participation in this enterprise as managers or full partners. While the activities required to manage the tree are similar to any other tree crop or agricultural cash monocrop, the cultivation of the *Acacia* was treated exclusively as "forestry". The identification of the Forest Enterprise Project as a male activity drew heavily on outsider perceptions of forestry as a professionalized, male domain, rather than on regional and local practices of gendered labor in tree cropping. This has both social and economic consequences in the lives of women and men farmers and in the distribution of plant and tree species in the landscape.

Gendered livelihood strategies could undergo major changes if timber cash cropping in monocropped blocks takes root in Zambrana-Chacuey largely without women's participation. First, women would lose access to and control over whole classes of plants if they are reconfigured into separate monocrop enterprises – some of the plants would disappear from the household production repertoire and some would become cash crops under men's control in increasingly specialized production domains. In other cases women would have less production space in cropland or patio, would lose access to intercropping opportunities or partnership roles with their husbands in commercial croplands (cocoa, coffee, fruit), and would retain less control over management of cash crop finances at the household level.

The ecological impact of women's exclusion from timber cash cropping activities, as in the case of gendered tenure, might include partial or complete replacement of their diverse patio plots and/or cropland with monocrop timber blocks. The exclusion of women as partners in the production process could also thwart changes in the new timber enterprise, which women might otherwise render more diverse or adapt to intercropping.

Class and gender differences in organizational affiliation

In addition to landholdings and livelihood strategies the membership is also differentiated by the strength and structure of household connections to the Federation. Most Federation-affiliated households (71 percent) are linked by two or more memberships, and the number of household affiliations ranges from one to four. The strength of a household's affiliation is reflected in the landscape as evidenced by the planting of timber tree blocks at farm level. Sixty percent of the households sampled had planted *Acacia* blocks by 1993 (compared to 87 percent planting some trees), most of them connected to the Federation by two or more memberships.

The Federation is also a distinctly gendered organization as reflected in the patterns of affiliation of men and women. The gender differences in Federation membership occur both within and between households and between different associations. Household linkages to the Federation are structured by the gender of the members connected as well as their choice of organization (Figure 10.5). As of 1993, women in Federation-affiliated households were mainly members of community Women's Associations (60 percent) and a small percentage were members of the local Farmers' Association (4 percent) and the Wood Producers' Association (4 percent) (Figure 10.5). Nearly a third (32 percent) were not members of any association as individuals. While a similar number of men were nonmembers (38 percent), nearly half of the men (44 percent) were members of both Farmers'

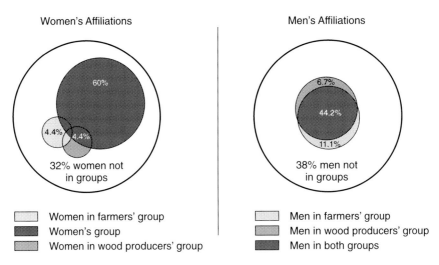

Women's Affiliations

Men's Affiliations

Figure 10.5 Comparisons of men's and women's affiliation with the Federation

Source: Rocheleau *et al.* (1995) and Ross (1995)

Note: Based on survey findings, 1992, 7% sample

Associations and the Wood Producers' Association. A small percentage of the men affiliated with only one of these groups (Farmers, 11 percent and Wood Producers, 7 percent) (Rocheleau and Ross 1995; Ross 1995).

Men's and women's ability to control the *Acacia* and gain access to the Forestry Enterprise Project differed substantially due to the selective affiliation of the project with the predominantly male Farmers' Association, and later the Wood Producers' Association. The patterns of household and project connection to the Federation determined in large part what knowledge entered the household, who controlled it and who used it, as well as whose interests were represented in various activities. Over 20 percent of all Federation-affiliated households were linked solely by women, primarily through the Women's Associations (Figure 10.6) putting them at a distinct disadvantage for access to timber producer services channeled through the Farmers' Associations and the Wood Producers' Associations. The women's groups received commercially marginal "auxiliary projects" such as small livestock production and household vegetable gardens instead of the timber production project.

The existence of women's groups has allowed women to organize and act independently on their own agendas, which in some cases provides a vehicle for women leaders or members to promote the Forestry Enterprise Project among women's groups. Yet, even in such cases the project's response to the women's groups' interest in and demand for services varied markedly among the project staff in each community and sub-region. The formal structure of the project services provided no direct linkages to women's associations, unless initiated by field staff.

The individual membership criteria for the Wood Producers' Association biased membership toward men with blocks of land amenable to monocropping, and tended to deter both women and the near-landless. The individual

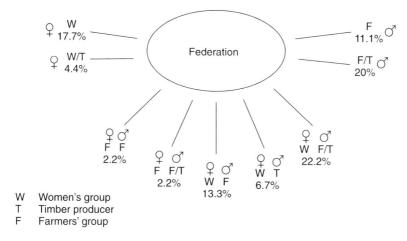

W   Women's group
T   Timber producer
F   Farmers' group

Figure 10.6 Gender structure of household linkages to the Federation

Source: Rocheleau *et al.* (1995) and Ross (1995)

membership fee discouraged separate membership for both spouses in any given household. The minimum entry requirement of fifty trees planted in 0.05 hectare on their own land also restricts the ability of women and near-landless men to qualify for membership.

The lack of institutional linkages did not prevent women from planting the *Acacia* – many women planted it as part of group efforts or as individuals. Rather, these barriers limited the *scale* at which the women could plant timber trees, which in turn affected their eligibility for WPA membership and the degree and quality of support that they received. Near-landless men, as well as those not interested in forestry faced a different problem. The Wood Producers' Association was replacing the commercial and technical support role of the Farmers' Association in some communities, threatening ties to the Federation of those not producing timber.

Federation membership is open to all smallholder farm residents of the region. However, the practical exclusion of most women and the smallest landholders from the Wood Producers' Association contradicted the founding principles of the Federation. In addition, the male orientation of the Forestry Enterprise Project conflicted with local practices in the gender division of tree cropping. These mismatches between local conceptions of subsistence and multi-species forestry and those introduced with the *Acacia* created new divisions within households, communities, and the Federation as a whole.

TRANSFORMING THE FORESTRY ENTERPRISE PROJECT: TECHNOLOGICAL DIVERSIFICATION, TENURE INNOVATIONS, AND ORGANIZATIONAL STRUCTURE

The current trends within the project toward simplification or replacement of plant associations in the regional landscape and the marginalization of women and near-landless households could be counteracted by deliberate interventions of the Federation, ENDA, the DGF, and the people of the region. The future opportunities for women wood producers, near-landless families and more remote communities will hinge in part on the ability of the Wood Producers' Association (WPA) and the Rural Federation (RF) to reconcile their respective mandates of production (WPA), support of farmer/foresters (WPA and RF) and advocacy for the poor (RF) and to cooperate to provide support for the diverse population of existing and potential wood producers in Zambrana-Chacuey. Our collaborative field research with the Federation, ENDA, and the Instituto Superior de Agricultura (ISA) identified several specific courses of action grouped under technology, policy (tenure and credit), and organizational structure.

Technology

The Federation, ENDA, and the Wood Producers' Association could better serve the interests of the near-landless and of women in general by diversifying the repertoire of species, spaces, and products. This strategy would address land constraints and displacement of food crops as well as transportation bottlenecks and farmer dependence on fixed arrangements for timber sale and transport. This would provide farmers with greater flexibility of planting choices as well as harvest and marketing options. There is a clear need for trees that can produce timber, small poles, specialty woods, and fruit and which can be safely intercropped with coffee, cocoa, and food crops in gardens, croplands, patios, and riparian forests. Farmers have identified (and experimented with) several species of exotic and indigenous trees which could serve this purpose[15] (field interviews 1993). For example, observations during field research yielded one clear possibility among the exotic trees amenable to intercropping. Several farmers had planted *Cordia alliodora* as part of early, informal project research efforts. By 1993, some of these trees had far surpassed the *Acacia* in stature and growth rate. Prior experience in Costa Rica has already established the high value of *Cordia* timber and the feasibility of intercropping in multispecies stands of coffee (Beer 1988).

Households and whole communities located far from motorable roads also need tree species that produce nontimber products (small poles, specialty wood, fruit) which can be transported by pack animals to markets or collection points. The participating organizations could better serve the full membership of the Federation by broadening the terms of technical assistance to include processing and marketing of a variety of tree products and to creating portable sawmills in communities unable to transport logs to market. The diversification of products and the use of tree species amenable to intercropping would have a positive effect on women's participation in forestry and agroforestry, as well as their decision-making power in landscape change.

Tenure and credit innovations

Tenure and credit innovations to support farm forestry can also transform the conditions of production for smallholders. While planting *Acacia* for

Plate 10.4 Cristobalina Amparo meeting with the Wood Producers' Association of Las Cuevas. She was the only woman among the 12 forest promoters in 1993

Source: Dianne Rocheleau

timber clearly increases land value, the dilemma of smallholders is that they must choose between food crops and a small block of timber. In addition to intercropping one obvious solution is to lease land for tree farming, which 62 percent of those surveyed are willing to do under group auspices. Legal support would be required to draft secure agreements as to land and tree ownership and management. Credit for land purchases by groups might also facilitate group planting of timber by the near-landless and by women who cannot secure more household land to plant trees. Two Farmers' Associations already have timber plots, and many groups have cultivated shared plots of food crops (such as cassava) and garden vegetables for sale or for home use. If this practice were extended to tree plots for women's groups then many women (regardless of class) could invest in a commercial timber enterprise.

Organizational innovations

Organizational changes could dramatically change the terms of women's and near-landless households' participation in forestry activities, particularly their share in the profits, the products, and the decision-making process. The Federation and the Wood Producers' Association could diversify so as to link their activities to those of the women's groups. Women could contribute their insights and skills in forestry activities as planners, participants, and critics. A formal liaison with the Federation and WPA could provide women and men without adequate land with legal, social, and technical advice as well as financial support for forestry production.

One possible change is to introduce group membership in the Wood Producers' Association for women's groups or other community groups whose members are unable or unwilling to join separately. Family or household memberships could also help to bring in women whose husbands are already members as individuals. Either the Federation or the Wood Producers' Association could also appoint special technical and planning liaison people to meet with women's groups and other Federation groups about forestry activities and to solicit their opinions on decisions pending before the Association.

Even if women are not interested in the timber project, its progress affects their interests in land, land cover, and land use options at household, community, and regional level for decades to come. Beyond women's membership in the Wood Producers' Association, there is a related but distinct need for the women's group to be represented within the Wood Producers' Association. There is clearly a broader need for a Federation-wide or other regional forum for discussion and coordination of such projects across distinct interest groups, including non-participants (whether supportive, neutral, or opposed to the project). A complementary plan of action between ENDA, the Federation, and the Wood Producers' Association could specifically address the forestry concerns of women and the near-landless households and beyond that, could transform the ongoing forestry activities to serve all Federation members.

CONCLUSION

The encounter between the women and men of Zambrana-Chacuey and a regional forestry initiative embodies both the promise and the pitfalls of social forestry programs throughout the world that attempt to combine reforestation with smallholder or community-level production of tree products for home use and profit. By almost any standard measure the first decade of the ongoing forestry project has been a resounding success. Over 85 percent of the more than 500 Federation-affiliated households in the region have planted timber trees on their farms. While some farmers have planted just a few experimental trees, many of them have converted tobacco plots to timber, while others have purchased new plots specifically to plant commercial timber for processing at the new cooperative sawmill.

Many farm families, however, have neither the land nor the cash to buy more land for monocropped timber trees and are unable to participate fully in commercial timber production. Additionally, some men farmers have converted women's multi-species patio gardens to timber trees, while other men have prevented their wives and children from planting trees with the project. Yet other women have received almost no information about the program. "Whose trees, in whose space, under whose control?" have emerged as key questions in this initiative, along with the concern over the widespread promotion of a timber tree monocrop in a landscape characterized by both ecological and economic diversity.

Yet it would be a mistake to treat the success of smallholder timber production as a monolithic threat – to women, the poor, and the regional ecosystem; it is crucial to consider the alternatives, with and without the tree project. The same tree that may replace women's patio gardens, coffee and cocoa stands, and remnant forests, may also protect the land and people from other less desirable options. Timber is currently competing with tobacco, citrus, and pineapple for land in the region and the latter two are controlled by large agribusiness corporations that displace smallholder farmers. Moreover, the timber cash crop option need not be limited to monocropped blocks of one species of timber tree under the exclusive control of men, as discussed in the body of this chapter. The evaluation of the experience to date and the exploration of possible future directions must address the complex, gendered, and class-divided realities of the Zambrana-Chacuey region.

The outcome matters not only for the region itself but has taken on a broader significance. Zambrana-Chacuey has become the model for a social and ecological experiment likely to be replicated at the national level and perhaps internationally. The story of the region and its recent experience with timber cash cropping also provides more general insights into the social and ecological dynamics of land use and land cover change at global, national, and local levels. Women's and men's experience and interpretation of "environment" and their use of "forest" and "trees" as sites, symbols, objects, and tools of political struggle in Zambrana-Chacuey challenge the prevailing theory and practice of "sustainable development" as well as its pre-eminent critics. The history and the stories of the social forestry project provides insight into alternative visions of gendered environmental science and politics in Zambrana and elsewhere.

NOTES

1   Also see the chapter in this volume by Connie Campbell about the women in the rubber tappers' defense of the forest in Xapuri, Acre, Brazil.

2   Edible forests are fruit and tree crops mixed with indigenous trees.

3   See also Rocheleau (1991), Schroeder (1993), and Schroeder and Suryanata (1996) for discussion of similar issues in Africa and Asia.

4   "Nontraditional agricultural exports" refers to any crop not previously exported including local staples such as cassava and other roots.

5   According to the 1981 census there were 10,671 residents in Zambrana-Chacuey.

6   The Federation is a grassroots organization that was created out of, and remains a part of this peasant movement.

7   ENDA-Caribe's headquarters are in Santo Domingo. We worked closely with their field office in Cotuí. The staff in Cotuí consisted of technical experts and social promoters. Aside from one of the social promoters, who was from Ecuador, and the head of the medicinal plant program, who was from Germany, the staff were all Dominican males.

8   Clark University (ECOGEN Project funded by USAID and the Gender and Environment Project funded by the Ford Foundation), the Instituto Superior Agricola (ISA), ENDA, the Federation, and the Wood Producers' Association all collaborated on this effort. A core team (Dianne Rocheleau and Laurie Ross, Clark; Julio Morrobel, ISA) was joined at various phases of the field work by Memerto Valerio (ENDA Forester/Project Manager), Daniel Zavallos (ENDA social promoter), Ricardo Hernandez (independent researcher), and Cirilo Brito and Cristobalina Amparo (both Federation leaders and wood-producer promoters). Contributions from a host of Federation members and ENDA staff complemented these efforts.

9   The Federation was the original institutional home of the project and represents a large proportion and a wide cross-section of the smallholding farming population in Zambrana-Chacuey.

10 These methods reflect a turn toward imagery and narrative in feminist field research and theory, particularly in poststructural and standpoint approaches. Donna Haraway (1991) encourages feminist scholars to reclaim vision and imagery, to project the multiple perspectives of situated subjects and to engage in an explicitly social project of scientific research on "nature." For more discussion of the theoretical and epistemological context for this methodology see *The Professional Geographer*, November 1995 issue, special section on feminist research methods (Rocheleau 1995; Mattingly and Falconer 1995).

11 While the random sample and formal survey was a good deal of trouble to implement (it required the compilation, from scratch, of a membership list of more than 700 people in fifty-nine groups and it forced us into a rigid and inflexible schedule), it did broaden our understanding of the federation membership, the region, and the effects of the project across a wide range of circumstances. These results were shared and discussed first with the Federation in a formal half-day workshop and later with ENDA in a similar format.

12 Medium or even large landholders within the Federation would still be considered smallholders by regional and national standards.

13 Once the new role of *Acacia* was established, Federation members and ENDA staff sought and eventually obtained tree titles – for home use – for more than twenty other tree species planted by various farmers under the auspices of the forestry project. This may yet open up the possibility for future negotiation about the availability (and economic and tenurial value) of whole forest stands and previously planted or protected trees (Rocheleau and Ross 1995).

14 The Federation conducted its own census of one section of Zambrana-Chacuey in order to justify a request for electricity in the region. Twenty percent of the people named as heads of household on this list were women. Other formal surveys and the formal census of the zone, including ENDA's, state that women-headed households account for 3–6 percent of the population. The Federation's definition of head of household is based on who manages daily work and decisions.

15 Among these are *Cordia alliodora* (Capa Prieto), *Simaruba glauca* (Juan Primero), *Guarea guidonia* (Cabirma), *Didimopanax morotoroni* (Sable), *Colubrina arborescens* (Corazon de Paloma/Cuerno de Buey), *Acrocarpus fraxinifolius* (Cedro Rojo), *Grevillea robusta* (Grevillea/Roble), *Catalpa longissima* (Roble).

There is also widespread interest in several *Citrus* species and *Musa* species (Platano, Guineo, Rulo), as well as *Carica papaya* (Lechosa), *Passiflora edulis* (Chinola), and *Annona muricata* (Guanabana).

REFERENCES

Arizpe, L., Paz, F., and Velazquez, M. (1993) *Cultura y Cambio Global: Percepciones Sociales sobre la Deforestacion en la Selva Lacandona*, Mexico City: Centro Regional de Investigaciones Multidisciplinarias.

Beer, J. (1988) "Litter Production and Nutrient Cycling in Coffee (*Coffea arabica*) or Cacao (*Theobroma cacao*) Plantations with Shade Trees," *Agroforestry Systems* 7, 2: 103–14.

Behar, R. (1993) *Translated Woman: Crossing the Border with Esperanza's Story*, Boston: South End Press.

Betances, E. (1995) *State and Society in the Dominican Republic*, Boulder, Colorado: Westview Press.

Escobar, A. and Alvarez, S. (eds.) (1992) *The Making of Social Movements in Latin America*, Boulder, Colorado: Westview Press.

Flora, C. and Santos, B. (1986) "Women in Farming Systems in Latin America," in J. Nash, H. Safa *et al. Women and Change in Latin America*, New York: Bergin and Garvey.

Guzman, V., Portocarrero, P. and Vargas, V. (eds.) (1991) *Una Nueva Lectura: Genero en el Desarrollo*, Santo Domingo, Dominican Republic: Ediçiones Populares Feministas.

Haraway, D. (1991) *Simians, Cyborgs and Women: The Reinvention of Nature*, New York: Routledge.

Katz, C. (1993) "Growing Girls/Closing Circles: Limits on the Spaces of Knowing in Rural Sudan and U.S. Cities," in C. Katz and J. Monk (eds.) *Full Circles: Geographies of Women over the Life Course*, London: Routledge.

Katz, C. and Monk, J. (eds.) (1993) *Full Circles: Geographies of Women over the Life Course*, London: Routledge.

Katz, E. (1992) *Intra-Household Resource Allocation in the Guatemalan Highlands: The Impact of Non-Traditional Agricultural Exports*, Ph.D. Dissertation, Department of Agricultural Economics, Madison: University of Wisconsin,; Ann Arbor, Michigan: UMI Dissertation Services.

Lernoux, P. (1982) *Cry of the People*, New York: Penguin Books.

Lynch, B. (1994) "State Formation, Public Lands, and Human Rights: The Case of the Dominican Republic and Los Haitisses," unpublished paper prepared for the XVIII International Congress of the Latin American Studies Association, Atlanta, Georgia, March 10–12.

Mattingly, D. and Falconer-Al-Hindi, K. (1995) "Should Women Count? A Context for the Debate," *Professional Geographer* 47, 4: 427–36.

Menchú, R. (1983) *I, Rigoberta Menchú: An Indian Woman in Guatemala*, London: Verso.

Momsen, J. (1993) "Women, Work and the Life Course in the Rural Caribbean," in C. Katz and J. Monk (eds.) *Full Circles: Geographies of Women over the Life Course*, London: Routledge.

Moore H. (1988) *Feminism and Anthropology*, Minneapolis: University of Minnesota Press.

Raynolds, L. (1994) "The Restructuring of Third World Agroexports: Changing Production Relations in the Dominican Republic," in P. McMichael (ed.) *The Global Restructuring of Agro-food Systems*, Ithaca, New York: Cornell University Press.

Rocheleau, D. (1991) "Gender, Ecology and the Science of Survival: Stories and Lessons From Kenya," *Agriculture and Human Values* 8: 156–65.

—— (1995) "Maps, Numbers, Text and Context: Mixing Methods in Feminist Political Ecology," *Professional Geographer* 47, 4: 458–66.

Rocheleau, D. and Edmunds, D. (1995) "Gendered Property and Forests: Women, Men, Trees and Tenure," paper prepared for the GEN-PROP E-Mail Conference on Gender and Property, International Food Policy Research Institute, June–December 1995.

Rocheleau, D. and Ross, L. (1995) "Trees as Tools, Trees as Text: Struggles over Resources in Zambrana-Chacuey, Dominican Republic," *Antipode* 27: 407–28.

Rocheleau, D., Ross, L., Morrobel, J. and Hernandez, R. (1995) *Farming the Forest, Gardening with Trees: Gendered Landscapes and Livelihoods in Zambrana-Chacuey, Dominican Republic*, ECOGEN Working Paper, Worcester, Massachusetts: Clark University.

Ronderos, A. (1992) *Toward an Understanding of Project Impact on Gender Negotiation: Forestry, Community Organization, and Women's Groups in Guanacaste, Costa Rica*, Master's Thesis, Worcester, Massachusetts: Clark University.

Ross, L. (1995) *Overcoming the Barriers that Prevent Local Control over Modern Technology in Zambrana-Chacuey, Dominican Republic*, Master's Thesis, Worcester, Massachusetts: Clark University.

Schroeder, R. (1993) "Shady Practice: Gender and the Political Ecology of Resource Stabilization in Gambian Garden Orchards," *Economic Geography* 69, 4: 349–65.

Schroeder, R. and Suryanata, K. (1996) "Case Studies and Class Power in Agroforestry Systems," in R. Peet and M. Watts (eds.) *Liberation Ecology*, London: Routledge.

Sharpe, K. (1977) *Peasant Politics: Struggle in a Dominican Village*, Baltimore: University of Maryland.

Silva, Paola (1991) "Mujer y Medio Ambiente en America Latina y el Caribe: Los Desafios Hacia a el Ano 2000," in Fundaçion Natura and CEPLEAS (eds.) *Mujer y Medio Ambiente en America Latina y el Caribe*, Quito, Ecuador: Fundaçion Natura – CEPLAES (Centro de Planificaçion y Estudios Sociales).

Townsend, J. (1995) *Women's Voices from the Rainforest*. London: Routledge.

Urban, A.M. and Rojas, M. (1994) *Shifting Boundaries: Gender, Migration and Community Resources in the Foothills of Choluteca, Honduras*, ECOGEN Working Paper, Worcester Massachusetts: Clark University.

Valerio, M. (1992) *Agroforesteria y Conservación de Suelos*, unpublished *AGROSIL* document, Santo Domingo: ENDA-Caribe.

Vargus-Lundius, R. (1991) *Peasants in Distress: Poverty and Unemployment in the Dominican Republic*, Boulder, Colorado: Westview Press.

Veras, J. (1984) "Indice Cronologico de la Legisación sobre Asuntos Forestales en la Republica Dominicana," in B. Santos (ed.) *Foresta: Alternativa de Desarollo*, Santiago: PUCMM.

11

WHERE KITCHEN AND LABORATORY MEET

The "tested food for Silesia" program

Anne C. Bellows

Numerous innovations, challenges, and contradictions are inherent in the work of a local grassroots environmental movement composed largely, but not exclusively, of women in the region of Upper Silesia in Poland. The activism is placed in the broader context of women's political participation under the impact of massive structural changes in central and eastern Europe. The so-called "sulfur triangle," of which Silesia is a part, spans four countries: Poland, the Czech and Slovak Republics, and (the former East) Germany. The devastating pollution suffered there reflects conditions in many other industrialized regions in central and eastern Europe and beyond. Therefore, the unique strategies developed by the activists, as well as the challenges they face, have immediate relevance beyond their own locality.

The Silesian activists have experienced personal loss in their families and communities from the extraordinary rates of cancers, tumors, and respiratory illness in their communities. The founders of the "Tested Food for Silesia" program, all professional engineers and all women, had the technical skills to begin to address what has often been accepted as an intractable pollution problem and even been labeled a necessary by-product of industrial development. Formalizing themselves into the "Gliwice Circle of the Polish Ecological Club" (PEC-G) during the post-1989 upheaval, they have taken advantage of the changes in economic and political life to develop new strategies to minimize or avert one aspect of local environmental health risk: locally grown and contaminated food supplies. Inevitably, by inserting themselves proactively into the midst of an environmental nightmare, the group's "Tested Food for Silesia" project became enmeshed in the structural dynamics of production, accumulation, and confrontations over local, central, and foreign state power. An investigation of the objectives to improve environmental health must consider those changing social forces as well as the resultant new forms of marginalizing women's political participation. Not surprisingly, the group's organizing strategies have reflected a need to and interest in forming coalitions with sympathetic political groups – including women's networks – to advance their cause.

CENTRAL AND EASTERN EUROPEAN WOMEN'S ORGANIZATIONS AROUND ENVIRONMENTAL HEALTH ISSUES

Many women in environmental movements, in central and eastern Europe[1] as elsewhere, have organized to protect their homes and families and the resources to sustain them. Some of their names even reflect this identity, for example, the Ukrainian *Mama 86*, the Czech *Prague Mothers*, and the Slovak *South Bohemian Mothers*.[2] Women within these movements understand the connection between bureaucratic ineptitude or abandonment and the hardships of everyday life in terms of daily experience. Slavenka Drakulic points out that "every mother [in Bulgaria] can point to where communism failed, from the failures of the planned economy and the consequent lack of food, milk . . . " (Drakulic 1991: 18). Recognition of the impact of structural order on local life and health is not a function of taking ideological sides.

The "Tested Food for Silesia" activists address the legacy of polluted local soils and foods from a past regime against a contemporary backdrop of what Drakulic calls "a form of imperialism," the incursion of western food into local stores, impairing the market of traditional local goods and threatening self-sufficiency (Drakulic 1991: 13). Temma Kaplan writes that "[a] government that fails to guarantee women their right to provide for their communities according to the sexual division of labor cannot claim their loyalty" (Kaplan 1982: 560).[3] Much of central and eastern European women's environmental organizing does not contest the social roles of women as mothers and keepers of households. Rather it grounds political activity in this role. This is not to say that all central and eastern European women environmentalists are mothers or form "motherist" groups. However, the prevalence of this orientation does signify a widespread identification with traditional roles. Anna Titkow and Henryka Domanskiego note the double-edged nature of this position:

> A traditional vision of women's positions in society is remarkably popular among Polish women. The question remains however as to whether or not this is only a stereotype implied by the fact that the majority of polled working women claim that they would rather continue to work than to remain at home even if not forced to by their financial situation. Despite this interesting reaction, women continue to accept traditional gender roles, even if only in rhetorical statements and this acceptance will continue to sanction the existing division of gender roles and all of its attendant consequences.
>
> (Titkow and Domanskiego 1995b: 299)

The activism of motherist and other women's environmental initiatives is framed in the context of work experience specific to each country. Political activity fits in and around other commitments and must be considered both in terms of levels and venue of employment and household responsibilities. In central and eastern Europe, 1989 became the threshold year between full employment and a phenomenon unknown since before World War II, unemployment. Women generally, and everyone over forty years of age (especially women), have experienced rejection as "surplus labor" in the new market-oriented central and eastern European states. In Poland, the number of women in paid jobs has fallen steadily from 78 percent in 1985, to 71 percent in 1990, to 57 percent in 1994 (Knothe *et al.* 1995: 16–20). Women in Poland, as elsewhere in central and eastern Europe, are also more vulnerable to long-term unemployment and are (re-)hired half as often as are men.[4] Of course both before and after 1989, traditional roles placed unpaid household work squarely in the hands of women.

Not only the amount of labor, but the type of labor is specific to the region. Shana Penn (1993), Director of the Network of East–West Women, points out that in central and eastern Europe women work extensively in the natural sciences, constituting the majority of medical doctors in a number of countries. This gives them primary access as professionals to statistical and experiential changes in public health. Women are integrated as journalists in the media and can track environmental health stories and the degree to and manner in which they are conveyed to the public. To a greater extent than their western counterparts, central and eastern European women are well represented in the engineering fields, leading to the kind of concentration of technical expertise in the Gliwice "Tested Food for Silesia" project.[5]

At the paid workplace, scientific data corroborate the home-based knowledge of pollution dangers. Many have implicated governmental intransigence in the face of appeals against censoring this information. Because women are the predominant caretakers of the sick[6] and are vulnerable in pregnancy and childbirth to early warning signs of environmental contaminants, Dimitrina Petrova and others believe that governments, both pre- and post-1989 withhold information out of an assumed right to exploit both women and the environment (Petrova 1993).[7] Given their employment in medical, scientific, and technical fields, women activists have access to this incriminating data. They can and have participated in its disclosure, and they are cognizant of the tides of official openness and censoring controlled by government authorities.[8]

Malgorzata Fuszara (forthcoming) states that all Polish women's movement activities, even at the turn of the century, were connected to other movements. Indeed it is often difficult to recognize the ubiquity of women's organizing because it is so often entwined in other struggles, many of them relatively conservative. As mentioned above, many environmental organizers even mobilize as *mothers*, not as *women* per se. Given that the identity of struggle is often linked to occupation, the place and forms of protest in women's environmental organizing may be as much the household as their place of paid employment; and the two places can merge to leverage political advantage against a recalcitrant state.

Ewa Charkiewicz relates a story of a well-respected geographer, Professor Zofia Odechowska from Warsaw, who had developed a research specialty in the area of water quality with links to health and who had worked extensively on the environmental "Mazurian Landscapes" project.[9] Odechowska had also been trying, to no avail, to lobby the government in the early 1980s to address the problem of, among other things, chloride contaminants in the Warsaw drinking water. It was on the basis of her status and appearance as a retired grandmother that Odechowska forced her way past guards to demand clean water of General Jaruzelski on behalf of city children with heightened allergy symptoms *and also* to present scientific evidence of links between local health and environmental quality. As an outcome of this woman's efforts, Jaruzelski demanded a report from the city, and later ordered the development of new wells in response to the findings.[10] The point is not that Warsaw's drinking water is now safe; it is not. Rather, it was the moment that the retired activist professor donned the robes of the grandmother and the hallowed role of *Matka Polka*[11] that she obtained results where others, women and men alike, had been unsuccessful in convincing officials to act.

This special space that mothers demand, or that may in some ways be reserved – even with resistance – for them, is a phenomenon that recurs in various forms and in numerous places in the central and eastern European countries. However, this special "maternalized" access to protest space is actually narrow and serves as much to control activism as to provide a distinctive channel for it. More socially acceptable is the public work that promotes social welfare beginning with one's own family, rather than a movement on behalf of rights, autonomy, or equity. Barbara Einhorn cites a 1989 survey conducted by Renata Siemienska (1991), which "demonstrated that only 42.6 percent of respondents strongly approved of the women's movements, as opposed to over 70 percent who supported ecology, peace, and human rights movements, with 9.3 percent maintaining that there should not be a women's movement at all" (Einhorn 1991: 16–36).[12]

There are a number of cases in central and eastern European literature where motherist groups publicly confronted the central authoritarian state over environmental issues, broadening in the process access to protest space for the larger population. Several authors cite the "Prague Mothers" group that organized in the early 1980s, campaigning in streets – closed to activism since the end of the Prague Spring – against pollution and for women's right to choose motherhood rather than the dual burden and expectation of women's paid and unpaid labor contribution (Siklova 1992; Einhorn 1991; Jancar-Webster 1992). Jirina Siklova notes that members were persecuted for their action on these issues. Svetlana Kupryashkina suggests that in the Ukraine, the Chernobyl disaster gave rise to *all* political movements. Women acted first because they felt the responsibility of their children's health more directly and because men were running the institutions that came under attack.[13] Liliana Maslarova (1993) notes that Bulgaria did not have a history of activism and underground opposition like the western edge of central and eastern Europe, especially Poland. "The first bells went off" in Bulgaria, as in the Ukraine, with the Chernobyl nightmare. Women immediately took to the streets with baby carriages to protest. Maslarova reports that the unusual and radical activity in the streets contributed to a fervor of dissent that led to the use of public space to promote nationalist and ethnic rights issues as well.

However, the integration of paid work and household occupations, and the self and public identification of women activists as mothers and partners, compromises women's entrance into mainstream political and economic life. The right of the mother to protest on behalf of the family in public space does not equate with equal access to that same public space for work and political representation.

Looking at Polish women's role in the pre-1989 opposition movement, Penn points out that the contribution of women's activism was denied to history or trivialized when *Solidarity* achieved its goal of political change. The women, such as Helena Luczywo, editor of *Tygodnik Mazowsze*, who ran the underground presses, were not acknowledged in international reports of Solidarity activities, nor were the regional Solidarity directorships of Barbara Labuda, Danuta Winiarska, and Anka Grupinska (Penn 1994). Joanna Regulska argues that despite women's activism, "their involvement during the opposition years did not translate into an invitation to the negotiating table" (Regulska 1994: 40). Siemienska points out that in the 1989 Round Table talks where the transition from communist rule was negotiated, neither Solidarity nor the communists included a woman representative (Siemienska 1994: 72–3). In Poland's 1990 parliamentary elections, women's represen-tation dropped from 20.2 percent to 13.5 percent (Enloe 1993: 23).[14] Parliamentarian Malgorzata Fuszara has demonstrated how post-1989 polit-ical dialogue at the national parliament, the *Sejm*, alienated women as equal social actors (Fuszara 1993: 244–8).[15] Elzbieta Pakszys and Dorota Mazurczak note that an early effort of the new Solidarity government was to eliminate the government's "Plenipotentiary for Women's Affairs," estab-lished in 1986 by the Department of Labor and Social Policy, and that:

> [i]n March 1992, Anna Popowicz was fired from her position as head of the first cabinet level office on women, youth, and family for questioning discrim-inatory public policies arising from the alliance between Solidarity and the Catholic Church.
>
> (Pakszys and Mazurczak 1994: 147)

In Poland, women's lack of political representation as well as the govern-ment's close collaboration with the Church has restricted women's control over their reproductive lives, most notably through a ban on abortion access, but also through the loss of health and childcare services. This interferes not only with women's access to employment, but also with the ability to control their physical interaction with a polluted environment. Krystyna Slodczyk (1994) documents the relationship of higher mortality rates and environmentally degraded regions in Poland as well as a higher sensitivity to environmental pollutants among women as compared to men. Referring to the impact of the ecological crisis in Poland, Wanda Nowicka writes, "[i]n Silesia, women having pregnancies where the fetus is deformed are obliged to carry the pregnancy to term. Such cases of deformity happen more often in Silesia than in other parts of Poland as a result of serious pollution" (Nowicka 1994: 152–3).

In summary, women in central and eastern Europe have organized polit-ically against environmental pollution at least in part when their traditional rights and responsibilities to protect their families and communities are impaired and their families suffer. In Poland as elsewhere, a combination of unpaid home-based and paid employment-based knowledges molds this moti-vation into distinct expressions of activism. Women have unique access to public political space as mothers and guardians of family welfare, for as long as they are perceived as victims, and as activists without salaries or political backing. As their radical projects become mainstreamed, however, the right of access is reclassified as a presumption, making it difficult for women to maintain their leadership within the new stable institutions they helped form.

HEALTH PROBLEMS AND THE PHYSICAL ENVIRONMENT IN UPPER SILESIA

The *voivod* of Katowice, an administrative unit about the size of several counties in the United States, is the heart of Upper Silesia.[16] While it represents only 2 percent of Poland's land area, 18 percent of Poland's total industrial production occurs in Upper Silesia. Of Upper Silesia's 3,000 factories, 300 are considered environmental hazards. Within this 2 percent of Poland, the population experiences 30 percent of the national dust emissions; 40 percent of national non-dust air pollution; and 60 percent of the total national waste disposal. Upper Silesia has the highest population density in the country (900 persons per square kilometer). Ten percent of the Polish population live in this region, which continues to attract more people "looking for work, better pay, and a higher standard of living" (Sokolowska and Migurska 1993: 6).

In 1983, Poland began to classify its environmentally degraded regions. According to the "Polish Institute of Tuberculosis and Lung Diseases," the death rate in all of the twenty-seven designated degraded areas is 21.6 percent higher than in non-degraded areas. The *voivod* of Katowice is the most degraded area. It has three times the total pollution indicators of the next worst-rated *voivod*, Legnica (*Wiadomosci Gliwickie* 1993).[17]

Of this densely populated, highly industrialized, and immensely polluted land of Upper Silesia, over half is used for agricultural production. Forty percent of all locally consumed vegetables come from these soils despite warnings of related health risks.[18]

Food products are one form of human ingestion of locally based pollution. Along with toxic levels of farm inputs used in agricultural production, high concentrations of lead, cadmium, zinc, and other metals lodged in the soils and in the air constitute the most critical health risks in locally grown foods. The World Health Organization and Polish experiments show that 60–80 percent of all metal toxins get into the human body through food. These toxins affect the nervous, digestive, and circulatory systems, particularly threatening the health of young children (*Wiadomosci Gliwickie* 1993; Sokolowska and Migurska 1993).[19]

In other countries and under different circumstances, capitalism has just as easily cornered the blame. Many writers have been quick to blame the former communists for Poland's severe environmental problems. For example, Peterson describes the legacy of Soviet designs of industrial development: "[b]ecause the state was the ultimate property owner in the system, it assumed liability for environmental mishaps and thereby encouraged high-risk and hazardous development" (Peterson 1993: 15). As Lebowitz states, "capitalism . . . appropriates the natural conditions of production without regard for their requirements for reproduction" (Lebowitz 1992: 92).

In reality, both socialist and market-driven economies are built on the logic of "scientific development" principles that champion exploitation and change, not resource or environmental stability.[20] Governments survive best in strong economies and have every reason to support resource exploitation. They will, therefore, rationalize the value of national economic growth over local ecological and human health tragedies. And although central states, both socialist and capitalist alike, might want to decentralize financial responsibilities (Lake and Regulska 1990; Regulska 1993a and b), they have little incentive to develop local institutions critical of production processes. This suggests that those who shoulder the burden of production's toxic by-products have few allies among those who do not directly experience the pollution. And among themselves, they will find resistance to criticism, especially if they benefit from employment in some of the polluting industries.

Local populations in Upper and Lower Silesia are aware of the health risks that have resulted in a death rate significantly higher than the rest of the country. A 1993 local Gliwice report stated, "Every 10th child is born with physical defects. More dangerous than the physical defects are the hidden defects that cause genetic code disorders" (*Wiadomosci Gliwickie* 1993). In Wroclaw, Lower Silesia, a 1994 community survey showed that most people believe that locally grown food is unsafe (Bellows and Regulska 1995: 9).[21] At the same time they will consume this food and even reject data presenting the problem.[22] There is a distrust of nonlocal knowledge and solutions that contradicts statements that locally grown foods are perceived as unsafe. This is explained first of all by the need to maintain a level of local self-sufficiency under challenging economic conditions. Most people depend on their extended family gardens for free, and on local farmers for low-cost, produce. Second, after years of censored public data on the environment, the sudden influx of crisis-level information together with the new unemployment and diminishing services arouses apathy or anger, but not necessarily immediate action. Finally, people are wary of marketing claims of "safe" or "healthy" food, particularly from imported sources. They will trust sources from clean areas in other parts of Poland, and even their own produce, before they will trust imports on the international "healthy food" market.

Most people in Upper and Lower Silesia believe that the local pollution problem is beyond their local control. This reflects little confidence in local power. Contradicting their distrust in nonlocal knowledge and solutions, they rely instead on their national state out of a combination of national pride, trust in the national state's domestic oversight and monitoring power, active or passive discouragement of activist initiatives, and the simple desire to shift the burden for responsible action. Given the commitment of states to economic development, passive trust in any state is undoubtedly misplaced. In the Silesia region, activism is the exception to the rule, rare and special in its potential to demonstrate to others the possibility of achieving local power and creating change.

MOTIVATION AND STRATEGIES FOR ORGANIZING: PERSONAL, POLITICAL, AND SPATIAL

A handful of local mid-career women, all chemical and/or civil engineers and activists in Gliwice, with the technical expertise to carry out their vision, initiated the original "Tested Food for Silesia" project in 1989. Their motivation was fueled by concern for the failing health in their communities – especially that of the children – and grew from personal experience of loss in their families and impairment of their own health. The transition from concern to practical action resulted in part from identifying a problem and developing local strategies to solve it. Some of the activists were employed at the public "Regional Agricultural Research and Chemical Testing Station" in Gliwice and, therefore, knew of existing data incriminating locally grown garden and farm produce. With the support of the station, they also had access to the technical facilities necessary to continue and to refine this research.

The group grew slowly in the beginning. The homogeneity of the engineer mothers with grown and growing children reflected common interests and friendships. The group was never closed to men but initially attracted only one, a retired engineer. He had participated in the Polish conservation movement in the first half of the twentieth century and was gratified to find another group to work with.[23]

The group had few resources beyond their dedication and expertise. They held meetings in homes and kept their files in washbaskets. For a short time in 1993, the group had a one-room office, albeit unheated. This was critical to house a newly acquired possession, the photocopier. The decision to buy the photocopier was much debated against the purchase of an overhead projector, both machines critical to the work of educational outreach.[24]

To address local food safety as one aspect of environmental management, the activists built a complex base of support through collaboration with many groups. This work has effectively reorganized local political alliances, expanded them spatially, and as a result earned the group local credibility and authority. After 1989, the activists readily explored new kinds of possibilities for local engagement and related mechanisms such as lobbying, publication, and markets to influence public affairs. The activists first pursued formal legal identity by aligning themselves with the national environmental NGO, the Polish Ecological Club, becoming the local PEC-Gliwice Circle. In this capacity they lobbied the Gliwice City Council for funds to support a vegetable-testing project in December 1991 (Heler 1992).[25] In June 1992, the Council passed Act # 214 that commissioned the PEC-Gliwice and the Regional Chemical and Agricultural Station to cooperate on the analysis of soil samples from 45 Gliwice municipal gardens as well as to identify safe food sources to test and import into Gliwice (Jagielski and Sokolowska 1992).[26] The sum of 105 million zloty, about US$9,000 in 1992, was appropriated for the project. The PEC-Gliwice also lobbied, though unsuccessfully, for money to subsidize nursery and kindergarten purchases of the more expensive organic vegetables.

By formalizing themselves into an NGO structure in 1991, the group broke new political ground in Gliwice and contributed to a national effort to establish a legal private nonprofit sector in society. The group also established a relationship with a local municipality to research, publish, and educate the populace about the risks of industrial and agricultural mismanagement. This too was a profound and even radical step, considering previous official censorship on these issues. Although the national leadership had changed, the government was still centralized and most of the pre-1989 local–national political links were still well entrenched. The management of

Plate 11.1 Staff and volunteers with map of more/less polluted agricultural
and garden areas

Source: Anne C. Bellows

most industrial centers has continued in the hands of former communists
(known as *nomenclatura*). Relinquishing data they could formerly suppress was
naturally challenging for them, given their past and the unknown and somewhat threatening future. Inquiries into the environmental records of local
industrial production are still not welcome.

Undeterred, the PEC-Gliwice has used the newly legal private market
sector as a strategy to develop a producer-to-consumer system for the introduction and distribution of tested foods in polluted urban centers. They
initiated and are expanding a network of certified organic farmers[27] in *voivods*
north and northeast of Gliwice, of wholesale food distributors, and of retail
stores in Gliwice and elsewhere in Upper Silesia. Arrangements for direct
distribution from wholesalers to hospitals, nurseries and kindergartens have
been accomplished in some local *gminy*. These institutional consumers usually
require some kind of government subsidy that PEC-Gliwice members help
negotiate.[28] PEC-Gliwice members have provided the technical services of
vegetable testing for heavy metals, pesticides, and nitrates on a regular and
"spot check" basis at farms, wholesaler points, and in retail stores. Testing
has accomplished two objectives. It builds data for trend analyses of regional
pollutant levels. It also serves to monitor the integrity of participant links in
the distribution system. Exchanging tested for non-tested produce before the
goods reach consumers presents a temptation because tested foods cost more.
Retailers (and others) violating the program are threatened with dismissal
and in some cases participating stores have been dropped from the program.[29]

Community education has provided the basic foundation for the "Tested
Food for Silesia" project. PEC-Gliwice members have spoken locally with

Plate 11.2 Staff and one daughter and a Peace Corps volunteer in main office with processed organic food exhibited behind them

Source: Anne C. Bellows

parent and community groups, media, schoolchildren, and administrators of schools, hospitals, and local vocational training schools since they first organized. They have also developed materials for dissemination through these groups. Members have attended national and international conferences to share information with groups interested in their work. In recognition of these efforts, the PEC-Gliwice received a grant from the "Katowice Voivod Environmental Protection Fund" in September 1994 to support and expand their activities.

In December 1993, the PEC-Gliwice initiated the idea of a Federation of Polluted Silesian Towns[30] to promote education about locally grown food risks and to expand their market distribution system for safe and organic foods to the regional scale. The nascent organization serves a number of objectives. It expands knowledge about food risks and safer alternatives to a larger area and it brings together the local governments, retailers, health and education service providers, and consumers to address a severe local environmental health problem. Regional cooperation has helped to encourage local governments to prioritize food safety issues. It also encourages Polish farmers in safer regions, who want to convert to organic methods, by demonstrating the existence of a regional market. The overall objective is to increase the supply and variety of safer foods.

The group knows that local inhabitants will continue to grow and consume food from contaminated soils. The present declining standard of living compels inhabitants to garden to supplement their incomes, a tradition that historically provided a survival buffer against the ineptitude of communist food dispersal mechanisms and the uneven distribution of community wealth and poverty under capitalism. Yet soil remediation costs are high and its effectiveness questionable in an environment still subject to severe pollution.

PEC-Gliwice activists, therefore, provide education on risk minimization to family gardeners who still plant and harvest locally. For example, the metal absorption ratio in plant parts is [fruits and seeds]:[leaves and roots] = 1:10, showing that fruits and seeds are ten times safer to grow and consume than leaves and roots. Celery, parsley, leek, lettuce, spinach, carrots, beets, and radishes are discouraged. Legumes, gourds, onions, garlic, tomato, as well as fruit trees and shrubs offer lower risks (Polish Ecological Club, Gliwice Circle, Handout 1993). Local farmers are encouraged to consider non-edible plant harvests and those needing to produce for profit are encouraged to switch cropping to flowers, landscaping products, and industrial-use products. However, the return on such crops is lower and the markets less well understood and trusted. Blaming farmers without providing reasonable alternatives is pointless as their poor economic status inhibits taking financial and production risks.

The diligence of the group and their broad base of community and regional support, as well as their international reputation, have increased their local profile and their membership has grown exponentially. The city of Gliwice has promised to make available a building to the PEC-Gliwice and to contribute to its renovation.[31] Nonlocal funders also identified the innovative project as an organizing model. Between 1994 and 1995, paid staff on the "Tested Food for Silesia" project expanded to six and the number of volunteers to twenty; the number of participating retail shops has increased from seven to twenty-five. Maria Staniszewska, President of the Gliwice Circle of the Polish Ecological Club, described twenty individuals being trained to help expand the project in Upper Silesia:

> They include three students, nine women who are 30–40 years of age, and eight participants in the 50–60 year age range. There are only four men. Most of them [all 20] are graduates from the Gliwice Polytechnic School. Their professions are various, and include for example chemical and environmental engineers, students, economists and housewives. I want the participants to learn how to promote and market our tested food project.[32]

The "Tested Food for Silesia" group did not think of themselves as a "women and environment" group when they began, nor do they now, but they wanted to both spread word about themselves and participate in a variety of supportive networks. Their first outreach activity to a women's network, to write a description about themselves for the *Women and Environment* newsletter, was actually suggested from inside the national Polish Ecological Club organization.[33] Subsequently, they have participated in a number of regional and international women and environment conferences and consultations.[34]

The sense of taking the first steps, of starting somewhere, underscores the PEC-Gliwice philosophy. Fully acknowledging that the program's foremost priorities are food safety and not anti-pollution, chemical engineer Janina Sokolowska states:

> Taking away chemicals from agricultural production is a gradual process, which may take 10 years and even more. Everyone is aware of that. One must make the first step. We are mostly concerned about the children who

live in Gliwice and in Silesia. Children absorb [up to] 80% of heavy metals from the foods they eat. Availability of healthy foods is crucial in the early stages of children's development.

(Lodlin 1992)

This underscores the primacy of preventive health strategies in the short term. Local pollution reduction requires more time, labor, and financial resources than the PEC-Gliwice can afford alone, particularly given their short-term priorities.

CHALLENGES AND CONTRADICTIONS

The effectiveness of the "Tested Food for Silesia" program challenges many social assumptions. Against the common understanding that the environmental and human health tragedy in Upper Silesia is immutable because of the costs of clean-up and the imperative for industrial production, the PEC-Gliwice has taken the initiative and gained allies to work for short-term and long-term solutions. This challenges the conviction that local action is a waste of time.

This work has been accomplished during a period of social upheaval and political change. The PEC-Gliwice group has grasped the implications of these changes and evaluated them critically. They have taken advantage of the market system concept by making it easier for farmers to risk introducing organic methods. For example, the PEC-Gliwice group will contract harvest sales with organic farmers in safe areas in Poland at the beginning of the growing season. The certainty of future sales not only provides relief, but improves farmers' chances of applying for commercial credit. The PEC-Gliwice has developed a new commodity, "tested foods," based on local needs and demand instead of a centralized government production schedule that was designed for the entire country. A small select market distribution system of producers, wholesalers, retail shop owners and consumers has been built to facilitate the movement of the new tested foods to the new consumers. Working with shop owners, they have increased consumer choices by making tested vegetables available for purchase, and by presenting shop owners and consumers with information about the tested food. The ability to sell different food products at variable prices encourages the farmer, wholesaler, and grocer to take the risk to grow, distribute, and sell a new product. Selling a variety of food (and other) products at variable prices was always a function of the illegal underground, or "black" market. The difference now is legal sanction.

The PEC-Gliwice also anticipated the dismal limitations of the new market system and its reformed structure and developed strategies to address those failings. Local and regional governments are lobbied for funds to subsidize the research and testing costs to bring down the higher price of tested foods for everyone, and for purchase subsidies for institutions dedicated to the populations most vulnerable to pollution: kindergartens, nursery schools, and hospitals. Local inhabitants who continue to garden receive information on minimizing their risks by planting select food products. The PEC-Gliwice forms networks to share data and strategies in ways that can strengthen efforts to actually address the power of capital – now in private instead of public hands – to reduce industrial pollution at the source. Finally, they have championed pro-environmental candidates, including some from their own ranks, in local government elections.[35]

The public message in Poland is that women, particularly mid-career and older women, are the least powerful in society. The successors to the communists in Poland denied women's participation in the underground and shut women out of the new society. Women are the last to be hired. They have minimal political representation. Control over their reproductive lives has been curtailed making entry into paid and political work more difficult. Yet the public message is contradicted by the exceptional effectiveness and ongoing activity of members of the "Tested Food for Silesia" program.

But what are the drawbacks in the program? Public education about risks from local food production may threaten the existence of land in agricultural use within the Gliwice local center. In the post-1989 period, local governments have been strapped for money at the same time that central governments have wanted to relieve themselves of the burden of public services. Local government autonomy allows flexible municipal management with programs like "Tested Food for Silesia," but autonomy is expensive and connected to fiscal independence. Local governments are under pressure to maximize economic development and augment income. Communal gardens are government property. They can and have been sold since 1989 to yield to the development of residential property and commercial and industrial expansion.[36]

Therefore, activism on behalf of the quality of food production can affect the total production available to Upper and Lower Silesian inhabitants who depend on their garden sources. Similarly, if local farmers are "redlined" and not permitted to produce for sale, the long-term regional self-sufficiency of food production, albeit now toxic, will decline. John Ragland and Seweryn Kukula argue that small farms are similarly making life bearable in the transition to a market economy:

> It is common for single-industry towns [throughout Poland] to have 25–30 percent unemployment. However, few people have become destitute, because they are part-time farmers or have relatives with land who are helping them through this distressing but not devastating period
> (Ragland and Kukula 1995: 193)

Once converted from agricultural to commercial and industrial use, land is almost impossible to reclaim for farming purposes.

Because of their primary and immediate goal to address health concerns, activists direct their efforts toward local government actors and those in their tested foods market distribution system. Stopping polluters and cleaning contaminated land takes more time and money than the members can afford. As a result, the industrial polluters escape the direct scrutiny of the PEC-Gliwice's activities. The PEC-Gliwice's focus on "importing" healthy and safe foods from safe regions in Poland can actually shield local industry from public attention. Local industrial managers might also benefit from the condemnation of local garden and farm land they polluted, by being able to buy it for industrial expansion or residential space for themselves.

In considering possible drawbacks, one must consider what happens to an activist group that is built on the priorities and experiences of women when their reputation wins them new volunteer recruits, some very young, some who are men. Existing evidence suggests that after radical projects are mainstreamed, as in the case of the Polish underground movement of the 1980s, women's representation and leadership diminish. In any one situation, a change of leadership might not seem threatening because the group's goal is environmental health, not women's leadership. However, when seen as part of a trend, and given the public message of women's powerlessness, one can view such an eventuality as an appropriation of the unpaid community labor of women. It would represent their eclipse from public record and memory as soon as their work received public acceptance, a stable institutional structure, and paid management positions. This would bode a tremendous loss of human resources and independent vision for a community and for ongoing social change.

In June 1995, the PEC-Gliwice organized a training program for local government officials from Upper Silesia interested in learning more about the "Tested Food for Silesia" program. Toward the end of the two-day workshop, a male participant offered a comment on the group's activities: "It is very good that women start projects like the 'Tested Food for Silesia' program; they do it very well. However, once established it should be turned over to the work of men." The workshop presenter at the time, Halina Kacprzak, a founder of the "Tested Food for Silesia" project, responded in a soothing and nonplussed way, that "all people are necessary to address pollution problems."[37]

The PEC-Gliwice's work of addressing food safety has taken hold in Upper Silesia and will not disappear. The group has created an environmental management model of great and immediate relevance. The question remains not only whether the model, as it matures, will continue to benefit from its founders' priorities and leadership, but whether the form(s) in which it is replicated elsewhere will reflect those experiences. To maintain the essence of the program, whether in Upper Silesia or as a model for replication, requires inclusion and respect for the knowledge of and experience in: kitchens, cooking, family and community health, and education, as well as laboratory expertise in chemical testing. This necessitates at least three things. First, a consciousness must be developed by women and men of the unique priorities, organizational strategies, and public contributions that can develop out of women's lives. Second, a consciousness must develop among women in activist groups so that they protect themselves and their own interests while they work for their families and communities. Third, the combined spheres of knowledge that gave rise to programs like "Tested Food for Silesia" must be part of the expansion and replication of the model, not only by "women and environment" networks but by all environmental networks. In the cooperative effort necessary to address source polluters as well as human health, women's contributions have always been critical. Women's work deserves this acknowledgement and they must participate in the leadership of mainstreamed programs they once introduced as radical.

NOTES

1 Real differences between countries in central and eastern Europe exceed commonalities and demand caution in a discussion of "regional" phenomena. Here, most of the literature reflects the pre-1989 political cohesion between the countries and certain similarities in political opposition.

2 See, for example, reports on central and east European women's environmental groups in the newsletter *Women and Environment*, funded by the Dutch NGO MilieuKontakt and edited by Ewa Charkiewicz and Judith Bucher. The newsletter published several issues in 1992–3. Unlike their West European counterparts, the majority of central and east European groups that contributed articles about themselves were self-identified motherist groups. Co-founder of the "Tested Food for Silesia" program, Janina Sokolowska, introduced the group's work in the newsletter's third issue (June 1993).

3 Based on early twentieth-century riots in Barcelona, Kaplan's analysis of women's activism transcends time and region.

4 See, for example: Regulska (1994); Titkow (1994); Knothe (1993); Dane GUS (1994).

5 Note that the Technical College of Opole in Silesia introduced a new Faculty of Environmental Engineering in the fall of 1993 after years of lobbying by faculty member Dr. Krystyna Slodczyk. Interestingly, the majority of students in this new department are women. The women students also form the majority of a new activist organization, the Club of Eco-Engineers.

6 "According to Polish law, only women can take some days off in order to take care of sick child [*sic*]" (Kalinowska *et al.* 1995: 35).

7 See confirmation of this also in Hauserova (1993) and Mirovitskaya (1993).

8 Interviews in the summer of 1994 indicated that technical and administrative workers at the national (Warsaw) and local (Silesia) levels as well as scientists at regional Silesian universities have experienced growing government and industrial collusion in censoring pollution and health data. The initial window of openness after 1989 has been closing. Researchers, activists, and local government officials alike find data difficult to obtain, and the media reluctant to publish it. Legal requirements to make such information available are obfuscated by, for example, providing aggregate data for large geographic areas that frustrates the identification of specific pollution sources, plotting toxic migration, and linking these factors to community health statistics.

9 Mazuria is part of the beautiful and almost pristine region of lakes and forest in northeast Poland.

10 Interview with Ewa Charkiewicz, April 1995. Charkiewicz is a founding member of the Wole Byc activists who integrated the need for environmental protection with larger structural political changes. The group began in the early 1980s and continues. She also co-edited *Women and Environment* (see Note 2). In the newsletter's first issue (December 1992), Charkiewicz introduced Odechowska's research and activism. It is also noteworthy that the chloride contaminants in question originated in Silesia, the industrial home of the "Tested Food for Silesia" project.

11 *Matka Polka* is the proverbial "Polish Mother" figure embodying all the virtues of motherhood, caring, loving, protecting, etc.

12 Unfortunately, Einhorn does not specify if the respondents in Siemienska's report are all women, women and men, or indeed, all men. The context suggests they may all have been women.

13 Svetlana Kupryashkina, November 1993. Presentations at Douglass College, New Jersey and personal discussion. She notes that there had been several more minor accidents at Chernobyl before the meltdown. These received limited press coverage and the citizenry accepted public assurances without protest. See also Kupryashkina (1993).

14 Elections in the following central and eastern European countries in 1990 resulted in the percentage of women in parliament dropping: from 29.3 to 8.6 in Czechoslovakia; 26.6 to 7.2 in Hungary; 34.4 to 5.5 in Romania; and from 33.6 to

20.5 in the pre-reunification election of March 1990 in East Germany (Enloe 1993: 23).

15 Fuszara (1993: 250) also notes that whenever parliament gained power, after 1956, in 1989 and 1991, women candidates always had trouble in the elections.

16 A *voivod* is a large administrative unit encompassing numerous *gminy*, or municipalities. Katowice is both a *voivod* and the name of the major city within the *voivod*. Gliwice is a city, a large *gmina*, within the Katowice *voivod*. Silesia is a geographic entity referring to a region without contemporary administrative borders that was formerly under the control of Prussia and Germany. Upper and lower divisions represent higher and lower elevations within the Oder River watershed. With minimal variance, the geographic region of Upper Silesia falls within the borders of the Katowice *voivod*.

17 The twenty-seven polluted areas are

recognized as particularly hazardous for ecological or human health . . . [and] feature loss of resistance, elimination of self-regeneration processes, and degeneration of biosystems as well as enhanced health hazards and incidence of diseases. These areas are concentrated near the biggest and most onerous sources of environmental pollution, chiefly, wastes, dust, gases, and sewage. . . . Highly polluted areas with the highest death rates, located in southwestern parts (particularly the Katowice voivodship), represent the highest threat to human life

(Potrykowska 1993: 255–6).

For a central and eastern European regional perspective, compare World Bank (1993):

Since the mid-1960s *a life expectancy gap* [WB emphasis] has emerged in which the countries of Central and Eastern Europe have fallen behind Western Europe, North America and Japan by approximately 5 years. This gap is primarily attributable to relative increases in mortality from chronic diseases in mid-life. However, the reasons for this pattern of relative increase are not yet clear. The explanation must involve some combination of factors in the socioeconomic and physical environments, behavioral and lifestyle choices such as smoking and diet, and shortcomings in the delivery of health care.

(World Bank 1993: Sec. 2.50, p. 17)

18 In addition to the research and education of the Polish Ecological Club of Gliwice, "Tested Food for Silesia" project, Ragland and Kukula (1995) suggest that 600,000–800,000 hectares (1.7–2.3 percent of Poland's land area) should be taken out of leafy vegetable and root crop (such as cabbage, cauliflower, and potatoes) production.

Interviews (summer and fall, 1991 and summers of 1993 and 1994) with Professor Mieczyslaw Gorny, Warsaw Agricultural University and founder of EKOLAND, an NGO serving organic farmers and agricultural researchers. According to Gorny, no food in Silesia, with a few exceptions in the Czechstochowa *voivod*, should be grown for human consumption.

19 Among the health hazards affecting central and eastern European health are, "exposures to chlorinated hydrocarbons and pesticides, etc. . . . food pollution. . . . Data on the chemical and industrial contamination of food has been analyzed from four countries – Bulgaria, CSFR, Lithuania, and Poland" (World Bank 1993: Sec. 2.52/vii, p. 18).

20 See, for example, Shiva (1989: 88). She calls this "maldevelopment."

21 N = 460; 78 percent of respondents were from Wroclaw, 22 percent from the surrounding area; 70 percent were between 30 and 49 years of age. Over 75 percent of respondents were female.

22 Conversations, summer 1993, with Ewa Prawelska, Deputy Director of the Regional Training Center in Cracow of Foundation in Support of Local Democracy, and Mieczyslaw Gorny, Professor of Nutritional Sciences, Agricultural University of

Warsaw. Both stated that in their public work experience, literature, and public meetings on the subject of environmental health are rejected because the information is so depressing and not accompanied with strategies that residents might use to address the health problems.

23 Group interview, January 1994, with "Tested Food for Silesia" program members Maria Stanisiewska, Kazimierz Byrka, Janina Sokolowska, and Barbara Migurska.

24 Group interview, August 1993, with "Tested Food for Silesia" program members Janina Sokolowska, Halina Kaprzak, and Barbara Migurska, and Tomasz Terlecki from the Krakow office of the Polish Ecological Club.

25 See also the original Gliwice Municipal Act (1992, Act # 214).

26 The soils and leafy plant and tuber produce of both nonlocal Polish farmers marketing through PEC-Gliwice and local grower-producers have been tested regularly since 1992 for levels of heavy metals, including lead, cadmium, and zinc, and also for nitrates, pesticides, and radioactivity. Testing for nickel stopped early on as only negligible traces were found. Radioactivity also appears to present little or no concern at this time in the Katowice *voivod*.

27 "Certified organic" is an identity or "label" available to foods grown on farms passing a site inspection usually based on standards set by the International Federation of Organic Agricultural Movements (IFOAM), though sometimes adjusted according to different regional or national criteria. The produce grown and/or the animals raised on the land must conform to strict limitations and rules regarding feed, what can be applied in the way of green manure, other organic or biodynamic applications, etc. Conversion to organic production can take up to three years before past inputs of pesticides, nitrates, and other poisons have been washed out of the soil. Some land – next to major highways, for example, or in the vicinity of major pollution sources, such as in the Gliwice area – can never become certified due to continual deposition or longevity of heavy metals in the soil. Cadmium, for instance, has a surface soil life of 250 years. In Poland, certification review is handled by EKOLAND, a nonprofit organization established first in 1982 and recognized as a formal NGO since 1989. IFOAM standards are not strict enough for the PEC-Gliwice standards because they rely on soil tests, not produce tests. In the *voivod* of Czechstochowa, for example, there are IFOAM-certified farmers whose produce the PEC-Gliwice rejects. In Upper Silesia much pollution is airborne and quickly absorbed by fast-growing leafy vegetables. Adequate sensitivity to this contamination is not possible according to the guidelines establishing "certified organic."

28 For example, Tarnowski Gorny is a *gmina* within the Katowice *voivod* that cooperates closely with the "Tested Food for Silesia" program. Believing the produce will contribute to a local preventive health policy, the municipality of Tarnowski Gorny has helped local kindergartens purchase PEC-G tested foods by subsidizing purchase costs. Interviews, June 1994, with Teresa Wylag, Regional Director of twenty-six kindergartens and nursery schools in the Tarnowski Gorny area; Mr. Burzka, Mayor of Tarnowski Gorny; and Dr. Kowalczyk, Chief of the Environmental Council of Tarnowski Gorny.

29 Interviews, summers 1994 and 1995, with "Tested Food for Silesia" members, especially President Maria Staniszewska and Janina Sokolowska.

30 By summer 1995, cooperating and interested towns and cities included Katowice, Tychy, Raciborz, Zabrze, Tarnowski Gorny, Bielsko Biala, Chorzow, Bedzin, Bytom, Gostyn, Myslowice, Olkusz, Sosnowiec, and Zwonowice.

31 The building is a historical site protected by the city and managed by the Sanitation Department where the PEC-Gliwice has many allies as a result of their other projects related to recycling. Interviews and site visit, July 1994, with "Tested Food for Silesia" members and Henryk Dylong, Director of the Gliwice Sanitation Department.

32 Correspondence from Maria Staniszewska, President of the PEC-Gliwice, to the Local Democracy in Poland Office of Rutgers, The State University of New Jersey on cooperative project "Environmental Health and Tested Food in Silesia: Cooperation through a Federation of Polluted Silesian Towns," May 1995.

33 Interview, August 1993, with article author Janina Sokolowska (1993).

34 See, for example, Sokolowska (1993); Sokolowska and Migurska (1993); Bellows (1994); Bellows and Regulska (1995); Kalinowska *et al.* (1995).

35 The last local government elections were in June 1994. Barbara Migurska of the PEC-Gliwice also ran for office, though unsuccessfully, for the first time.

36 Interviews, January 1994, with Dr. Krystyna Slodczyk, ecologist and faculty member, Technical College of Opole (Lower Silesia) and Professor Janusz Slodczyk, Dean of Economic Planning, Opole University. They stated that public support of communal gardens has been withdrawn in Opole, most notably close to the city center. The land has been or will be sold and most probably moved out of agricultural production. Certainly it will not be available for communal kitchen gardens and public recreation.

37 Seminar-training program "Ecological and Tested Food for Silesia: Cooperation with Local Municipalities," June 8–9, 1995, Gliwice, Poland. Part of the project described in Note 32.

REFERENCES

Bellows, A. (1994) "Polish Project Links Environment, Health and Food Supply," *Surviving Together: A Quarterly on Grassroots Cooperation in Eurasia* 12, 3: 5.

Bellows, A. and Regulska, J. (1995) "'Setting the Agenda': Environmental Management and Leadership Training for Women in Silesia," New Brunswick, New Jersey: Center for Russian, Central and East European Studies, Rutgers, the State University of New Jersey.

Charkiewicz, E. (1992) "Women and Environment: A View from Poland," *Women and Environment* (Amsterdam: Milieukontakt), No. 1 (December): 2–4.

Dane GUS (1994) *Aktywnosc Sawodowa i Bezrobocie w Polsce, XI*, Warsaw: Main Statistical Office.

Drakulic, S. (1991) *How We Survived Communism and Even Laughed*, New York: Harper Perennial.

Einhorn, B. (1991) "Where Have All the Women Gone? Women and the Women's Movement in East Central Europe," *Feminist Review*, 39 (Winter): 16–36.

Enloe, C. (1993) *Sexual Politics at the End of the Cold War: The Morning After*, Berkeley and Los Angeles: University of California Press.

Fuszara, M. (1993) "Abortion and the Formation of the Public Sphere in Poland," in N. Funk and M. Mueller (eds.) *Gender Politics and Post-Communism: Reflections from Eastern Europe and the Former Soviet Union*, New York and London: Routledge.

—— (forthcoming) "Women's Movements in Poland," in C. Kaplan and J. Scott (eds.), volume of edited papers from the "Transitions, Environments, Translations: The Meanings of Feminism in Contemporary Politics," joint conference of the Institute for Research on Women, Rutgers University and the Institute for Advanced Study, Princeton, April 28–30, 1995, New Brunswick, New Jersey.

Gliwice Municipal Act (1992) "Foundation for Environmental Protection: 'Tested Food for Silesia'," Municipal Act # 214, Gliwice, Poland: City Council.

Hauserova, E. (1993) "Bronchitis, Laryngitis, Otitis . . . and Acute Hysteria," *Yazzyk Magazine* (Prague), No. 2.

Heler, J. (1992) "Attested Carrot," Collection of the Polish Ecological Club, Gliwice Circle, trans. J. Tyrpa.

Hrebenda, Z. (1992) "Safe Product," *Dziennik Zachodni* (Western Daily), 1 (January 2): 3, trans. K. Losos and J. Tyrpa.

Jagielski, J. and Sokolowska, J. (1992) "Map of Polluted Vegetable Gardens," *Nowiny Gliwickie* (News Gliwice) 44 (1811), October 29, trans. J. Tyrpa.

Jancar-Webster, B. (1992) "Technology and the Environment in Eastern Europe," in J. P. Scanlan (ed.) *Technology, Culture, and Development: The Experience of the Soviet Model*, Armonk, New York and London: M. E. Sharpe.

Kalinowska, E., Kozinska-Balyga, A., and Moskalewicz, B. (1995) "Women's Health in Poland," in Polish Committee of NGOs organized for the United Nations

Women's Summit in Beijing, September 4–15, 1995, *The Situation of Women in Poland* (English version).

Kaplan, T. (1982) "Female Consciousness and Collective Action: The Case of Barcelona 1910–1918," *Signs* 7, 3: 545–66.

Knothe, M. A. (1993) "Job Market for Women in Warsaw," in *Women & Business*, 2.

Knothe, M. A., Lisowska, E. and Wlodyka, E. (1995) "Women and the Economy," in Polish Committee of NGOs organized for the United Nations Women's Summit in Beijing, September 4–15, 1995, *The Situation of Women in Poland* (English version).

Kupryashkina, S. (1993) "Women Seek New Roles, Rights in Ukraine," *Surviving Together: A Quarterly on Grassroots Cooperation in Eurasia* 11, 1: 5.

Lake, R. W. and Regulska, J. (1990) "Political Decentralization and Capital Mobility in Planned and Market Societies: Local Autonomy in Poland and the United States," *Policy Studies Journal* 18, 3: 702–20.

Lebowitz, M. A. (1992) "Capitalism: How Many Contradictions," *Capitalism, Nature, Socialism* 3, 3: 92–4.

Lodlin, L. (1992) "Attesting for Health," *Wiadomosci Gliwickie* (Gliwice News), trans. J. Tyrpa.

Maslarova, L. (1993) "Political Action and Environment in East and Central Europe," paper presented at the Association of Women in Development biennial conference, October 21–24, 1993, Washington, D.C.

Mirovitskaya, N. (1993) "Women and Environment in Russia: Two Variables of Reforms," paper presented at the annual meeting of the International Studies Association, March 22–28, 1993, Acapulco, Mexico.

Nowicka, W. (1994) "Two Steps Back: Poland's New Abortion Law," *Journal of Women's History* 5, 3: 151–5.

Pakszys, E. and Mazurczak, D. (1994) "From Totalitarianism to Democracy in Poland: Women's Issues in the Sociopolitical Transition of 1989–1993," *Journal of Women's History* 5, 3: 144–50.

Penn, S. (1993) "Political Action and Environment in East and Central Europe," paper presented at the Association of Women in Development biennial conference, October 21–24, 1993, Washington, D.C.

—— (1994) "The national secret," *Journal of Women's History* 5, 3: 55–69.

Peterson, D. J. (1993) *Troubled Lands: The Legacy of Soviet Environmental Destruction*, Boulder, Colorado: Westview Press.

Petrova, D. (1993) "The Winding Road to Emancipation in Bulgaria," in N. Funk and M. Mueller (eds.) *Gender Politics and Post-Communism: Reflections from Eastern Europe and the Former Soviet Union*, London: Routledge.

Polish Committee of NGOs (1995) *The Situation of Women in Poland* (English version), report to the United Nations Women's Summit in Beijing, September 4–15, 1995.

Potrykowska, A. (1993) "Mortality and Environmental Pollution in Poland," *Research and Exploration* 9, 2: 255–6.

Ragland, J. and Kukula, S. (1995) "Balancing Agriculture with Physical and Economic Environment in Eastern and Central Europe with Special Reference to Poland," in A. S. R. Juo and R. D. Fried (eds.) *Agriculture and Environment: Bridging Food Production and Environmental Protection in Developing Countries*, ASA Special Publication No. 60, Madison, Wisconsin: American Society of Agronomy.

Regulska, J. (1993a) "Local Government Reform in Central and Eastern Europe," in R. J. Bennett (ed.) *Local Government in the New Europe*, London and New York: Belhaven Press.

—— (1993b) "Self-governance or Central Control? Rewriting the Constitutions in Central and Eastern Europe," in A. E. D. Howard (ed.) *Constitution Making in Eastern Europe*, Washington, D.C.: The Woodrow Wilson Center Press.

—— (1994) "Transition to Local Democracy: Do Polish Women Have a Chance?" in M. Rueschmeyer (ed.) *Women in the Politics of Postcommunist Eastern Europe*, Armonk, New York and London: M. E. Sharpe.

Shiva, V. (1989) *Staying Alive: Women, Ecology and Development*, London: Zed Books.

Siemienska, R. (1991) "Women's Issues in the Transitional Period in Poland," mimeo, Warsaw University.

—— (1994) "Women in the Period of Systemic Changes in Poland," *Journal of Women's History* 5, 3: 70–90.

Siklova, J. (1992) Unpublished paper, presented to S. Wolschik, Harvard University, Cambridge, Massachusetts.

Slodczyk, K. (1994) "Environmental Degradation and Mortality in Poland," paper presented at the panel on "Women and Environmental Health in Silesia" at the Women, Politics, and Environmental Action conference, June 1–4, 1994, Moscow.

Sokolowska, J. (1993) "Healthy Food for Silesia," *Women and Environment* (Amsterdam: Milieukontakt), No. 3 (June): 5–6.

Sokolowska, J. and Migurska, B. (1993) "Tested Food for Upper Silesia," in *Women and Environment in Central and Eastern Europe*, Report of the 10th East–West Consultation: Regional Conference October 1993, Lipnik nad Becvou, Czech Republic. Amsterdam: Milieukontakt Oost Europa, East–West Project.

Staniszewska, M. (1995) "Project Report Nr. 1: Environmental Health and Tested Food for Silesia: Cooperation through a Federation of Polluted Silesian Towns," New Brunswick, New Jersey: Local Democracy in Poland, Rutgers, the State University of New Jersey.

Titkow, A. (1994) "Polish Women in Politics: An Introduction to the Status of Women in Poland," in M. Rueschmeyer (ed.) *Women in the Politics of Postcommunist Eastern Europe*, Armonk, New York and London: M. E. Sharpe.

Titkow, A. and Domanskiego, H. (eds.) (1995a) *Co to Znaczy Byc Kobieta w Polsce*, Warsaw: The Polish Academy of Science, Institute of Philosophy and Sociology.

—— (1995b) "Instead of Summary," in A. Titkow and H. Domanskiego (eds.) *Co to Znaczy Byc Kobieta w Polsce*, Warsaw: The Polish Academy of Science, Institute of Philosophy and Sociology, trans. A. Kinecka.

Wiadomosci Gliwickie (Gliwice News) (1993) "Do You Know What You Eat: A Report About Food Pollution in Silesia," January 11, trans. J. Tyrpa.

Women and Environment (1992–3), Newsletter edited by E. Charkiewicz and J. Bucher, Amsterdam: Milieukontakt.

World Bank (1993) *Environmental Action Programme for Central and Eastern Europe*, report prepared for the European Council of Environmental Ministers meeting in Lucerne, April 28–30, 1993, Washington, D.C.: World Bank.

12

"HYSTERICAL HOUSEWIVES" AND OTHER MAD WOMEN

Grassroots environmental organizing in the United States

Joni Seager

Women are the backbone of virtually all environmental organizations in the United States. With few exceptions, women comprise 60–80 percent of the paid members in mainstream environmental membership groups, and a much larger percentage than that in grassroots and animal rights organizations. The "gender gap" in environmentalism also turns up in other measures. For example, many public opinion polls suggest that, almost everywhere in the world, women favor passing more stringent environmental protection laws and spending more money on environmental protection, and they express more alarm over the state of the environment than their male counterparts.[1] It seems evident that women feel connected to environmental issues and that many women see the environment as one of "their" issues. But despite this synergy between women's concerns and environmental issues, women continue to be excluded from the ranks of "ecomanagement." The management and policy-making ranks of most mainstream environmental organizations remain dominated by men, and women are under-represented among the support professionals – lawyers, scientists, and lobbyists – on whom environmental organizations increasingly rely. Most women in the United States who are active in the environmental arena (and there are many thousands of them) are taking on environmental issues in their own backyards, with their own energy and money, and in their own terms as leaders and workers in grassroots, community-based organizations that lie outside the formal structure of green politics and environmental organizations.

FERNALD RESIDENTS FOR SAFETY AND HEALTH

In the mid-1980s, Lisa Crawford was a self-described "average person" living and working in Fernald, a small community in rural Ohio, 20 miles northwest of Cincinnati.[2] She and her family lived in a rented farmhouse across the road from a "Feed Materials Production Center," a facility to which she had not paid much attention, and which she and many others in the community assumed was a factory producing animal feed. Lisa thought this was a great place to raise her son – fresh well water, fresh air, clean living. This bucolic vision changed in January 1985 when Lisa learned that the facility across the road was one of the US government's military nuclear production facilities, run by the Department of Energy (DoE), producing nuclear materials for weapons. It was called a "feed materials" plant because it processed various uranium compounds into uranium metal to "feed" into the other various nuclear weapons complexes across the country. Fernald also served as one of the major storage facilities for nuclear wastes from around the country, including "hot" material from the Manhattan bomb project.

The nuclear weapons complex in the United States, long shrouded in secrecy, consists of fourteen research, production, testing, and assembly facilities spread across twelve states. In the mid-1980s, the production of nuclear materials for all nuclear weapons was concentrated at four sites: Hanford, Washington; the Savannah River plant near Aiken, South Carolina; the Idaho National Engineering Laboratory near Idaho Falls, Idaho; and Fernald, Ohio. Warhead production occurred at six other sites. After a decade of activism and persistent investigation, and with the end of the Cold War, it is now widely known and acknowledged that all of these weapons sites are environmental disaster zones, among the most grossly contaminated in the country: militarized secrecy, combined with a sense of a higher purpose – national security – produced egregious environmental and safety abuses at these nuclear weapons facilities.[3]

What Lisa learned in January of 1985 was not only that she was living across the street from a nuclear facility, but that radioactive discharges – both intentional and accidental – from this plant had grossly contaminated the air, water, and soil in a wide zone surrounding the facility. Her landlord called to say that the well for their farmhouse water supply was contaminated; her husband immediately insisted that the family stop drinking the water. In her own words, Lisa was "really devastated" by this turn of events, and at a loss as to how to respond to the revelations about the plant across the street. She called on a friend, a nurse, who had earlier voiced suspicions about what was going on at the facility. Together, they decided that someone had to look into what was going on at the Fernald plant and they founded a community action group called Fernald Residents for Environmental Safety and Health (FRESH), now heralded as one of the country's most effective community grassroots groups. FRESH was among the first grassroots groups to bring public scrutiny to bear on environmental conditions at military facilities, and although they had no idea that they were about to take on the "military–industrial complex," the fact that they did so paved the way for subsequent investigations at other facilities. The core group of FRESH founders were all women and, although Lisa emphasizes that there are many men involved in FRESH, she reports that the group (especially the more active core) still consists of about 95 percent women.

The first challenge for Lisa and for FRESH was to force the DoE to disclose information about the state of the facility. The DoE was accustomed to operating its nuclear facilities behind a thick veil of secrecy, and resisted intrusions from "outsiders." Over the next few years, through countless community meetings, official hearings and official investigations, FRESH faced little but obstruction, delaying tactics, and obfuscation from the DoE and other government officials – the people who Lisa calls "the blue suits." FRESH never called for a closing of the facility; instead, they doggedly stuck to a list of demands which primarily called for information, truthfulness, cooperation, and accountability from the DoE. Several years later, Lisa says that one of the more painful personal transformations wrought by her involvement in the Fernald fight is that she has become a much less trusting person; as she says, "the process has made us all skeptics and cynics."

FRESH eventually did force full disclosure, and the picture that emerged was one of massive and egregious environmental violations. Over the operating life of the plant, the Fernald facility had released 394,000 pounds of uranium and 14,300 pounds of thorium into the air. Another 167,000 pounds of uranium were discharged into the nearby Great Miami River; Lisa's drinking well, and several others nearby, were contaminated with uranium; contaminated groundwater is migrating off-site toward an important source of drinking water, the Great Miami Aquifer (World Resources Institute 1992; Radioactive Waste Campaign 1988). Six waste pits on site hold 892,000 cubic yards of buried radioactive waste, and two on-site cement silos contain Manhattan Project waste "so hot," Lisa was told, that "you can't even get near it" – which Lisa says makes her wonder how they got it there in the first place. The waste pits and settling ponds at the Fernald site contain more than 12 million pounds of uranium in a "toxic soup" generated in thirty-four years of uranium processing. In terms of volume, Fernald is the third largest radioactive waste dump in the United States, behind Hanford and the Savannah River plant.

In 1989, the DoE stopped production at Fernald and closed the facility; it is now among the dozens of dangerous military facilities around the country slated for "clean-up." Fernald is listed as a "Superfund" site, as are the other three materials production centers. The most recent government estimate is that it will take fifty years and fifty billion dollars to clean up the Fernald site alone; many experts doubt that "cleanup" – full rehabilitation of the site – is really even possible. The role of FRESH has shifted from serving as a pressure group for disclosure to serving as a pressure group to monitor clean-up compliance.

NATIVE AMERICANS FOR A CLEAN ENVIRONMENT

By the late 1980s, concerns about "environmental justice" were transforming American environmentalism. The awareness that minority communities in the United States (as elsewhere) disproportionately bear the burdens of pollution and environmental degradation has reshaped the agenda of activists and policy-makers, expanding the range of what are considered to be environmental issues to include urban, public health, and community quality of life issues, and focusing attention on the "social frame" of the state of the environment.[4] The breakthrough in bringing these "new" environmental issues to the attention of a wider constituency was achieved only as a result of years of unnoticed and unheralded organizing by small grassroots groups fighting for bigger principles through local battles.

Native Americans played a key role in the emergence of the environmental justice movement, but – like other minority groups – organized Native American activism on environmental issues is surprisingly recent. "Native Americans for a Clean Environment" (NACE), one of the most influential national Native American environmental organizations, was started in 1984 by a Cherokee woman, Jessie Deer-in-Water, a "hairdresser living in rural Oklahoma."

In the 1970s, Jessie had been involved in a campaign against the opening of a nuclear power plant, and before that she had been "somewhat active" in Native American issues, but she did not consider herself to be particularly aware of or involved in environmental issues. The catalyst for her environmental activism was a simple question from her husband: as she recalls, one night in the early 1980s, her husband noticed a story in the local newspaper about a pending permit application for a nuclear waste "injection well" process. When he asked Jessie what the injection well permit application meant, she started down a path of environmental investigation that put her on a collision course with the national nuclear and defense industries.

The applicant for the waste injection well permit was Kerr-McGee, a large Oklahoma-based engineering and nuclear management conglomerate that had operated a uranium conversion facility near Vian, Oklahoma, since 1970. "Uranium conversion" is a process which refines raw, mined uranium into usable nuclear fuel. The Kerr-McGee plant, Sequoyah Fuels, located in eastern Oklahoma midway between the small communities of Gore and Vian, was built to supply converted uranium to domestic nuclear power plants, to nuclear plants in seven foreign countries, and to the Department of Energy, which used the fuel to make weapons-grade material for nuclear bombs. Sequoyah Fuels was strategically vital to both the nuclear and defense industries; environmental activists suggest that this position allowed Kerr-McGee to operate outside normal regulatory controls and led to a lengthy record of environmental violations (Bleifuss 1987: 12–13). The uranium conversion process creates high volumes of radioactive and other toxic waste; according to the Nuclear Regulatory Commission, in the early 1980s Kerr-McGee was producing about 7.8 million gallons of toxic sludge a year (Bleifuss 1987: 12). In 1982, Kerr-McGee applied for permission to dispose of this waste by shooting it into underground wells at high pressure.

In search of information about the injection well permit, Jessie called on her friends from the earlier nuclear-power-plant struggle. As she pieced together information on the Kerr-McGee operations, Jessie became more and more alarmed. Like the residents of Fernald, the families in Vian depended on private wells for their drinking water, and the prospect of Kerr-McGee shooting nuclear waste into the ground angered Jessie. That night, she typed up a petition opposing the permit and took it the next day to her workplace, a beauty shop; as she says, "this was a great organizing site because ten or twenty women came through each day" and were a perfect "captive audience." In a town of 1,500, Jessie managed eventually to get 1,100 signatures on the petition. When she sent copies of the petition to the elected representatives and the officials overseeing the permit process, she got back letters stating simply that there was no need for alarm and that Kerr-McGee had assured the officials of the safety of their process. As Jessie said, "then, I knew it was going to be a long fight."

Through the next months, Jessie sat through dozens of public hearings and official presentations on the merits of the Kerr-McGee plan. In her opinion, the Public Health Department was complicit with Kerr-McGee and acted as though its primary role was to facilitate the permit process. After several contentious meetings with Health Department officials, one woman said to Jessie, "If someone had come to me a year ago on this issue I would have thought they were a Communist. . . . I had to see it to believe it that the Health Department is not in fact working in my best interest at all." Jessie wryly notes that, as a Native American, she had the advantage of not having to overcome "any delusions about the government working for our best interest."

The permit hearings revealed new information about Kerr-McGee's oper- ating procedures at the Sequoyah Fuels facility. Nationally, Kerr-McGee had a long record of questionable environmental and safety practices at its nuclear plants (*Akwesasne Notes* 1985: 27). In Vian, questions about the proposed injection wells drew attention to the fact that Kerr-McGee had been routinely dumping wastes into the nearby Arkansas River for years. In the mid-1980s, Kerr-McGee was annually dumping 11,000 pounds of uranium waste into the river (Bleifuss 1987: 13). This was of particular interest to Jessie Deer- in-Water because a 1970 US Supreme Court ruling determined that while, by treaty, the river itself had been ceded to the state, Native Americans retained control of the riverbanks and bed. Jessie saw this ruling as an opening wedge against Kerr-McGee and decided that the best plan of attack against the company was to fight this as a Native land rights issue. She persuaded twelve other members of the Native American community, six men and six women, to join her as founders of Native Americans for a Clean Environment (NACE).[5] Like Lisa Crawford in Fernald, Jessie found that, generally, the most willing workers in this struggle were women. As she says, "It was mostly women who could be persuaded to take a stand on an issue that everyone thought was hopeless!"

The injection well permit was defeated after two years of community pressure spearheaded by NACE – a remarkable victory given the economic and political clout that Kerr-McGee wields in Oklahoma. NACE faced resis- tance and stonewalling on all official fronts, and mysterious incidents – lost mail, misfiled accident reports – dogged their campaign (*Environmental Action* 1990: 23).

By the mid-1980s, anticipating the closure of nuclear facilities in other parts of the country – including the facility at Fernald – the US govern- ment was anxious to get out of the business of uranium processing and was encouraging private industry to take over a bigger share of this work. Kerr- McGee applied to build a second nuclear facility near Vian. Despite renewed opposition to this plan, the facility was approved and opened in 1986. Although Jessie saw this as a setback, she never considered it a full defeat; community pressure resulted in significant safety modifications to the design of the second facility, including provisions for warning sirens to alert the surrounding community of accidents at the plant.

GRASSROOTS ACTIVISM

Although emotions ran high and tempers hot in both Vian and Fernald, neither FRESH nor NACE adopted civil disobedience or confrontation as a primary tactic. Rather, the procedural strategy of both community groups was shaped by organizational persistence. In both Vian and Fernald, small community groups were taking on large and powerful industrial and governmental entities; as Lisa said, "Our job is to drive DoE crazy . . . we want to bug them and irritate them and make them accountable every time they turn around, and make them realize we won't go away." Both NACE and FRESH depended on labor-intensive informational networking; both groups relied heavily on one-on-one contacts to build their constituency. While this quiet organizational strategy was highly effective in both Vian and Fernald, it also meant a high level of personal intrusion. For Jessie and Lisa, even informal, unplanned encounters – casual conversations in the grocery store line, for example – typically also became "business" conversations. Jessie and Lisa became "public property," and the environmental campaign infiltrated every aspect of their lives; "compartmentalization" was not possible, given the small-town setting and the community-level organizational strategy.

The experiences of Lisa Crawford and Jessie Deer-in-Water, while each distinctive and particular to their local situation, illuminate several themes that turn up repeatedly in the stories of women environmental grassroots activists.[6]

Many women who are now environmental leaders were, initially, reluctant activists; most grassroots women report little prior community activism, little environmental knowledge, almost no experience in public speaking or organizing; many describe themselves modestly as "mere housewives." For example, Lisa – now a woman who is a regular guest on national news shows and a seasoned witness at Congressional hearings – cheerfully admits that she had done no public speaking and no advocacy work before founding FRESH: "I was never really a political person, never really read the newspaper, never really watched the news . . . but, whew, you *become* very political very fast." Jessie Deer-in-Water did have a more substantial background in political activism before taking on the Kerr-McGee battle. The limited information available on women's activism suggests that for "minority" women in the United States, Jessie's personal history may be more typical – many women (and men) who are members of a racial minority have a prior commitment to activism on race, class, or social inequity issues that then blends into or leads into environmental activism.[7]

In the great majority of cases, women become involved in environmental issues because of their concern for the health of their family; they organize around "women's" issues of health and safety, family and children. When I directly asked Lisa why she thought it was mostly women who were the core members of FRESH (as in other groups), she did not hesitate before answering: "Mothering . . . women want to protect their children . . . we're trained to think that mothers should save the world." When DoE officials first attempted to calm the community around Fernald with smug reassurances of "don't worry . . . it won't hurt you," Lisa reacted with an immediate distrust, expressed as "don't try to tell that to a mother!" Jessie expresses a similar sentiment, but cast in a broader framework; women, she thinks, have an intuitive feeling that the earth is on a destructive path, and women have expanded their sense of themselves from giving life to saving it. Jessie has an expansive sense of the importance of environmental activism: "If you look anywhere in a city and see a neglected sidewalk, you'll see grasses and even trees trying to come through a little crack. The Earth will heal herself if we just let her, but if we give it a hand that's even better."

In both Fernald and Vian, women's participation not just *changed* the nature of the environmental activism – rather, it was *only* because some women were willing to take a stand that the environmental issue *became* an issue. But in a broader frame, it seems clear that women's participation often changes the nature of environmental discourse by introducing new concerns and ways of expressing them. Their sense of responsibility for saving lives and their concern for immediate family and personal health and safety means that women activists often express distinctive standards of environmental efficacy. For women in the midst of a grassroots environmental battle, the nature of expertise – and of whose voices "count" – is often defined in emotional, personal, and moral terms. Lisa says that her early response to the enormity of the violations at Fernald was, "I'm not a scientist, I don't have a Ph.D., but I say to you that this is not right; morally, ethically, this is wrong." This conviction that morality and emotionalism are legitimate platforms for environmental authority challenges the mainstream wisdom that disinterested, neutral observers are the best environmental experts. A recent letter to the editor of the *New York Times* from a woman in Washington, D.C. underscores the growing schism over what constitutes a legitimate basis for environmental evaluation:

> [T]his new environmentalism does not call for more expert verification of the dangers of toxics or more cost-benefit analysis of environmental protection measures. It calls for pollution prevention, to minimize the risk to ourselves and our children by looking for alternatives to toxic chemicals ... the residents of [a polluted site] don't need scientists to tell them toxic pollution is harmful. They see it every day.
>
> (*New York Times* 1993)

Because women activists are often emotionally connected to environmental issues, they typically see nothing wrong with expressing those emotions. Lisa Crawford suggests that this may be one of the reasons why men are often not in the forefront of community battles – men appear to be less willing to speak out, more afraid of appearing foolish, and less willing to show emotions about issues that are not "supposed to be" personal and emotional, such as the environment. This impression of women's willingness to go out on a limb when men hold back is reinforced by Jessie's observation that in her struggle it was mostly women who were willing to take up what was widely seen to be a "hopeless" cause.

As activists, women foreground issues of prevention, not just remediation, and of accountability, not just amelioration. As part of the Fernald cleanup, the DoE is shipping wastes from the Ohio facility to Nevada and Utah. The FRESH position is that this toxic waste shell game is an unacceptable

solution, and they are now fighting with the DoE to make them take responsibility for the toxic waste in situ – they want the DoE to stabilize, neutralize,
and store or destroy the toxic wastes safely at the Fernald site without creating
a new problem for another, perhaps unsuspecting, community. Lisa is
eloquent about this sense of a larger responsibility:

> There are hundreds of DoE facilities across the country. Ohio has five of
> them, one just as bad as the other, and there are sites many times bigger than
> ours. You have to look at your site as just one little piece that fits in with
> hundreds of other little pieces across the country. They tell us that they're
> trucking dozens of barrels of waste to Nevada every day. I say what good does
> that do? I picture Nevada as being this state of nothing but barrels. I feel
> guilty. All we're doing is taking crud out of my community and shipping it
> off to someone else's community and contaminating someone else's land. And
> I have a real problem with them trucking anything along anyone's highways
> and roads. We have elementary schools along the routes to the major express-
> ways and so do a lot of other communities. They can't move a barrel from
> one side of a building to another without dropping it and, my God, do you
> want them to truck this stuff along the roads?! I don't!

In Oklahoma, Kerr-McGee was perfecting a process to convert nuclear
waste into fertilizer, which it was test-marketing around the country. Referring
to this dubious practice, Jessie echoes Lisa's sentiments: "There's no way
that what happens in Oklahoma does not affect people everywhere else . .
. networking is the only way around this."

Local grassroots organizations are often criticized for their parochial
tendencies – the "not in my backyard" syndrome. However, the sense of a
larger connectedness and of moral obligation to a wider community that
women so often express suggests a counterbalance to the localizing pull of
environmental issues. But for local environmental activists – especially
women's organizations – with limited resources, it is difficult to actually
create and sustain broader networks among the hundreds of small groups
scattered across the country. Because government agencies and large corporate organizations operate at a national (or international) scale, they have
tremendous advantages over the local groups that may oppose them – the
ability to move wastes across the country, for example, and the capacity to
play off one community against another. In the United States, one organization – the Citizen's Clearinghouse for Hazardous Waste (CCHW) – works
specifically to address this imbalance. CCHW serves as a national umbrella
networking organization to facilitate the flow of information among community-based grassroots environmental groups. Perhaps not surprisingly, CCHW
was founded by a woman grassroots activist – Lois Gibbs, the activist who
exposed the environmental toxic waste disaster at Love Canal (New York)
in the late 1970s.[8]

On several occasions, Lisa has been told by angry men to "go home,
little lady, and take care of your family – you shouldn't be out here messing
with things that aren't your business." For Jessie, similar personal attacks
were sometimes combined with racial slurs. Most women activists at some
time or another face great resistance to their activism, often especially intense
from men. For many men, the notion of a woman activist is an oxymoron.
Women activists are stepping outside the bounds of sanctioned female

behavior, and the techniques which men invoke to put women back in their place are often entirely based on sexist "policing" – there can hardly be a woman environmental activist in the world, for example, who has not been called a "hysterical housewife." It is clear that when women walk out of their homes to protest a clear-cutting scheme, toxic waste dump, or highway through their community, their gender and sex identity goes with them – in a way that is not true for male activists.

Lisa Crawford considers herself to be fortunate because she has a "wonderful, supportive" spouse who ungrudgingly assumed a large share of her domestic responsibilities when FRESH started to consume more and more of her time and energy. She says that this is absolutely crucial – without a supportive partner, she says, it would not be possible for women to throw themselves into the maelstrom of environmental activism. And while she does not feel that she faces this problem, Lisa acknowledges that many women she knows do have husbands who are resentful of their activism. For unknown numbers of women, the cost of activism is often the breakup of a marriage.

The experiences of women activists raise several larger issues. The first is that our knowledge of the environmental activism of women may contribute to the larger debate of whether there is a distinctive women's point of view or a "women's voice" on social issues. It would appear that the prominence and distinctiveness of grassroots environmentalism as a women's activity, when combined with the evidence of differences between women's and men's attitudes to environmental issues, does suggest this is the case. At a generalized level, women do seem to express environmental caring and concern differently than do men. Women speak of broader connectedness, of concern for family welfare, of an activism catalyzed by an assumed life-giving (or life-saving) role. If this is the case, if we are witnessing the emergence of a distinctive women's voice on the environment, does this suggest a universal female "ecoconnectedness?"

While this possibility is a source of empowerment for some women, others find that raising the notion of a special "women's voice" is complicated and fraught with contradictions. Many Western feminists are leery of ascribing "special" and inherent characteristics to women (or men) and are uncomfortable with the idea that there may be a "women's perspective" or a "women's activism" on the environment. Given the obvious fact of class, race, income, and status differences among women within and between countries, it is difficult to talk about a distinctive "women's" contribution to environmentalism without falling into simplistic essentializing and universalizing. Furthermore, many women argue, the notion of a distinctive "women's role" has been at the heart of patriarchal oppression of women for centuries, and feminists should be the last ones to embrace such a stance.

Neither Jessie nor Lisa would say that there was an overarching distinctive "woman's position" on the environment, and both emphasize the extent to which their environmental work involved a coalition of women *and* men. However, both activists are clear about the fact that women took on environmental issues when men did not ... that women use arguments of morality and emotionality when men do not ... and that women take up environmental issues out of a sense of "life-giving" or life-saving obligation and outrage.

From this vantage point, the emergence of distinctive women's activism on the environment does not necessarily imply the workings of biological destiny, nor does it necessarily suggest a universalized gender-specific behavior. Rather, if there is a universalism of a "women's voice" on the environment, this may reflect a certain universalism in women's social location. Women's environmental activism occurs within the context of and often as a result of their particular social roles, especially as mothers or caring guardians, and as people generally assigned responsibility for cleaning up other people's messes. These roles in many key ways do transcend boundaries of race, ethnicity, and class. At the international "Global Assembly of Women for a Healthy Planet" in Miami 1991, Peggy Antrobus – one of the founders of "Development Alternatives with Women for a New Era" (DAWN) – outlined the common consciousness that women bring to (and that brings women to) environmental activism:

> We are different women, but women nonetheless. The analysis and the perspectives that we get from women are certainly mediated by, influenced very profoundly by differences of class, and race, and age, and culture, and physical endowment, and geographic location. But my hope and my optimism lies in the commonalities that we all share as women – a consciousness that many of us have, if we allow ourselves to have it, of the exploitation of our time and labor in unremunerated housework, subsistence agriculture and voluntary work. Our commonality lies in the often conflicting demands of our multiple roles as caretakers, as workers, as community organizers. Our commonalities lie in our primary responsibility for taking care of others. Our commonality lies in our concern about relationships; the commonality that we share is the exploitation of our sexuality by men, by the media and by the economy. The commonality that we share is in our vulnerability to violence. Our commonality finally lies in our otherness, in our alienation and exclusion from decision-making at all levels.

Antrobus's remarks highlight the issue of the nature of environmental awareness. The experience of women on the environmental front lines should help us change our notion of what environmental destruction looks like. Women are usually the first to notice – or to anticipate – environmental problems in their communities. Typically, what women "notice" is pretty mundane. Because women, worldwide, still have primary responsibility for feeding, housing, and childcare, they are often the first to notice when the water smells peculiar, when the laundry gets dingier with each wash, when children develop mysterious ailments – or they are the first to worry that these assaults on family safety and health are imminent. This may appear to be a humble entry point for environmental awareness, but in fact it has catalyzed a powerful environmental challenge at the grassroots. Moreover, environmental degradation *is* typically mundane: it occurs in small measures, drop by drop, well by well, tree by tree. Degradation of environmental quality usually does not occur in big, flashy, or global ways.

This might suggest the necessity for a reassessment of the conduct and priorities of environmental research which has, in the United States, always favored "big science." The notion that local people (and, often, especially local women) who are close observers of their local environment may be the most reliable narrators about environmental problems is discomfiting to big scientific and environmental organizations whose prestige depends on solving "big" problems in heroic ways.

In both Vian and Fernald, there was limited environmental expertise initially available to the community groups. Mainstream environmental groups in the United States devote surprisingly little attention to nuclear issues and there is virtually no leadership at all from the big groups on military and defense-related pollution issues; wildlife and natural area protection dominate the agenda of most mainstream environmental groups, and these issues receive virtually all of the staff time and financial resources available. While NACE and FRESH consulted with local and national environmental groups to the greatest extent possible – as Jessie said, "we tried to milk all the help out of them we could" – both groups, ultimately, were virtually on their own and had to assemble their panels of scientific and environmental "experts" piecemeal. Much of the expertise was provided by members of the community groups who educated themselves (very quickly) about matters that were, in many cases, highly technical. In the United States, most grassroots environmental groups form around issues of toxic waste management and disposal, nuclear facilities, or military-facility pollution. These are areas of concern that are low on the list of priorities of most of the large mainstream environmental organizations.[9] This gap in interests and priorities is a key factor contributing to a widening schism between the mainstream and grassroots environmental movements.

A related issue prompted by women's experiences is the nature of environmental knowledge. The value of the environmental work being done by women grassroots activists suggests the importance of expanding our notion of what constitutes appropriate environmental "expertise." Increasingly, science is considered to be the primary tool of mainstream environmental analysis and the arbiter of environmental concerns. Science-based environmentalism, however, has a number of drawbacks, including the fact that "science" takes environmental assessment further and further away from the realm of lived experience. Instead of listening to people with intimate, local knowledge, reliance on "scientific fact" pushes amateurs to the fringes, and undercuts the valuable environmental knowledge that local observers may have accumulated over the years. This describes a process that when exported on a grand scale defines colonialism, a point made by environmentalists in the developing world, most eloquently by Vandana Shiva (1991). Mainstream environmentalism appears to be increasingly in the thrall of science.[10] Nonscientific expertise, and especially expertise grounded in daily experience, is increasingly dismissed and devalued within mainstream environmental circles. Since a higher proportion of environmental "amateurs" are women, and women are still very much a minority in the ranks of scientists, an integral byproduct of the shift toward science-based environmentalism is the increasing marginalization of women.

Many women activists – especially those who have been belittled by male "experts" – are especially leery of the role of science and scientific expertise in arbitrating environmental disputes. Grassroots women activists have their priorities in an order that makes sense to them, and that seems relevant to the local situation, and they resist the notion that they should contort their environmental arguments to fit a framework imposed by outside credentialed experts. Lisa Crawford's comment that, while she did not have a Ph.D., she knew what was right and what was wrong, resonates broadly throughout the women's grassroots community. Many grassroots activists echo similar sentiments. A Canadian activist, for example, recently said much the same thing:

> I didn't know anything about chemical waste then, but I sensed that it wasn't right. It was a gut feeling. . . . I gradually became aware that you didn't have to be an expert. There are very few experts for one thing. If you have 12 scientists looking at a report, for instance, you'll probably have 12 different interpretations; and depending on who the scientists or experts worked for, that determines the policy.
>
> (Scott 1987: 9)

As grassroots environmental watchdogs, women often see it to their advantage that they have been socialized to listen to their "gut feelings." Grassroots activists know how to tap specialized expertise when they need it, but they do not let the presence of "blue suits" on either side of the fence dictate their agenda.

NOTES

1 See, for example, UNEP (1989); Schahn and Holzer (1990); Hill (1993).

2 The information on Lisa Crawford and Jessie Deer-in-Water derives from personal interviews with each, and from the taped record of an environmental conference "Engendering Environmental Thinking," held at the Massachusetts Institute of Technology, Cambridge, Massachusetts, May 20–22, 1992.

3 For information on the network of nuclear weapons facilities, see: Ehrlich and Birks (1990); Shulman (1992); Radioactive Waste Campaign (1988).

4 On environmental justice, see: Russell (1989); Bullard (1990).

5 A brief announcement of NACE's founding was printed in *Akwesasne Notes* 1985: 27.

6 For more descriptions of the work of women activists in the United States, see: Garland (1988); Bullard (1993); Newman (1994); Seager (1993). For good international comparisons, see: Dankelman and Davison (1988); Sontheimer (1991); Caldecott and Leland (1983).

7 This possibility is suggested in a number of recent books on the environmental justice movement. See, for example, Bullard (1993).

8 For more information on Lois Gibbs and Love Canal, see Seager (1993) and Gibbs's autobiography (1982).

9 For a recent survey of priorities of US environmental organizations, see Snow (1992).

10 For a fuller discussion of this, see Seager (1993).

REFERENCES

Akwesasne Notes (1985) Summer: 27.

Bleifuss, J. (1987) "Kerr-McGee Lays Waste to Eastern Oklahoma," *In These Times*, Aug. 19–Sept. 1: 12–13.

Bullard, R. (1990) *Dumping in Dixie: Race, Class and Environmental Quality*, Boulder, Colorado: Westview Press.

—— (ed.) (1993) *Confronting Environmental Racism: Voices from the Grassroots*, Boston: South End Press.

Caldecott, L. and Leland, S. (eds.) (1983) *Reclaim the Earth: Women Speak Out for Life on Earth*, London: The Women's Press.

Dankelman, I. and Davison, J. (1988) *Women and Environment in the Third World*, London: Earthscan.

Ehrlich, A. H. and Birks, J. W. (eds.) (1990) *Hidden Dangers: Environmental Consequences of Preparing for War*, San Francisco: Sierra Club Books.

Environmental Action (1990) "Jessie DeerInWater," Jan./Feb.: 23.

Garland, A. Witte (1988) *Women Activists: Challenging the Abuse of Power*, New York: Feminist Press.

Gibbs, L. (1982) *Love Canal: My Story*, Albany, N.Y.: State University of New York.

Hill, K. (1993) "Gender, Moral Voices and the Making of Environmental Policy: A Case Study in Norway's Ministry of Environment," unpublished paper, Harvard Graduate School of Design, Cambridge, Massachusetts.

Newman, P. (1994) "Killing Legally with Toxic Waste: Women and the Environment in the US," in Vandana Shiva (ed.) *Close to Home: Women Reconnect Ecology, Health and Development Worldwide*, Philadelphia: New Society Publishers.

New York Times (1993) "Letter to the Editor," March 29.

Radioactive Waste Campaign (1988) *Deadly Defense: Military Radioactive Landfills*, New York: Radioactive Waste Campaign.

Russell, D. (1989) "Environmental Racism: Minority Communities and their Battle Against Toxics," *Amicus* 11, 2: 22–32.

Schahn, J. and Holzer, E. (1990) "Studies of Individual Environmental Concern: The Role of Knowledge, Gender and Background Variables," *Environment and Behavior* 22: 767–86.

Scott, A. (1987) "Margherita Howe, Environmental Watchdog," *Women and Environments* 9, 2: 8–10.

Seager, J. (1993) *Earth Follies: Coming to Feminist Terms with the Global Environmental Crisis*, New York: Routledge; London: Earthscan.

Shiva, V. (1991) *The Violence of the Green Revolution*, London: Zed Books.

Shulman, S. (1992) *The Threat at Home: Confronting the Toxic Legacy of the US Military*, Boston: Beacon Press.

Snow. D. (ed.) (1992) *Inside the Environmental Movement*, Washington, D.C.: Island Press.

Sontheimer, S. (1991) *Women and the Environment: A Reader*, New York: Monthly Review Press.

UNEP (1989) United Nations Environment Program, *Public and Leadership Attitudes to the Environment in 14 Countries*, New York: Lou Harris/UNEP.

World Resources Institute (1992) *1993 Environmental Almanac*, Boston: Houghton Mifflin.

ACKNOWLEDGMENTS

I wish to acknowledge and thank Lisa Crawford and Jessie Deer-in-Water for their generosity in giving me the time and attention to conduct the formal interviews that provide the basis for this paper. My thanks also to the University of Vermont for a faculty research support grant to assist the completion of this project.

Part V

CONCLUSION

13

FEMINIST POLITICAL ECOLOGY

Crosscutting themes, theoretical insights, policy implications

Barbara Thomas-Slayter, Esther Wangari,
and Dianne Rocheleau

AN EMERGING FEMINIST POLITICAL ECOLOGY

Feminist political ecology brings into a single framework a feminist perspective combined with analysis of ecological, economic, and political power relations. It does not simply add gender to class, ethnicity, race, and other social variables as axes of power in investigating the politics of resource access and control and environmental decision-making. The mutual embeddedness of these hierarchies forbids this simplistic approach. Instead, the perspective of feminist political ecology builds on analyses of identity and difference, and of pluralities of meanings in relation to the multiplicity of sites of environmental struggle and change (Hart 1991; Ghai and Vivian 1992; Tsing 1993; Pankhurst 1992; West and Blumberg 1990).

In this volume, authors of eleven case studies from around the world analyze gendered experience of and responses to environmental changes, problems, and hazards. The environmental issues explored include: managing forests and trees, strengthening resource management in agricultural systems, reduction and safe disposal of hazardous wastes, and protecting the integrity of food and water supplies in both rural and urban communities.

The chapters present a diversity of situations along a continuum from urban to rural and from industrial to agrarian. Mega-cities are represented by the experiences of West Harlem Environmental Action (WHE ACT) in New York City, while the women of extractive reserves in Acre in the Brazilian Amazon represent the most rural, land-based situation among the cases. The Polish, United States, Austrian, and Spanish examples all address the gendered experiences of and responses to industrialization, though some of these impacts are visited upon rural spaces and others on densely populated urban areas. The gendered knowledge, tenure, and organizational issues presented in the Philippines, India, Kenya, Zimbabwe, Dominican Republic and Brazilian Amazon cases all have in common an agrarian or forest setting, with increasing commercialization of production and rapidly changing gendered livelihoods, landscapes, property regimes, and social relations.

The conditions in all eleven cases reveal the many distinct forms and responses to the restructuring of relations between rural and urban spaces and agrarian and industrial sectors. In each instance people are affected by the reintegration of cultural, economic, ecological, and political spheres on new terms within regional, national, and international systems. Gender relations both influence and are shaped by all of these global processes. In every case, the gender divisions of knowledge and authority, of property and decision-making rights, and of organizational affiliation are shifting.[1]

Women (and in some cases men) in each of these places have acted in response to some combination of threats to health, livelihood, quality of life and social justice. In Austria, the Hainburg Forest protection movement focused on quality of life and justice for other species and their habitats/ecosystems, for women and for the entire citizenry of an imperfect democracy. In Silesia, health was the overriding concern, with a justice component focused on women and children. The rubber tappers union in Brazil addressed livelihood, justice, and quality of life issues, all related to forest protection and management. Concerns about justice in the extractive reserves of Xapuri shifted from an exclusive focus on class to include gender. In Kenya, women's environmental interests rested on access to land and other forms of capital and on resources for livelihood security, in both short- and long-term contexts. For the women of Embu and Machakos access to land was a priority, although protection of forest and water resources for future use was a key issue and just distribution of resources by class and gender figured prominently among women's concerns.

The cases in this volume introduce us to a broad range of organizational situations in which women are increasingly involved. They may be formal associations or informal networks, registered nongovernmental organizations or movements; they may operate within a local, regional, national, or international context; they focus on a range of issues including health, resources, development, environmental protection, or related justice concerns. Sometimes the organizations are oppositional, arising in response to conflicts with local authorities or to larger state or corporate entities; in other instances, they formulate a new agenda around which the members rally. In many cases, local-level groups involve a delicate meshing of old styles, traditions, and systems with new objectives.[2] Occasionally, the organization is one of long standing with a number of other agendas. In most of the cases presented in this volume, the impetus for action is an environmental issue or concern which has mobilized the energies of a group, whether established, emergent, or nascent.

The gender basis of the organizations varies. In some instances, the women are working within a predominantly men's association (Xapuri, Brazil); or they may function within a joint women's and men's association (Zambrana-Chacuey, Dominican Republic). Sometimes the association is exclusively for women (Machakos and Embu, Kenya), and sometimes the organization is largely, but not exclusively, composed of women (Gibraleon or Catalonia, Spain). Within the organizations women may be founders, joiners, casual or loyal supporters. Overall, women have demonstrated that the issues which concern them are broader than the family or the household and involve collaborative effort with others who share their objectives.

All the cases draw attention to both local agency and local creativity, as well as to issues of scale and the interconnections among local, regional, national, international, and global phenomena. They demonstrate ways in which women on the periphery of state or local power are defining and redefining their situations, albeit under formidable constraints. They portray women actively engaged in protesting, interpreting, and reinterpreting their needs, their environmental concerns, and the issues pertaining to the security and well-being of their families, households, and communities.

All the stories encourage us to focus on the importance of gendered human agency and social structure in shaping pivotal environmental policies and decisions. They raise key questions about the relationship between knowledge and power as local communities and grassroots groups and movements struggle to find ways to resist dominant world views of the environment and resource use (Escobar and Alvarez 1992). Local communities may be disempowered by outsiders who extract what they need (and dump what they do not want) to the detriment of local residents. Their experience raises further questions about the relationships among knowledge, power, action, and resistance.

Feminist political ecology provides a valuable framework for analyzing and comparing the stories of these women from around the world. It offers an approach which derives theory from practical experience, avoiding the pitfalls of maintaining a strict distinction between theory and practice. It links an ecological perspective with analysis of economic and political power and with policies and actions within a local context. Feminist political ecology rejects dualistic constructs of gender and environment in favor of multiplicity and diversity, and emphasizes the complexity and interconnectedness of ecological, economic, and cultural dimensions of environmental change. It recognizes the relationship among global, national, and regional policies and local processes and practices.

CROSSCUTTING THEMES

There are numerous crosscutting themes which emerge from these eleven case studies. We select seven which are of particular relevance to the perspective of an emerging feminist political ecology.

Linking environment and survival

Environmental matters *are* survival matters. It is largely impossible to separate the two concerns; the issues are linked. Just ask a member of the West Harlem Environmental Action (WHE ACT) fighting the North River Sewage Treatment Plant; a Filipina from Siquijor struggling to find *romblon*, a plant once widely available for weaving sleeping mats and supplementing income; a Brazilian woman rubber tapper seeking protection of extractive reserves; or a member of the Bilbao Residents' Association fighting the toxic fumes from an abandoned mine threatening to engulf her neighborhood.

In some instances, these issues are related to livelihoods and the conditions necessary to meet basic subsistence needs, as clearly delineated in the

Himalayan community in the Saklana Valley of Uttar Pradesh. In others, the concern is for health and safety, as, for example when Lisa Crawford of Fernald, a town in rural Ohio, learned that the "feed production center" across the street from her house was actually a United States Government nuclear weapons production facility discharging radioactive materials into the groundwater that was the source of her domestic water supply.

While the links between environment and survival are evident, the local meanings vary. Women and men must be situated in the context of particular ideas, actions, and practices; analysis must proceed within the localized situation as it is linked to the broader context. In every one of the eleven cases, the actors have dispelled the myth that environmental concerns are a luxury for the wealthy or an "amenity" to consider as a postscript to economic development. Rather these concerns are central to the lives, livelihoods, survival, and well-being of the residents of these communities.

The impact of large economic and political systems on localities

Inexorably, residents of all the local communities presented in this volume are being drawn into broader ecological, political, economic, and social systems, propelled by their increasing needs for and reliance on cash, as well as the spread of communications and technology. In Kenya, for example, the coffee and corn farmers of Machakos or the cotton farmers of Embu find that the globalization of the market shapes their local production systems. In the Philippines, the expanding market has affected men's and women's ability to build systems of social exchange, an established mechanism for creating social capital. Women's social capital, used to sustain their distinct livelihoods, has suffered and declined in the face of a growing commercialization which has strengthened opportunities for men in the broader market economy.

In the Saklana Valley, Mehta observes the consequences of the Himalayan region's growing involvement in the market economy. Men are increasingly involved in off-farm employment while women, as agriculturalists, are left behind in a world which does not offer them the resources, information, and tools to carry out their work effectively. In West Harlem, the community is singled out as a site for citywide waste handling and disposal services and facilities deemed unacceptable in wealthier and whiter neighborhoods. The Northeast Regional Sewage Treatment Plant was shifted to West Harlem from the location originally planned which is now part of the Trump Towers complex.

International pressures are beginning to affect Austria as the nation joins the European Union. Despite the problems with the Danube Power Company delineated by Wastl-Walter, Austria has ecological standards which are comparatively high in relation to the rest of Europe. Ironically, Austria is experiencing pressure from the European Union to modify some of its laws for greater environmental leniency in response to international economic interests. Moreover, nuclear power facilities and other pollutants just beyond Austria's borders are of grave concern to the women involved in Austria's grassroots environmental movement.

Throughout this volume, we have been introduced to the women and men in local communities whose lives are changing dramatically as larger economic and political systems continue to transform their communities. They struggle to find the economic, political, and environmental resources to resist, to develop alternatives, and to deal more effectively with externally induced changes.

Gender-based asymmetrical entitlements

Asymmetrical entitlements to resources – based on gender – constitute a recurring theme. Access to resources – whether by de facto or de jure rights, exclusive or shared rights, primary or secondary rights, ownership or use rights – proves to be an important environmental issue for women virtually everywhere. Most of the cases are explicit on this point. In Kenya's Embu and Machakos districts, for example, most women depend on the use of local land and water resources to produce food and energy and to earn income, yet women lack legal rights and control over land and their rights of use and access are insecure. Gender inequities already extant in customary law have been exaggerated and new sources of inequity have been introduced by modern land tenure and legal reforms.

In the Indian Himalayas, Mehta notes the gender-based asymmetries in women's exclusion from certain kinds of property rights in rapidly changing farming systems. For example, men own and control tools based on animal energy. Women depend on men both to initiate the agricultural season by plowing and also to market the product of their efforts. In Saklana, with the increasing specialization, out-migration, and dependence on the market, women are increasingly responsible for agricultural production, yet gender entitlements to resources are becoming more asymmetrical than ever.

Analysts of resource access and tenure in the Philippines also observe this asymmetry, though Filipinas begin with a more egalitarian ownership and inheritance structure. In Agbanga, gender-based specialization in the market economy and the prioritization of men's earning ability are sharpening the gender differences in resource tenure, and creating an atmosphere of competition among residents, along the lines of class and gender, for scarce productive resources. Similarly in Zambrana-Chacuey, in the Dominican Republic, the introduction of a timber tree (*Acacia mangium*) as a men's cash crop contributes to the asymmetries in access, ownership, and control over resources. In each case, women have a disproportionate share of responsibilities for procuring resources for the household, and for maintaining the environment, with very limited formal rights.

The value of local knowledge

Feminists take a strong stance on the necessity for recognizing all kinds of knowledge. The contributors to this volume are no exception. The cases fall under the umbrella of "perspectives from the margins" in which the views, knowledge, and perspectives of ordinary people are respected and valued.

All these cases recognize local, gendered knowledge not only for its intrinsic worth but also for its valuable contribution to sustainable environments and economies. Women, as is evident throughout this volume, hold much of this knowledge especially in the arenas of environment and health. This knowledge is based on their experience, responsibilities and daily practice rather than an intrinsic or essential "feminine" quality.

The Zimbabwe case reflects most clearly a theme important in a number of the contributions: validating local women's knowledge. Working with a group of women on tree tenure and tree use, Fortmann makes a strong argument for:

1 recognizing that knowledge is gendered,
2 generating and sharing knowledge and expertise,
3 acknowledging ideas and giving credit to local colleagues, and
4 local ownership of a research project.

She documents the use of such an approach and the resulting empowerment of women in two Zimbabwe villages.

The issues surrounding local knowledge, its use, recognition, and transference are prominent in the other cases as well. In Machakos district, Kenya, in 1985, women coping with drought, famine, and the large-scale outmigration of men, found themselves relying on gendered knowledge of wild plants for food and medicines as well as reconstructing men's prior knowledge of rangelands and fodder to feed starving cattle. In the Himalayas, women's knowledge systems are not only eroding between generations, but are also being devalued, as men go off to the towns and cities seeking new opportunities and new forms of employment. Implicit in much that is happening in Saklana is the assumption that even though women's labor may now predominate in agriculture, their knowledge and experience are not particularly useful or valuable. This viewpoint is often reflected by agricultural extension officers and male elders who ignore women's knowledge of farming systems and their environment.

Some of these same conditions are occurring in the Philippines. Movement from horizontal social exchanges to hierarchical exchanges and market relations poses a threat to women's incomes and women-centered work with consequences for environmental sustainability and the continued transmission of locally based knowledge of farming and resource management. In the United States, women experience some of these same phenomena as demonstrated by the cases from West Harlem, New York; Fernald, Ohio; and Vian, Oklahoma. Women confront health, environmental, and political bureaucracies, all of which demonstrate a clear lack of respect for local knowledge possessed by women grassroots activists and their "housewife" constituencies.

Gendered space

Many of the cases focus on "gendered space," spaces that are socially constructed as appropriate and suitable for men and those that are domains for women. In the Himalayan community, Saklana, men's spaces are expanding, both literally as they move to and from home and new markets and figuratively, in terms of the importance attached to male activities. Moreover, as men become more tightly linked to the outside world, women's relative isolation and their dependence on men's income-generating activities increase.

In Xapuri, as in Saklana, there is clearly defined gender appropriateness for different kinds of spaces. The "union space" was a social and political space for men. Until women began to organize within the union it was not appropriate as a place to be or as an activity for women. Only on occasion – for instance when a woman was a de facto head of household – was she allowed the space and authority to participate in union activities. One of the major achievements of women's participation in union activities and women's groups has been to create legitimate public political space for women in rubber tapper communities.

In Zambrana-Chacuey, Dominican Republic, the *Acacia mangium*, a timber cash crop tree, has become a site of struggle over domains on the farm between men and women. Women's inability to control the *Acacia* (whether to plant it themselves or to resist their husbands' plans to plant timber lots) has made their authority over all farm land, including the patio (homestead garden), increasingly vulnerable. The new tree – a commercial monocrop largely under the control of men – could also replace women's plants, including a wide variety of food, tree, and medicinal crops and affect their access to a whole range of medicines, condiments, and foods for household use and sale.

Command over space and gendered rights of control and access are sources of social and political power. New initiatives by a small core of activist women opened space in the union and made public space more available for all women in Xapuri. Similar initiatives "opened up spaces" in community organizations and local politics in West Harlem and in Ohio and Oklahoma, in regional politics in Spain and national politics in Austria. In West Harlem, women's longstanding community work became more visible, explicitly political, and public, while in the other cases many women entered community and national politics for the first time. Indeed, in Austria and West Harlem, organizing against big government and big construction provided a base for accommodating women's wider political ambitions – even though they were not successful at first. Each of these movements has inscribed new women's spaces into the political landscape at local and national levels.

Realignment of rural–urban spaces and production systems

Widespread stereotypes of isolated rural communities and the separation of urban and rural environmental problems into industrial versus agrarian domains distort our views of the highly interconnected realities of women and men throughout the world. Extractive industries (mining, logging, and commercial export agriculture) have long connected rural communities and major urban centers. Global restructuring of economies, ecologies, and polities is changing the shape and terms of those connections, including the gendered social relations within and between urban and rural places. Each of the cases illustrates the importance of rural–urban linkages, their changing nature and their effects on both gender relations and human–environment interactions.

The people of Fernald, Ohio, and Vian, Oklahoma, and several communities in Spain have experienced the conversion of their rural homes into hazardous waste disposal sites for both local and faraway industries. For the rubber tapper communities in Brazil, whose livelihoods are based on forest extraction, the impetus for their problems and their best solutions have come from changes in rubber markets, Brazilian land policy, the role of the military in national life, and the rise of national and international environmentalism as a powerful political force – all changes external to the community. In the Himalayas, the Philippine Islands, and the savannas of Kenya, migration of men to urban workplaces has transformed the gender division of labor and authority as well as gendered resource management and the landscape itself.

These cases offer glimpses of the shifting roles of rural and urban spaces and the impact of changing production systems on local communities. The realignment of urban and rural places, as well as manufacturing and agrarian production, is often at the heart of a gendered struggle to address uneven relations of power and to reverse destructive trends in situations of environmental change and, in some cases, crisis.

Women's collective struggles

In many parts of the world, women are becoming involved in collective struggles to address problems of resources, the environment, and economic survival, most often in women's organizations or those dominated by women, and sometimes jointly with men, whether as equals, members of women's groups within men's organizations, or as informal affiliates. The contributors to this volume recognize women's collective struggles, ranging from informal social networks for exchange of resources, to local groups, grassroots movements, and the formation of a political party.

In the Philippines, women from Napo and Tubod continue to build their gender-based exchange networks. In Brazil, women from Xapuri are involved in the base communities of the Catholic Church, a valuable avenue for social interaction among women who live under quite isolated circumstances. They have expanded their participation to include the national Rubber Tappers Union and national women's organizations.

Kenyan women in Embu and Machakos engage in collective efforts to diminish risk, uncertainty, and insecurity and to create new opportunities for themselves, their families and their households. Their self-help groups constitute a key strategy for meeting the challenges of a declining and unevenly distributed resource base and labor scarcity. They also provide or improve access to a variety of market and nonmarket exchange resources, public goods, and official connections to political processes.

Several cases – the Philippines, Kenya, Dominican Republic, Zimbabwe, Brazil – explore women's responses to exploitation and inequitable access to resources. Others identify the ways women have organized to address specific problems or hazards. In Fernald, Ohio, Lisa Crawford founded a community action group called Fernald Residents for Environmental Safety and Health (FRESH). FRESH is now regarded as one of the country's most effective community grassroots groups resisting hazardous waste dumping. In Oklahoma, Jessie Deer-in-Water started an organization called Native Americans for a Clean Environment (NACE). In both situations, the core workers of the struggle were women, who organized and mobilized grassroots groups. Both groups later affiliated with the national Citizens Clearinghouse for Hazardous Wastes and joined other communities in local and national struggles to prevent and control the production and disposal of toxic and hazardous wastes.

In Spain and Poland, as well as in the United States examples, the initial impetus for struggle against a specific hazard or danger at the grassroots level arose with women in the community. Later the struggle was joined by small numbers of men. Among the rubber tappers in Brazil and in the Dominican Republic's Rural Federation, the struggle involved both men and women. In Austria, the destruction of the Hainburg forest involved both men and women equally but carried greater symbolic significance for women as a double debut for feminist environmentalism and women leaders in national politics.

Within an overall framework of male dominance, women find a means of expression, organization, and rebellion (Pankhurst 1992). Their collective action, their engagement in struggles over resources, and their resistance to the implementation of environmentally destructive projects constitute elements of "oppositional" practice as defined by postmodern feminist cultural theory (Butler and Scott 1992). Confronting gender-based injustice and the relations of hierarchy, subordination, and domination, as the women activists described in this volume do, places them and their activities squarely in the political realm. Whether they are working at the local, regional, national, or international level, they seek significant political access and change for themselves, their families, their communities, and in some cases, women as a group.

THEORETICAL IMPLICATIONS

What, then, are the theoretical implications of these cases and their common themes? What can we learn from these in-depth explorations of gender, environmental issues, economy and polity in these localities from around the world? How do these interpretations of diverse local experiences contribute not only to global understanding, but to theoretical advancement of a feminist political ecology perspective?

Analysis must establish and improve understanding of the connections among environmental concerns, local knowledge, gender-based responsibilities and opportunities, and community organizations or grassroots movements. The emergent focus on women and the environment has occurred within several diverse schools of thought:

1 ecofeminism, including both essentialist and social constructivist ratio-
nales for the link between women and nature (Merchant 1980; Mies and
Shiva 1993; Shiva 1989);
2 feminist environmentalism, based on the material and economic linkages
of women's interests with "nature" (Agarwal 1991);
3 feminist and poststructuralist critiques of science, environmentalism,
and/or sustainable development (Butler and Scott 1992; Haraway 1990;
Harding 1991; Hynes 1989, 1991; Seager 1993; and
4 feminist alternatives for sustainable development (Sen and Grown 1987;
WEDO 1992; Thomas-Slayter and Rocheleau 1995b).

The history and relevant debates about these perspectives are summarized
and elaborated in recent works by Braidotti *et al.* (1994), Jackson (1993) and
Plumwood (1993).[3]

We reflect on four ways in which the cases and their common themes
build the perspectives of feminist political ecology and have implications for
modifying theory.

Recognizing the interconnectedness of all life

A feminist political ecology recognizes the interconnectedness of all life and
the relevance of power relations – including gender relations – in decision-
making about the environment. It analyzes the powerful underlying structures
which operate to the benefit of certain classes and groups, both locally and
across international boundaries. Feminist political ecology focuses on the ways
in which site-specific ecological and livelihood systems are linked into national
and global environmental, economic, and political systems which shape,
enable, and limit the opportunities and constraints occurring at the local level.

Whether in Silesia, Xapuri, West Harlem, or Zambrana-Chacuey, women
clearly make the unequivocal connection between the immediate environmental or economic issue and the broader survival, political, and philosophical concerns. The cases reveal that women's organizations and grassroots
movements utilize a direct problem-solving approach for addressing a specific
set of issues. Often they demonstrate effective use of non-violent methods.
Examples are found in the civil disobedience of the Austrian environmental
activists and the peaceful demonstrations and confrontations of the Brazil rubber tappers trying to protect their forests. In both these cases and others as
well, women recognize that these issues are indeed part of a larger set of concerns. They raise important questions about the nature of the particular society and its political life. They raise questions about the relations of human
beings to "nature" as manifested in various contexts, to other species, to "the
land," to "place," and to the resources that support both lives and livelihoods.

Questioning the presumption of technological progress and domination of nature

Throughout the world, many women are questioning the presumption of a
victory of technology over nature, thereby offering a critique of the technol-
ogy-driven mainstream trajectory of development in most national politics.

Women described in these chapters question the nature of environmental knowledge and expand the notion of what is considered environmental expertise. Women are asking questions about who defines environmental problems and who measures them.

These issues are explored vividly by Seager in "Hysterical Housewives." Seager identifies the different social location of men and women and argues that women, who are responsible for environments of daily life, are the most reliable narrators for observing and assessing environmental change. In the United States, as Seager observes, grassroots groups form around issues of toxic waste management and disposal, nuclear facilities, or pollution from military facilities. Mainstream environmental organizations are not focused on these issues. There is a widening gap between mainstream organizations focusing on high technology and nature preservation and those that are grassroots environmental movements, such as Fernald Residents for Environmental Safety and Health (FRESH) or Native Americans for a Clean Environment (NACE).

These gaps and different perceptions are explored in several of the cases: in Austria with regard to the hydroelectric power plant to be placed in the Hainburg forest; in West Harlem with the struggle against contamination of the community from a sewage treatment plant; in Catalonia as people observed their own landscape designated as a waste disposal site; in the Philippines as the technologies of dynamite fishing destroy the long-term viability of the coral reefs; in Silesia, as Polish women take back the prerogative to define what constitutes safe, healthy food; in Amazonia, as the rubber tappers – women and men – struggle to maintain rainforests as their home and workplace in alliance with international environmental organizations focused on preserving ecosystems, wildlife habitats, and "gene pools."

These communities – and these authors – are examining the tradeoffs as societies opt for ever more complex, sophisticated ways of negotiating industrial and post-industrial life. They suggest that other values, other issues, other perspectives emanating from the grassroots are flourishing and must be heard.

Recognizing that ideologies shape relationships among gender, knowledge, environment, and development

Ideologies – particularly those formulated within a patriarchal mode – create gendered access to information, knowledge, resources, and the technologies for improving livelihoods. A patriarchal model, which situates women in the private sphere, conditions both men's and women's responsibilities, and determines the social value assigned to each. The cases offer considerable variation in their analysis of the ways in which ideology shapes the convergence of gender, environment, and development. For example, in the Himalayas, seclusion, though not practiced as fully as in the plains, is a component of the patriarchal ideology widespread in that region. Mehta explores the ways in which it constrains women's freedom of movement and limits their access to the highly valued monetized domain, thus preventing women from broadening their roles in land and water management through access to new systems of knowledge and spheres of exchange.

This phenomenon, in a moderated form, is presented in the Brazilian, Kenyan, and Philippine cases as well. Women are perceived in terms of their roles as providers of labor and tenders of the household, rather than as farmers and managers of the land or of other significant means of livelihood. In effect, their work is devalued. Implicit in this ideology is the notion that women's labor is infinitely expandable, and that their domestic responsibilities (whether of a productive or reproductive nature) are not particularly significant.

Such values and norms shape individual identities and behavior as well as customary practice and political culture. They have practical consequences for women as they deny access to the resources needed for women's work and continue the process of their marginalization and social devaluation. Moreover, this ideological orientation lends itself to political uses of women. Discussing Poland's region of Silesia, Bellows emphasizes that women "fall into political institutions of motherhood defined for them in terms of patriarchy and nationalism." While Polish women were, for decades, fully integrated into industrial production and the scientific professions, they are once again being relegated to the home and the kitchen. The women engineer founders of "Tested Food for Silesia" found themselves forced to invoke their authority as mothers to gain political legitimacy, only to face an increasing trend to staff successful environmental organizations with male "professionals."

In all these cases, women are pressing against the boundaries of the ideological boxes into which they are placed. They may, as in Gibraleon, Spain, carry out their domestic chores on the street while marching and organizing. They may, as in Austria, determine that a feminist environmentalist is needed in politics. They may, as in India, assert the right to grow a commercial crop (peas) instead of a grain for home consumption, or, as in Brazil, seek to join the rubber tappers union. They may, as in the Dominican Republic, organize to promote women's interests under the banner of "Housewives Associations." Wherever, they are struggling against marginalization and powerlessness within an ideological framework which has historically supported and promoted unequal gender relations. In each of the cases presented here – and increasingly throughout the world – they do so within the context of livelihood, environmental safety, health, and conservation issues.

Addressing the different structural positions occupied by women and men

It is commonplace to find that men specialize and women pursue integrative roles in economic activity and in resource management. This situation is both shaped by, and in turn influences, men's and women's distinct structural positions within society. Gendered structural positions – both a cause and a consequence of gendered identities, ideologies, and practice – pose important theoretical questions. Authors in this volume differ in their analysis and interpretation of this phenomenon. Some suggest the causes are related to economic variables, to colonialism, and to material patterns of change foisted on the South by the North. Others focus on philosophical and psychological models of behavior differentiated according to gender, which most agree is a social construct. Whatever the cause of gendered social structures, the outcome is an uneven relation of power which characteristically disadvantages women.

Brú's analysis of communities in Spain draws on her perception of a male inclination to compartmentalize and specialize whereas she characterizes women as acting on the basis of a feminine model of simultaneity and versatility. She specifies that we can recognize in social mobilization ways of acting that are specifically feminine, related to gender-based roles, functions, and abilities.

The discussion of Agbanga, a community in the Philippines, reveals that gender-based specialization within the market economy has brought about the prioritization of men's earning capacity and their social capital. In fact, the increasing specialization of labor, combined with natural resource degradation and privatization experienced by households in Agbanga has put women's livelihoods at risk.

The same process is occurring in India, where highly gender-differentiated encounters with productive resources, markets, and money underscore the structurally dissimilar positions occupied by women and men within rural households and landscapes. There, specialization and monetization of the local economy are tying women (and their daughters) to the land in ways that are qualitatively different from previous patterns of livelihood and resource management. Men are gaining access to new forms of specialized knowledge and roles; women are not.

The residents of Zambrana-Chacuey, Dominican Republic, experience this kind of specialization as a result of programs and projects introduced from outside the community. Historically, gendered patterns of livelihood management and community participation in Zambrana-Chacuey have been relatively flexible. The Forestry Enterprise Project has led to a specialization by men in timber production, and the de facto exclusion of women from this lucrative activity during the early stages of the initiative. Women are ideologically, politically, and economically defined as "helpers" and household beneficiaries of this new income-generating, market-oriented activity. In practical terms, some women would like equal access to timber production while others would resist the encroachment of timber and its displacement of diverse food and tree crops in their gardens and croplands.

Throughout this volume, women can be observed questioning the socially determined modes of specialization. Individually and through their networks and organizations they are raising concerns about both the customary and newly imported ways of organizing economic production, access to resources, and environmental management.

USING FEMINIST POLITICAL ECOLOGY TO INFORM POLICY ISSUES AND DEBATES

To create effective and equitable policies related to the environment requires both broad and specific understanding across levels of analysis. Policies must be attentive to local social contexts, to local perceptions of issues, and to local concerns. Policy must also reflect understanding of the impact of global systems on the nation and on the local community in the context of blurring boundaries among all these levels. Moreover, both formal and informal structures are relevant to policy change and link the state, the organizations of civil society, the community, household, and individual.

Incorporating a feminist analysis can illuminate the ways in which gender positions both men and women vis-à-vis institutions that determine access to land, to other resources, and to the wider economy. An ecological approach allows us to see environmental management, resource use, and technological change as a dynamic, interactive process, rather than one of incremental and unilinear movement to subsume "nature" under economic progress and political control. An emphasis on "politics" recognizes the social and political context in which national and international governments and development agencies, operating at all levels, make policy. Linking gender and political ecology allows us to focus on the uneven distribution of resource access and control by gender, as well as according to other social variables such as class and ethnicity (Thomas-Slayter and Rocheleau 1995b; Rocheleau 1995). We believe that the perspective of feminist political ecology can further the design of policies for a more sustainable environment and for overcoming deep-rooted economic and political causes of poverty and ecological crisis. We suggest four policy issues which crosscut these contexts and are relevant to policy analysis from the perspective of feminist political ecology.

The convergence of feminist agendas, environmental agendas, and other social movements

An environmental problem is usually embedded within a set of broader issues and concerns. Time and again, local communities start with one problem and move on to deal with the political, social, and economic ramifications of the issue at hand. Many of the cases in this volume reveal that a feminist agenda and an environmental agenda often converge within broader concerns about ethics, values, and the nature of political life.

In West Harlem, as Vernice Miller points out, WHE ACT benefited from a strong legacy of political organization from the civil rights struggle. Yet the specific question of the sewage plant has galvanized the community to fight for greater control over planning decisions and to participate in a broader struggle against environmental racism, especially the siting of hazardous waste facilities in communities of color. Women have assumed a new level of leadership in the West Harlem community to engage in collective advocacy and to respond in a collective voice to those who have degraded and contaminated their neighborhood.

Similarly, in Catalonia, rural residents perceive as a threat – even an act of aggression – their government's plans to put a waste disposal site in their community to accommodate hazardous by-products generated in urban-industrial areas. These residents sense danger and feel anger at the prospect that their community has been designated to serve as the recipient of wastes. Women in Catalonia have led the resistance to the government's plans.

In Xapuri, in Brazil's westernmost state of Acre, the *mulheres seringueiras*, or female rubber tappers, are challenging stereotypes about women's roles as they seek improved economic opportunities, and preservation of the forests from which they earn their livelihoods. Similarly, the Austrian women who fought to save the Hainburg forest from a hydroelectric project also confronted the state's power through nonviolent protest. The power plant faded to secondary importance and the real subjects of debate were democratic rights, power, violence, social values, and the place of women in national political life. For these women, and others like them around the world, it is difficult – even unnecessary – to separate their overlapping agendas as women, as feminists, as environmentalists, and as citizens.

Gender bias in organizations and programs

Planned programs of both governmental and nongovernmental organizations often reflect a gender bias arising from systems which are based on an inherently patriarchal ideology and political framework, thereby serving male interests. In Zambrana-Chacuey, Dominican Republic, the women and men in the forestry project introduced through the Rural Federation and ENDA-Caribe participated equally in the struggle for peasant rights. Yet a forestry project focusing on the *Acacia mangium* as a lucrative timber cash crop introduced gendered patterns of enrollment in the Wood Producers' Association. The association established guidelines for planting, harvesting, and marketing *Acacia*; these guidelines magnified gender and class-based inequities and transformed them into obstacles to participation in timber cash cropping. The Forestry Enterprise Project has been mainly channeled to men's groups. The project staff initially perceived forestry as men's work and timber lots as men's domain on the farm, and failed to provide women with information and technical support for timber production and marketing.

As observed on Leyte and Siquijor Islands (Philippines), technology introduced by the government has had both a class and a gender bias, oriented toward the wealthier members of the community and toward men. In Saklana (India) development interventions have helped to reinforce and expand activities which are predominantly male-dominated, whether locally or related to off-farm migration.

These policies, programs, and projects have specific gendered impacts, even though they may seem to lack a specific gender content. They begin to weave together women's subordination through the interdependent relations of capitalism and patriarchy within the context of class and male domination. As such, interventions based within these institutions may increasingly alienate women, reinforcing their subordination.

The cases in this volume demonstrate that women increasingly recognize these biases. In the process, they are becoming organized and politicized. They are endeavoring – sometimes on their own and sometimes with men – to open up political space for the articulation of a more progressive discourse linking gender, environment, and development.

The impacts of privatization

Privatization has had enormous implications for women around the world, especially poor, rural women, who have lost access to commonly held resources and who often have a contingent relationship to property through their relations with men. In Kenya, bestowing title deeds on men implies that men are the heads of households and major decision-makers. Women's roles in decision-making, as well as their contribution to the production, reproduction, and maintenance of the families, are ignored, not to mention their work in resource management at household and community levels. Legally, women are not prohibited from land ownership in Kenya, but only about 5 percent actually own land. Customary laws and religious beliefs that support the subordination of women in property ownership have been exaggerated and distorted through land tenure reform to the point that women have also lost many of their previously recognized rights of land use and access within men's property. Where women are not legal owners of land, their access to land and water resources as well as credit is limited. This is particularly problematic in arid and semi-arid regions where other forms of non-farm income (other than migrant men's wages) are scarce.

Moreover, consolidation and the development of strictly defined boundaries, as well as the reduction of common lands and the creation of titled lands from common lands have reduced the flexibility and diversity of farming systems. The reduction of open access lands has affected women all over the world, particularly poor women, as they are often highly dependent on forest and grazing resources for meeting their daily needs and responsibilities. In Machakos and Embu (Kenya) women note the increasing distances they must travel for fuelwood. In Napo and Tubod (Philippines) they lament their loss of access to many of the raw materials previously available from the local natural resource base, which they use in household income-generating enterprises. In Saklana (India), they travel to increasingly distant forests to collect fodder for the buffaloes which they now keep to produce both milk and fertilizer. Moreover, the commons once provided places for young and old to meet and to pass along local environmental knowledge, a decreasing opportunity in the case examples of Kenya, the Philippines, and India.

Access to public space

Access to public space – and particularly to public space that is political – often presents a significant challenge for women. This fact proved to be an issue for the female rubber tappers in Brazil who were not supposed to attend the union meetings in any capacity other than preparing meals in the kitchen, or acting as symbolic innocents in confrontations with ranchers and authorities. In Catalonia, women mobilized vigorously on a short-term basis to deal with the danger posed by the threatened waste disposal facility. They did not view themselves as part of a larger political effort to affect policy and did not demand public space as such. On the other hand, in Austria, the confrontation over the hydroelectric power plant on the Danube became an opportunity for both women and men to question the role and policies of the state more broadly. In effect, women converted the issue into demand for a public and political space, particularly in their efforts to enable a Green feminist to run for the Presidency.

Bellows observes that the environment is not formally considered political space in Poland. Activism in regard to the environment connotes less anti-government censure than activism in other arenas. In the cases presented in this volume, issues which started out as defined in a fairly narrow environmental framework had a way of popping out of the box. A concern for access to trees for rubber became an endeavor to prevent certain kinds of lands from becoming privatized – a highly political act indeed. An effort to demand cleanup of toxic facilities in Fernald, Ohio, led to a confrontation with the established military and political power structures. In Bilbao, Spain, as in West Harlem, New York, concern about real health and environmental risks led beyond those specific risks to a deep understanding of processes of marginalization of neighborhoods and communities, on the basis of ethnicity and race, respectively. In these two situations, women continued to struggle not just against the sewage plant and the burning abandoned mine, but against the poverty, marginality, unemployment, drugs, and other issues confronting their families and communities. They struggled to achieve the dignity which they wanted for the neighborhoods they cherished.

In the cases explored in this volume, women find means of expression and rebellion, to engage in oppositional practice and to protest a range of injustices based on selective access to or degradation of resources for production, survival, and cultural continuity. These cases help us to understand how women construct collective identities, how these gendered identities differ across cultures and polities, and what kinds of political space and circumstances enable them to flourish.

MAKING A DIFFERENCE: LINKING THEORY AND ACTION IN WOMEN'S ORGANIZATIONS AND MOVEMENTS FOR A SUSTAINABLE ENVIRONMENT

The gendered roots of activism

Activism for women most often arises in connection with their social roles, particularly in defense of family and community, and most often for reasons of livelihood security, health and safety, as well as a sense of place. Bellows asserts that the motivating principle for protest is clearly associated with the protection of children and future generations. In fact, in a curious twist, women professionals in Upper Silesia (Poland) don the respectable public garb of mothers and grandmothers to advocate and organize the tests of soils and foods for toxic chemicals and other contaminants. Leaders of grassroots organizations bring to bear full moral and emotional persuasion on the issues confronting them. The arguments they use reflect their concerns about livelihood security, health, and life-threatening circumstances. Only secondarily do they couch their concerns in more broad environmental and economic arguments. Brú suggests that women's perception of environmental factors is closely linked to gender roles, especially in healthcare and family well-being. She too draws on the notions of roles and functions which are gender-based, making an argument similar to Seager's in this volume who states, in regard to the environmental activism in Oklahoma and Ohio, "the universalism of women's voice on the environment may reflect the universalism of their social location."

On the front lines: women, social action, and conflict

In a number of instances, women have been out in front on conflict issues or "on the front lines" in situations of great tension and potential violence. Their actions belie the myths existing in some cultures of women's passivity and unwillingness to take risks. This observation is clearly demonstrated in the case of the rubber tappers in Xapuri. Women joined the *empates* or forest demonstrations and actually took a place on the front lines in face-to-face confrontation with armed police. The presence of unarmed women, and even children, in the front lines was a tactic used by the rubber tappers to ensure a peaceful outcome in a very tense and potentially violent situation. They believed, correctly, that the police would be unwilling to fire on unarmed women and children and their presence gave a time frame in which the men – rubber tappers and officials – could enter negotiations. However, the women, who were called upon to take to the front lines to defend the forest, were expected to leave the negotiations to the men! The political space was not considered their domain.

In Gibraleon, women were not only on the front lines in confrontations with the police and military, but they were also central to the organization of the protests over the hazardous waste site and to the follow-up negotiations. Women played a key role in uniting the community, in creating an "authentic mythology of the conflict" through story and song, and in linking the nature and purpose of the conflict with the life of the community. In Austria, in defense of the Hainburg forest, unarmed, peaceful women, as well as men, faced a brutal intervention by the police. In each of these cases, the constancy and determination of the women involved is unquestioned.

In these and other cases, women identified the connection between their own jeopardy in the immediate environmental or economic concerns and the broad questions of survival, politics, and morality. The analysis suggests that we need to use the perspective of feminist political ecology to, in the words of Alvarez (1990), "unpack the state" and society more broadly. We need to examine the multiple institutional and ideological instances for finding points of access for gendered participation, organization, resistance, and protest.

Activism, changing attitudes, and increasing autonomy

The activism analyzed in this volume, the accompanying attitudinal changes, and the increasing autonomy experienced by many women, are beginning to modify the stereotypes present in many corners of the world about women's roles and behavior. The Brazilian women rubber tappers, who struggled on the front lines of the *empates*, are slowly changing perceptions about what is an appropriate role for women as they begin to translate their defense of the forest into new livelihood strategies. The legitimacy of a handful of women who have assumed leadership roles within the rubber tappers movement, by virtue of their positions as de facto heads of household, is permitting them the space and authority to begin to include other women in the movement. New social roles are emerging for women in this process. A similar process is occurring in Zambrana-Chacuey at local and regional levels, and in Austria at the national level.

In Bilbao, new strength among women is arising from their clear under-standing that the abandoned mine is neither an accident nor an isolated incident, but part of a process which is marginalizing them. The Old Bilbao Neighborhood Association, which continues its struggle against the degra-dation of the neighborhood, is a manifestation of this understanding.

Thus, it is evident that a stronger voice for women is emerging not only in Spain, but in Kenya, in Austria, in the United States, in Poland, in Brazil, and across other parts of the world where women are asserting themselves in the face of adverse environmental circumstances and even aggression toward their families and communities. Indeed, the environment is an entry point for women into politics, and women's organizations are serving as points of departure for environmental action and political involvement.

Global perspectives from local experience

The UNDP's *Human Development Report* for 1994, assessing the last fifty years, emphasizes an arresting picture of unprecedented human progress and unspeakable human misery, of rich and poor nations alike afflicted by growing human distress.[4] Starting from a very different ideological base Peet and Watts (1993) note the ecological and economic perils facing people throughout the world in an age of triumphal capitalism. The cases in this volume demonstrate the growing consciousness and activism on the part of women around the world as they respond to environmental and economic distress and to very specific problems in various locales. They support a feminist critique of a mindless, growth-oriented, technology-based develop-ment, provide an analytical link among various crises of environment, economy, and polity, and contribute to the definition of an alternative para-digm of social change.

Gender shapes the opportunities and constraints women and men face in securing safe environments, viable livelihoods, and strong communities across cultural, political, economic, and ecological settings. Environment and gender issues are central to debates about the nature of society and the claims which women and men can make on their respective societies. The social relations of gender are significant in their influence on entitlements and on environmental sustainability. They are also relevant to the emerging grass-roots activism on environmental matters.

The eleven case studies presented in this volume offer ample evidence in support of these propositions. They provide insights which can contribute to the integration of gender, environment, community organization, and grassroots movements into policy and political change. The perspectives of feminist political ecology cut across these issues to build an approach toward power relations taking into full account not only male dominance, but main-stream and privileged attitudes and control over the environment.

Today, women all over the world – and the grassroots organizations in which they are involved – are more concerned than ever about the fate of their environment and the linkages among declining ecosystems, degraded resources, and their increasing poverty. Feminist political ecology offers a new perspective on structures and processes of social change. Through its recognition of threats to equity and diversity, and its promotion of social and environmental justice, it helps to strengthen the balance between men's and women's rights and responsibilities in local communities. It clarifies linkages among gender, environment, livelihoods, and poverty, in ways that benefit both women and men. In so doing, it addresses the economic and political barriers to environmental sustainability and social justice.

NOTES

1 We are using the terms knowledge and authority in the sense of power-knowledge as explicated by Foucault (1979, 1980) and elaborated by Stamp (1989).
2 For a discussion of local organizations, see: Thomas-Slayter (1994).
3 For a more elaborate discussion of the ecological and institutional challenges of "bringing gender in," see Thomas-Slayter and Rocheleau (1995b: 93–7).
4 For a discussion of these issues in relation to Kenya, see Thomas-Slayter and Rocheleau (1995a).

REFERENCES

Agarwal, B. (1991) "Engendering the Environment Debate: Lessons from the Indian Subcontinent," CASID (Center for Advanced Study of International Development) Distinguished Lecture Series, Discussion Paper 8, Michigan State University.

Alvarez, S. E. (1990) *Engendering Democracy in Brazil*, Princeton, New Jersey: Princeton University Press.

Braidotti, R., Charkiewicz, E., Hausler, S., and Wieringa, S. (1994) *Women, the Environment and Sustainable Development: Toward a Theoretical Synthesis*, London: Zed Books.

Butler, J. and Scott, J. W. (eds.) (1992) *Feminists Theorize the Political*, London: Routledge.

Escobar, A. and Alvarez, S. (eds.) (1992) *The Making of Social Movements in Latin America: Identity, Strategy, and Democracy*, Boulder, Colorado: Westview Press.

Foucault, M. (1979) *Discipline and Punishment: The Birth of the Prison*, New York: Random House, Inc.

—— (1980) *Power/Knowledge: Selected Interviews and Other Writings 1972–1977*, New York: Pantheon Books, Inc.

Ghai, D. and Vivian, J. M. (eds.) (1992) *Grassroots Environmental Action: People's Participation in Sustainable Development*, London: Routledge.

Haraway, D. (1990) *Simians, Cyborgs and Women: The Reinvention of Nature*, London: Routledge.

Harding, S. (1991) *Whose Science? Whose Knowledge? Thinking from Women's Lives*, Ithaca, New York: Cornell University Press.

Hart, G. (1991) "Engendering Everyday Resistance: Gender, Patronage and Production Politics in Rural Malaysia," *The Journal of Peasant Studies* 19, 1: 93–121.

Hynes, P. H. (1989) *The Recurring Silent Spring*, London: Pergamon Press.

—— (1991) *Reconstructing Babylon: Essays on Women and Technology*, Bloomington: Indiana University Press.

Jackson, C. (1993) "Environmentalisms and Gender Interests in the Third World," *Development and Change*, 24: 649–77.

Merchant, C. (1980) *The Death of Nature: Women, Ecology, and the Scientific Revolution*, San Francisco: Harper and Row.

Mies, M. and Shiva, V. (1993) *Ecofeminism*, London: Zed Press.

Pankhurst, H. (1992) *Gender, Development and Identity*, London: Zed Press.

Peet, R. and Watts, M. (1993) "Introduction: Development Theory and Environment in an Age of Market Triumphalism," *Economic Geography* 69: 227–53.

Plumwood, V. (1993) *Feminism and the Mastery of Nature*, New York: Routledge.

Rocheleau, D. (1995) "Gender and Biodiversity: A Feminist Political Ecology Perspective," *IDS Bulletin* 26, 1: 9–16.

Seager, J. (1993) *Earth Follies: Coming to Feminist Terms with the Global Environmental Crisis*, London: Routledge.

Sen, G. and Grown, C. (1987) *Development, Crises, and Alternative Visions*, New York: Monthly Review Press.

Shiva, V. (1989) *Staying Alive*, London: Zed Books.

Stamp, P. (1989) *Technology, Gender and Power in Africa*, Ottawa: International Development Research Center.

Thomas-Slayter, B. (1988) "Household Strategies for Adaptation and Change: Participation in Kenyan Rural Women's Associations," *Africa* 58, 4: 401–22.

—— (1994) "Structural Change, Power Politics, and Community Organizations in Africa: Challenging the Patterns, Puzzles and Paradoxes," *World Development* 22, 10: 1479–90.

Thomas-Slayter, B. and Rocheleau, D. (1995a) *Gender, Environment, and Development in Kenya: A Grassroots Perspective*, Boulder, Colorado: Lynne Rienner.

—— (1995b) "Research Frontiers at the Nexus of Gender, Environment, and Development: Linking Household, Community, and Ecosystem," in R. S. Gallin and A. Ferguson (eds.) *The Women and International Development Annual, Volume IV*, Boulder, Colorado: Westview Press.

Tsing, A. Lowenhaupt (1993) *In the Realm of the Diamond Queen*, Princeton, New Jersey: Princeton University Press.

UNDP (United Nations Development Program) (1994) *Human Development Report, 1994*, New York: United Nations.

WEDO (Women, Environment and Development Organization) (1992)*World Women's Congress for a Healthy Planet, Official Report*, Miami, Florida: WEDO.

West, G. and Blumberg, L. (1990) *Women and Social Protest*, Oxford: Oxford University Press.

NOTES ON CONTRIBUTORS

EDITORS/CONTRIBUTORS

Dianne E. Rocheleau is the author of *Agroforestry in Dryland Africa* (with Fred Weber and Alison Field-Juma) and *Gender, Environment and Development in Kenya* (with Barbara Thomas-Slayter), numerous papers and articles on gender, political ecology, and popular participation in forestry, agriculture, and "sustainable development." She has resided and conducted research and environmental work (1975–present) in the United States, Dominican Republic, and Kenya, with short-term assignments elsewhere. She is an Associate Professor of Geography at Clark University and Director of the Gender and Environment Project of the George Perkins Marsh Institute at Clark University. She also serves on the advisory board of the Land Tenure Center, the Board of Trustees of the Center for International Forestry Research (CIFOR), and the Policy Consultative Group on Africa (World Resources Institute and USAID). Her current research includes: the multiple histories of ecological, economic, and cultural change in the dry forests and savannas of Ukambani (Kenya); gendered knowledge, rights, and institutions shaping the landscape of farm and forest regions in the Dominican Republic and Kenya; "sustainable development" and the environmental and economic restructuring of local landscapes, livelihoods, and life-ways.

Barbara Thomas-Slayter is author of *Politics, Participation and Poverty: Development through Self-Help in Kenya*, and *Gender, Environment and Development in Kenya* (with Dianne Rocheleau), as well as numerous articles on rural organizations and development, popular participation, gender and development, and gender issues in resource management. She currently serves as Director and Professor at the International Development Program and the Ecogen (Ecology, Gender, and Community Organization) Project at Clark University. She has served on the Board of Directors of the Coolidge Center for Environmental Leadership and is currently on the board of the Women and International Development Joint Harvard/MIT Group and the Board of Directors of Oxfam America. During 1992–4, she conducted research at the African Studies Center of Boston University as a Research Scholar. She has resided and conducted research in Kenya and the Philippines, with shorter-term assignments in Zimbabwe, India, Nepal, and other countries. Her current research includes: gender, community organization, and resource management in Kenya, Nepal, and the Philippines; methods for gender analysis and community-based development; and gender, community organization, and state policy.

Esther Wangari is Assistant Professor of Women's Studies at Towson State University in Baltimore, Maryland.

Dr. Wangari received her Ph.D. in Economic Development and International Trade from the New School for Social Research in 1991. Her dissertation on *Effects of Land Registration on Small-Scale Farming in Kenya: The Case of Mbeere in Embu District* was funded by the Ford Foundation, the International Development Research Centre (IDRC), and the Rockefeller Foundation. In 1992–3, she conducted research on gender, development, and environment as a postdoctoral fellow at the George Perkins Marsh Institute, Clark University, supported by a research grant from the Ford Foundation. At St. Cloud State University, Dr. Wangari taught in the Department of Interdisciplinary Studies in 1993–4, and in 1995 served as a Distinguished Woman Scholar of Color working on Global Understanding. At Gettysburg College she taught in the Global Studies Program (1995–1996).

Dr. Wangari's current research interests are focused on culture, resource management, class, and gender as well as policies for sustainable society and sustainable development. She is also interested in the nature of knowledge and determination of knowledge, and a cross-cultural understanding of human beings. Dr. Wangari has published various articles and recently a book, *The Heritage Library of African Peoples: Ameru*, and a chapter, "Intersection of Gender, Race, and Environment: Colonial and Post Colonial Policies," in *Oppression and Social Justice: Critical Frameworks*, fifth edition, 1995.

CONTRIBUTORS

Cristobalina Amparo is a leader of the Rural Federation of Zambrana-Chacuey and technical promoter for its Wood Producers' Association. She is a recognized authority on the history of the Federation and women's interests within it, as well as about the use of medicinal plants and their integration into nutrition, household garden, and agroforestry projects. She served in 1984 as the Federation's representative to the Program for Applied Research on Traditional Medicine in the Caribbean (TRAMIL) regional conference in Port-au-Prince, Haiti. She is currently engaged in work on health, production and women's interests within the Federation.

Anne C. Bellows is a Ph.D. candidate in Geography at Rutgers, The State University of New Jersey and works at the Center for Russian, Central and East European Studies on environmental training and research projects. She has worked on community development, environmental, and organic agriculture projects in Poland since 1991. As a member of the Network of East–West Women (NEWW), she provides networking support and information sharing for activists and academics interested in environmental issues. She brings on-site labor experience in conventional and organic agriculture, cooperatives, and marketing to her present work.

Cirilo Brito is a member of the board of directors of the Rural Federation and the Wood Producers' Association in Zambrana-Chacuey, Dominican Republic, whose knowledge and experience in farm forestry aided research and implementation of an agroforestry project in his zone. He assisted Frans Geilfus in writing *Trees to Serve the Small Farmer: An Agroforestry Manual for Rural Development*. He serves as technical promoter for the Wood Producers' Association.

Josepa Brú-Bistuer is the author of many articles on environmental issues, including industrial waste facilities, social science perspectives on the environment, and gendered perspectives on environment and equity. She currently teaches Geography and Environmental Science at the University of Girona, with past teaching experience at the Universities of Barcelona and Llieda, as well as professional experience in urban environmental planning including as a consultant to the OECD. Her current research is on the gendered nature of environmental issues, from state policy to local mobilization against toxic waste generation, treatment, and disposal.

Gladys Buenavista recently received her doctorate in Sociology from the Virginia Polytechnic Institute and State University (VPI & SU). While at VPI & SU she also served as a researcher for the Ecology, Community Organization, and Gender Project (ECOGEN). She is currently site co-ordinator in Bukidon, the Philippines, for the Sustainable Agriculture and Natural Resource Management Collaborative Research Support Project. Her current research interests include sociological issues in natural resource management and the relationship between participatory research, gender, and development.

Cornelia Butler Flora is author of numerous publications on gender, community development, and sustainable agriculture in the United States, Latin America, Africa, and the Philippines, including *Rural Communities: Legacy and Change, Rural Policies for the 1990s,* and *Sustainable Agriculture in Temperate Zones*. She is currently Director of the North Central Regional Center for Rural Development, covering the twelve north-central states of the U.S. and based at Iowa State University, and is head of the Technical Committee of the Sustainable Agriculture and Natural Resource Management Collaborative Research Support Project. She has served as Professor and Head of the Department of Sociology at Virginia Polytechnic Institute and State University and has performed consulting work for the InterAmerican Foundation, InterAmerican Development Bank, UNIFEM, USAID, and other international organizations on gender and development issues.

Connie Campbell is a visiting Assistant Professor in the Center for Tropical Conservation and Development (TCD) at the University of Florida. She has worked on participatory research with women in the Rubber Tappers Union and related groups in Rio Branco, Acre (Amazonia, Brazil). She recently completed her dissertation research (with Professor Marianne Schmink) as well as case-study and methodology materials for field researchers and trainers. Her current research includes history of women's involvement in the rubber tappers' struggles to protect and control forest resources in communities within Acre. She is stationed at the PESACRE station in Acre and TCD in Gainesville.

Louise Fortmann is the editor (with John Bruce) of *Whose Trees?* and the author of numerous publications on land tenure, property rights, community-based resource management systems, and social forestry. Professor Fortmann has conducted field research in Botswana, Tanzania, Kenya, Zimbabwe and has had short assignments in South and Southeast Asia. She is currently Professor of Natural Resource Sociology at University of California – Berkeley. She has served on boards of the Land Tenure Center, University of Wisconsin, the Center for International Forestry Research and the Institute for Sustainable Forestry.

Moya Hallstein is a feminist scholar currently engaged in graduate studies in the Department of Geography at the University of Kentucky. Her research interests include gender, class, and race in resource management in the United States and Costa Rica.

Ricardo Hernandez holds a degree in history from the Autonomous University of Santo Domingo (UASD) and has worked with a research team from Cornell University studying the social and economic impact of plantation work on small farmers in the Central Valley of the Dominican Republic. He has published work on the effects of the Rosario Dominicana gold mine on the Sanchez Ramirez zone and has worked as a researcher for CEDEE.

Manjari Mehta holds a Ph.D. in anthropology from Boston University and is presently teaching at Massachusetts Institute of Technology. She has served as Acting Coordinator for South Asian Programs at Oxfam America and as a Research Associate at the International Centre for Integrated Mountain Development in Kathmandu, Nepal. She has published several articles on the gendered impact of the introduction of cash crops into a middle Himalayan valley in north India. Her current research includes cross-cultural indigenous knowledge systems, women's roles as resource managers, and efforts at collective mobilization in hill areas.

Vernice Miller is an environmental justice activist who served as northeast region facilitator for the First National People of Color Environmental Leadership Summit. She has also served as researcher for the United Church of Christ's *Toxic Wastes and Race in the United States*. She has coordinated more than 600 U.S. environmental organizations and activists in preparation for the Earth Summit. She is a founding member of the New York City Environmental Justice Alliance and co-founder of West Harlem Environmental Action. She currently directs Environmental Justice Initiatives for the Natural Resources Defense Council.

Julio Morrobel is a Professor of natural resources conservation and management at the Higher Institute of Agriculture (ISA) in Santiago, Dominican Republic. His current research includes the impact of farm forestry on soil fertility as well as the effect of agroforestry innovations on rural communities.

Susan Quass is a feminist political activist, organizer, and scholar with experience in anti-imperialist, social justice, and women's movements in the United States, Asia, and the Pacific. She has worked with the Nuclear-Free and Independent Pacific movement and the Boston Women's Health Book Collective, and has authored numerous articles on Palauan women and self-determination in the Pacific. She is currently pursuing a Ph.D. at Clark University, where her research includes a study of women, gender, and human rights in Palauan self-determination; analysis of the strategies of solidarity and alliance demonstrated by the Women and Health Documentation Network; and the application of feminist theory to modes of social analysis used in international development practice.

Laurie Ross recently received her M.A. from the International Development and Social Change program at Clark University. While at Clark she was a researcher with the Ecology, Community Organization, and Gender project (ECOGEN). She has done field work in Zambrana-Chacuey, Dominican Republic about the distinct gendered, class, and age-determined interests in agroforestry technology and resources. Her M.A. thesis is entitled, "The Partnership Between Grassroots Social Movements and Development Organizations: Identity Reinforcement and Reinvention in Zambrana-Chacuey, Dominican Republic."

Joni Seager is a scholar and activist in feminist and alternative environmental networks. She is the author of three books (*Atlas Survey of the State of the Earth*; *Atlas of Women in the World*; *Earth Follies: Making Feminist Sense of the Environment*) and numerous articles and papers on gender and environment. She is currently an Assistant Professor of Geography at the University of Vermont. She is also an occasional commentator on environmental and feminist topics for CNN, National Public Radio, and the BBC. Her current research includes feminist analyses of militarism, environmentalism, and women's community-based environmental activism.

M. Dale Shields holds an M.A. from the International Development and Social Change program at Clark University. She lived as a Peace Corps volunteer on Siquijor Island in the Philippines from 1987 to 1989. While at Clark, she conducted research on issues of gender, class, and environment on the Philippine islands of Siquijor and Leyte under the auspices of the Ecology, Community Organization, and Gender Project (ECOGEN). Her Master's thesis is based on field work completed in 1991 as an ECOGEN researcher on Siquijor and is entitled " 'Iya-Iya, Ako-Ako': Gender and Class in Resource Management on Siquijor Island, the Philippines." She has worked as a volunteer, organizer, lobbyist in the United States with Public Interest Research Groups (PIRGs), Habitat for Humanity and the Heifer Project International.

Doris Wastl-Walter is author of twenty-seven publications including five co-edited volumes on urban geography, feminist geography, political geography, regional research, methods for gender analysis, gendered science, and community structure and development. She is currently Professor of Geography at the University of Klagenfurt, Austria, with research specialty in Austria, Hungary, and Slovenia. Her current research includes the relationship between community organization and state administration in policy, "sustainable development," and the environmental and economic restructuring of small communities in Slovenia and Austria; and gender relations in political parties and grassroot movements in different areas of Austria.

Daniel Zevallos is a Social Promoter working with the Rural Federation of Zambian-Chacuey, Dominican Republic and other non-governmental organizations throughout the region. An Ecuadorian national, he holds a degree in sociology and has worked with rural peoples' organizations in Ecuador and the Dominican Republic since 1991, on questions of environment, development and social justice.

INDEX